Introduction to Quantum Mechanics

Introduction to Quantum Mechanics

2nd Edition, Revised and Expanded

S.M. Blinder
Chemistry and Physics
University of Michigan
Ann Arbor, MI, United States

ELSEVIER

ACADEMIC PRESS
An imprint of Elsevier

Academic Press is an imprint of Elsevier
125 London Wall, London EC2Y 5AS, United Kingdom
525 B Street, Suite 1650, San Diego, CA 92101, United States
50 Hampshire Street, 5th Floor, Cambridge, MA 02139, United States
The Boulevard, Langford Lane, Kidlington, Oxford OX5 1GB, United Kingdom

Library of Congress Cataloging-in-Publication Data
A catalog record for this book is available from the Library of Congress

British Library Cataloguing-in-Publication Data
A catalogue record for this book is available from the British Library

ISBN: 978-0-12-822310-9

For information on all Academic Press publications
visit our website at https://www.elsevier.com/books-and-journals

Publisher: Susan Dennis
Acquisitions Editor: Anneka Hess
Editorial Project Manager: Andrea Dulberger
Production Project Manager: Kumar Anbazhagan
Designer: Mark Rogers
Cover: The Wave Field, a landscape sculpture by Maya Lin on North Campus, the University of Michigan, Ann Arbor

Typeset by VTeX

For Frances, Michael, Stephen, Matthew, Amy, Sarah, Lea, Michele, Kristen, Brooklyn, David and Clara.

Contents

20. Quantum computers

Preface to the 1st edition

This book is the product of 40 years of distilled experience teaching quantum theory to juniors, seniors and graduate students. It is intended as a less weighty text for one semester of the physical chemistry sequence or for a stand-alone course in quantum mechanics for students of chemistry, material science, molecular biology or earth science—possibly even physics—if the student is willing to put up with chemical applications.

The Supplements appended to several Chapters contain optional material for more adventurous students but can be guiltlessly omitted without loss of continuity. I have purposely limited the number of problems after each Chapter and geared them toward conceptual understanding rather than numerical drill. We refer to the excellent problem collections such as Shaum's Physical Chemistry (McGraw-Hill, New York, 1989) for those desiring a larger selection of problems.

I gratefully acknowledge the years of inspiration provided by my many teachers, students, colleagues and family members, too numerous and varied to be cited individually.

Ann Arbor, Michigan
May 2003

Preface to the 2nd edition

In the 15 years since publication of the First Edition, I have received an immense number of messages from readers, expressing mostly (but not all!) favorable comments and suggestions for enhancements and additions. In addition, my own experiences during this time have been very rich. I began my second career as a telecommuting computer scientist with Wolfram Research, the creators of Mathematica and other scientific software. In particular, I have been associated with the Wolfram Demonstrations Project, which has given me the opportunity of authoring several hundred interactive modules on quantum phenomena and related subjects and editing of several thousand other contributions by outside authors.

I have made a large number of edits of material in the First Edition and have made some significant additions. Some of these are on a somewhat higher level than the original content, but all have been highly recommended by students and colleagues. As a result, this book can no longer be described as a compact, one-semester textbook. But it can still be used as such by judicious omission of some of the advanced topics. I have added Chapter 5 on Special Functions, to replace the separate fragmentary presentations on the quantum-mechanical applications of Hermite, Legendre and Laguerre functions and to place these in a more unified context. I have also added other mathematical topics significant to applications of quantum mechanics.

Chapter 14 is an expanded account of density functional theory, which has now become the dominant approach in practical applications of computational chemistry. Chapter 17 covers the statistical thermodynamics of molecular gases, which very naturally follows the account of the parameters derived from molecular spectroscopy. Many undergraduate physical chemistry courses in which the main topic is quantum chemistry also cover statistical thermodynamics.

Chapter 20 gives an introduction to quantum computers. These are, at present, largely in a stage which might best be described as unrealized potential, much like nuclear fusion power. Still, the consideration of quantum computation has proven to be very illuminating in enhancing our understanding of the fundamentals of quantum theory and quantum information. This final chapter demands a higher level of proficiency than the rest of the book and may, of course, be omitted.

Finally, I would like to renew my thanks of my many colleagues, students, readers, critics and family members, who have made this endeavor not only possible, but highly rewarding.

Ann Arbor, Michigan
November 2019

About the author

S. M. Blinder is Professor Emeritus of Chemistry and Physics at the University of Michigan. Born in New York City, he got his undergraduate degree at Cornell University, where he did independent study with both Hans Bethe in the Physics Department and Peter Debye in the Chemistry Department. He completed his PhD in chemical physics from Harvard in 1958 under the direction of W. E. Moffitt and J. H. Van Vleck (Nobel Laureate in Physics, 1977). Professor Blinder has nearly 200 research publications in several areas of theoretical chemistry and mathematical physics. He was the first to derive the exact Coulomb (hydrogen atom) propagator in Feynman's path-integral formulation of quantum mechanics. He is the author of several earlier books: *Advanced Physical Chemistry* (Macmillan, 1969); *Foundations of Quantum Dynamics* (Academic Press, 1974); *Introduction to Quantum Mechanics* (Elsevier, 2004); *Guide to Essential Math* (Elsevier, 2008, Second Edition 2013); *Twenty-First Century Quantum Mechanics*, coauthored with Prof. G. Fano (Springer, 2017). He was the editor of the volume *Mathematical Physics in Theoretical Chemistry* (Elsevier, 2018).

Professor Blinder has been associated with the University of Michigan since 1963. He taught a multitude of courses in chemistry, physics, mathematics and philosophy, mostly however on the subject of quantum theory. He is currently a telecommuting senior scientist with Wolfram Research (creators of Mathematica and other scientific software). In earlier incarnations, Blinder was a Junior Master in chess and an accomplished cellist. He currently lives with his wife in Ann Arbor.

Chapter 1

Atoms and photons: origins of the quantum theory

1.1 Atomic and subatomic particles

Leucippus

Democritus

The notion that the building blocks of matter are invisibly tiny particles called atoms is usually traced back to the Greek philosophers Leucippus of Miletus and Democritus of Abdera in the 5th Century BCE. The English chemist John Dalton developed the atomic philosophy of the Greeks into a true scientific theory in the early years of the 19th Century. His treatise *New System of Chemical Philosophy* gave cogent phenomenological evidence for the existence of atoms, suggested by the laws of definite and multiple proportions. The law of definite proportions is based on the observation that chemical compounds appear to contain their constituent elements in small whole number mass ratios. Dalton hypothesized that this was a consequence of the existence of discrete units of matter. This is further reenforced by the law of multiple proportions, whereby different compounds of the same elements combine in a series of related whole-number ratios. The classic example is provided by the oxides of nitrogen, of which we would now identify the compounds NO_2, NO, N_2O, N_2O_4 and N_2O_5.

After over 2000 years of speculation and reasoning from indirect evidence, it is now possible in a sense to actually *see* individual atoms, shown, for example, by Fig. 1.1.

Introduction to Quantum Mechanics. https://doi.org/10.1016/B978-0-12-822310-9.00009-4

John Dalton

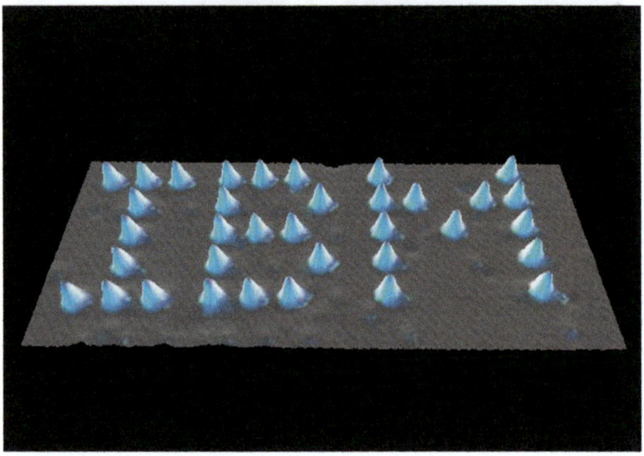

FIGURE 1.1 Image showing electron clouds of individual xenon atoms on a nickel(110) surface produced by a scanning tunneling microscope at (of course!) IBM Laboratories.

The word "atom" comes from the Greek *atomos*, meaning "indivisible." It became evident in the late 19th Century, however, that the atom was *not* truly the ultimate particle of matter. Michael Faraday's work had suggested the electrical nature of matter and the existence of subatomic particles. This became manifest with the discovery of radioactive decay by Henri Becquerel in 1896—the emission of alpha, beta and gamma particles from atoms. In 1897, J. J. Thomson identified the electron as a universal constituent of all atoms and showed that it carried a negative electrical charge. Its magnitude is now designated $-e$.

To probe the interior of the atom, Ernest Rutherford in 1911 bombarded a thin sheet of gold with a stream of positively-charged alpha particles emitted by a radioactive source. Most of the high-energy alpha particles passed right through the gold foil, but a small number were strongly deflected in a way that indicated the presence of a small but massive positive charge in the center of

J. J. Thomson

Ernest Rutherford

the atom (see Fig. 1.2). Rutherford proposed the nuclear model of the atom. As we now understand it, an electrically-neutral atom of atomic number Z consists of a nucleus of positive charge $+Ze$, containing almost the entire the mass of the atom, surrounded by Z electrons of very small mass, each carrying a charge $-e$. The simplest atom is hydrogen, with $Z = 1$, consisting of a single electron outside a single proton of charge $+e$.

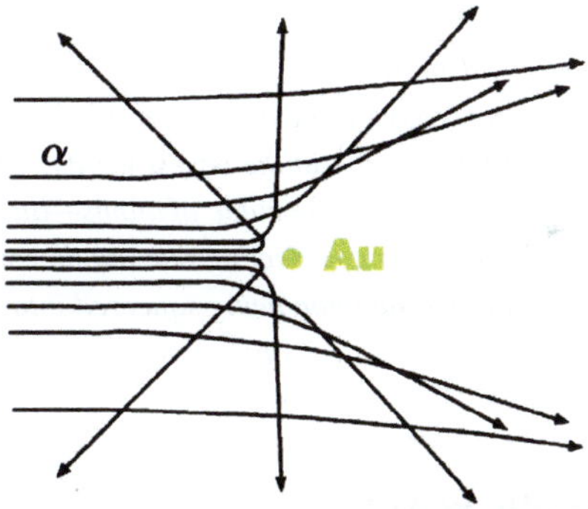

FIGURE 1.2 Some representative trajectories in Rutherford scattering of alpha particles by a gold nucleus.

With the discovery of the neutron by Chadwick in 1932, the structure of the atomic nucleus was clarified. A nucleus of atomic number Z and mass number A was composed of Z protons and $A - Z$ neutrons. Nuclei diameters are of the order of several times 10^{-15} m. From the perspective of an atom, which is 10^5 times larger, a nucleus behaves, for most purposes, like a point charge $+Ze$.

During the 1960's, compelling evidence began to emerge that protons and neutrons themselves had composite structures, mainly the idea of Murray Gell-Mann. According to the currently accepted "Standard Model," the protons and neutron are each made of three *quarks*, with compositions *uud* and *udd*, respec-

Murray Gell-Mann

Richard Feynman

tively. The up quark u has a charge of $+\frac{2}{3}e$, while the down quark d has a charge of $-\frac{1}{3}e$. Despite heroic experimental efforts, free quarks have never been isolated. By contrast, the electron maintains its status as an observable elementary particle.

We conclude this Section on atoms with a quote from Richard Feynman, in *The Feynman Lectures on Physics*, Vol I, p 1-2:

> *If by some cataclysm, all of scientific knowledge were to be destroyed, and only one sentence passed on to the next generation..., what statement would contain the most information in the fewest words? I believe it is the atomic hypothesis... that all things are made of atoms—little particles that move around in perpetual motion, attracting one another when they are a little distance apart but repelling upon being squeezed into one another. In that one sentence, you will see, there is an enormous amount of information about the world.*

1.2 Electromagnetic waves

Perhaps the greatest achievement of physics in the 19th century was the unification in 1864 of the phenomena of electricity, magnetism and optics by James Clerk Maxwell. An (optional) summary of Maxwell's equations is given in Supplement 1A. Heinrich Hertz in 1887 was the first to demonstrate experimentally the production and detection of the electromagnetic waves predicted by Maxwell—specifically radio waves—by acceleration of electrical charges. As shown in Fig. 1.3, electromagnetic waves consist of mutually perpendicular electric and magnetic fields, $\mathbf{E}$ and $\mathbf{B}$ respectively, oscillating in synchrony at high frequency and propagating in the direction of $\mathbf{E} \times \mathbf{B}$.

The wavelength λ is the distance between successive maxima of the electric (or magnetic) field. The frequency ν represents the number of oscillations per

James Clerk Maxwell

Heinrich Hertz

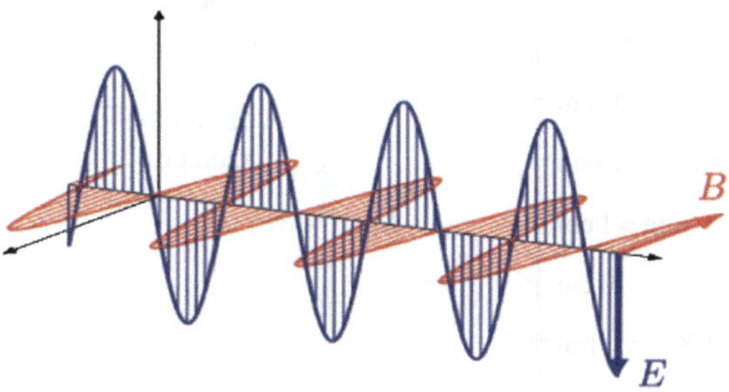

FIGURE 1.3 Schematic representation of monochromatic linearly-polarized electromagnetic wave.

second observed at a fixed point in space. The reciprocal of frequency $\tau = 1/\nu$ represents period of oscillation—the time it takes for one wavelength to pass a fixed point. The speed of propagation of the wave is therefore determined by $\lambda = c\,\tau$ or in more familiar form

$$\lambda\nu = c, \tag{1.1}$$

where $c = 2.9979 \times 10^8$ m/sec, the speed of all electromagnetic waves in vacuum equals. Usually c is referred to as the *speed of light*. Frequencies are expressed in hertz (Hz), defined as the number of oscillations per second.

Electromagnetic radiation is now known to exist in an immense range of wavelengths including gamma rays, X-rays, ultraviolet, visible light, infrared, microwaves and radio waves, as represented in Fig. 1.4.

1.3 Three failures of classical physics

Isaac Newton's masterwork, *Principia*, published in 1687, marks the beginning of modern physical science. Not only did Newton delineate the fundamental

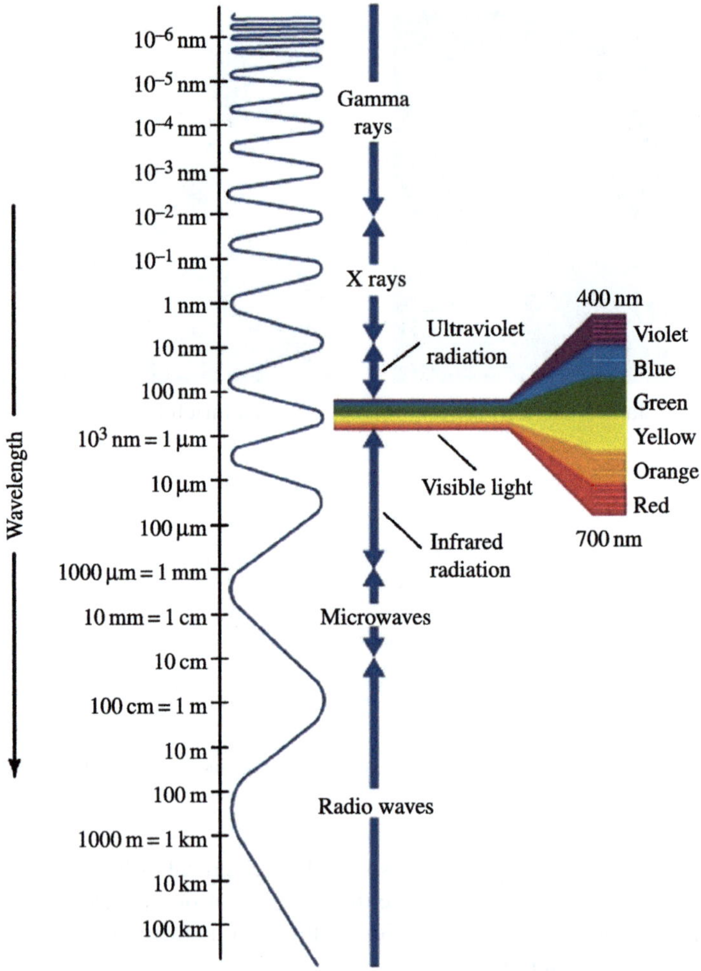

FIGURE 1.4 The electromagnetic spectrum, showing wavelengths of different types of radiation. Adapted from R. A. Freedman and W. J. Kaufmann III, *Universe* (Freeman, New York, 2001).

laws governing motion and gravitation but he established a general philosophical worldview which guided all scientific thinking for two centuries afterwards. This system of theories about the physical world is known as "Classical Physics." Its most notable feature is the primacy of cause and effect relationships. Given sufficient information about the present state of the Universe, it should be possible, at least in principle, to predict its future behavior (as well as its complete history.) This capability is known as *determinism*. For example, solar and lunar eclipses can be predicted centuries ahead, within an accuracy of several seconds. (But interestingly, we can't predict even a couple of days in advance if the weather will be clear enough to view the eclipse!)

Isaac Newton

The other great pillar of classical physics is Maxwell's theory of electromagnetism. The origin of quantum theory is presaged by three anomalous phenomena involving electromagnetic radiation, which could *not* be adequately explained by the methods of classical physics. First among these was blackbody radiation, which led to the contribution of Max Planck in 1900. Next was the photoelectric effect, treated by Albert Einstein in 1905. Third was the origin of line spectra, the hero being Neils Bohr in 1913. A coherent formulation of quantum mechanics was eventually developed in 1925 and 1926, principally the work of Schrödinger, Heisenberg and Dirac. The remainder of this Chapter will describe the early contributions to the quantum theory by Planck, Einstein and Bohr.

1.4 Blackbody radiation

It is a matter of experience that a hot object can emit radiation. A piece of metal stuck into a flame can become "red hot." At higher temperatures, its glow can be described as "white hot." Under even more extreme thermal excitation it can emit predominantly blue light (completing a very patriotic sequence of colors!). Josiah Wedgwood, the famous pottery designer, noted as far back as 1782 that different materials become red hot at the same temperature. The quantitative relation between color and temperature is described by the *blackbody radiation law*. A blackbody is an idealized perfect absorber and emitter of all possible wavelengths λ of the radiation. Fig. 1.5 shows experimental wavelength distributions of thermal radiation at several temperatures. Consistent with our experience, the maximum in the distribution, which determines the predominant color, increases with temperature. This relation is given by Wien's displacement law,

$$T\,\lambda_{\text{max}} = 2.898 \times 10^{6}\ \text{nm K} \tag{1.2}$$

where the wavelength is expressed in nanometers (nm). At room temperature (300 K), the maximum occurs around 10 μm, in the infrared region. In Fig. 1.5, the approximate values of λ_{max} are 2900 nm at 1000 K, 1450 nm at 2000 K and 500 nm at 5800 K, the approximate surface temperature of the Sun. The Sun's λ_{max} is near the middle of the visible range (380-750 nm) and is perceived by our eyes as white light.

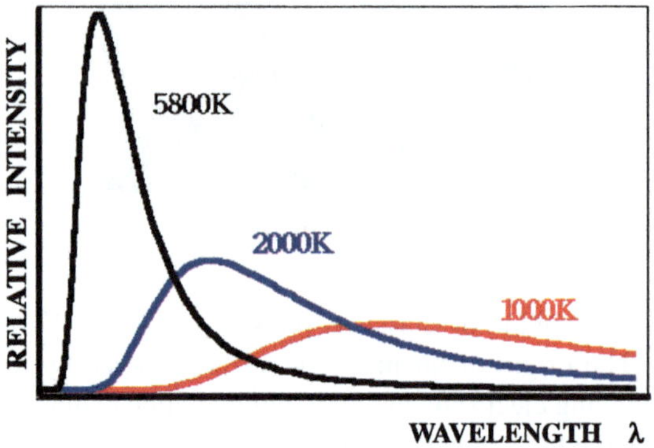

FIGURE 1.5 Intensity distributions of blackbody radiation at three different temperatures. The total radiation intensity varies as T^4 (Stefan-Boltzmann law) so the total radiation at 2000 K is actually $2^4 = 16$ times that at 1000 K.

John William Strutt, Lord Rayleigh

Max Planck

The origin of blackbody radiation was a major challenge to 19th Century physics. Lord Rayleigh proposed that the electromagnetic field could be represented by a collection of oscillators of all possible frequencies. By simple geometry, the higher-frequency (lower wavelength) modes of oscillation are increasingly numerous since it is possible to fit their waves into an enclosure in a larger number of arrangements. In fact, the number of oscillators increases very rapidly as λ^{-4}. Rayleigh assumed that every oscillator contributed equally to the radiation (the equipartition principle). This agrees fairly well with experiment at low frequencies. But if ultraviolet rays and higher frequencies were really produced in increasing number, we would get roasted like marshmallows by sitting in front of a fireplace! Fortunately, this doesn't happen, and the incorrect theory is said to suffer from an "ultraviolet catastrophe."

Max Planck in 1900 derived the correct form of the blackbody radiation law by introducing a bold postulate. He proposed that energies involved in absorption and emission of electromagnetic radiation did not belong to a continuum, as implied by Maxwell's theory, but were actually made up of discrete bundles—which he called "quanta." Planck's idea is traditionally regarded as the birth of quantum theory. A quantum associated with radiation of frequency ν has the energy

$$E = h\,\nu, \tag{1.3}$$

where the proportionality factor $h = 6.626 \times 10^{-34}$ J sec is known as *Planck's constant*. For our development of the quantum theory of atoms and molecules, we need only this simple result and do not have to follow the remainder of Planck's derivation. If you insist, however, the details are given in Supplement 1B.

1.5 The photoelectric effect

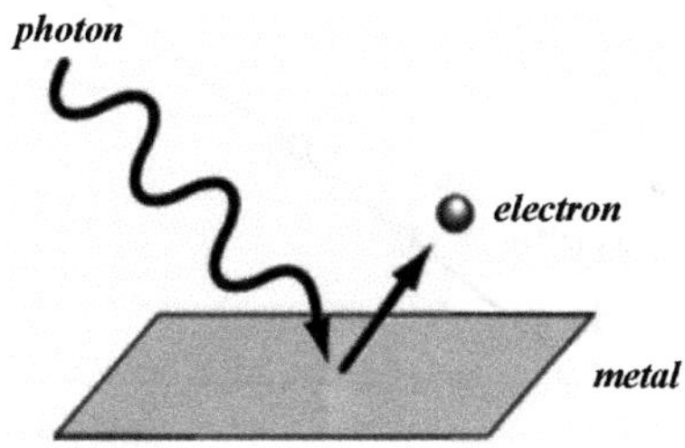

FIGURE 1.6 Photoelectric effect.

A familiar device in modern technology is the photocell or "electric eye," which runs a variety of useful gadgets, including automatic door openers. The principle involved in these devices is the photoelectric effect, which was first observed by Heinrich Hertz in the same laboratory in which he discovered electromagnetic waves. Visible or ultraviolet radiation impinging on clean metal surfaces can cause electrons to be ejected from the metal (see Fig. 1.6). Such an effect is not, in itself, inconsistent with classical theory since electromagnetic waves are known to carry energy and momentum. But the detailed behavior as a function of radiation frequency and intensity can *not* be explained classically.

The energy required to eject an electron from a metal is determined by its *work function* Φ. For example, sodium has $\Phi = 1.82$ eV. The electron-volt is a convenient unit of energy on the atomic scale: 1 eV $= 1.602 \times 10^{-19}$ J, corresponding to the energy which an electron picks up when accelerated across a potential difference of 1 volt. The classical expectation would be that radiation of sufficient intensity should cause ejection of electrons from a metal surface, with their kinetic energies increasing with the radiation intensity. Moreover, a time delay would be expected between the absorption of radiation and the ejection of electrons. The experimental facts are quite different. It is found that no electrons are ejected, no matter how high the radiation intensity, unless the radiation *frequency* exceeds some threshold value ν_0 for each metal. For sodium $\nu_0 = 4.39 \times 10^{14}$ Hz (corresponding to a wavelength of 683 nm), as shown in Fig. 1.7. For frequencies ν above the threshhold, the ejected electrons acquire a kinetic energy given by

$$\tfrac{1}{2}mv^2 = h(\nu - \nu_0) = h\nu - \Phi. \tag{1.4}$$

Evidently, the work function Φ can be identified with $h\nu_0$, equal to 3.65×10^{-19} J=1.82 eV for sodium. The kinetic energy increases *linearly* with frequency above the threshhold but is independent of the radiation intensity. Increased intensity does, however, increase the *number* of photoelectrons.

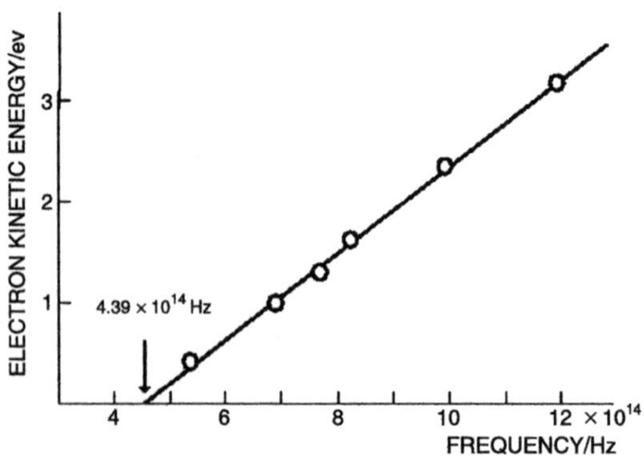

FIGURE 1.7 Photoelectric data for sodium (Millikan, 1916). The threshhold frequency ν_0, found by extrapolation, equals 4.39×10^{14} Hz.

In 1905, Albert Einstein proposed an explanation of the photoelectric effect (for which he received the Nobel Prize in 1921). Einstein's explanation of the photoelectric effect appears trivially simple once stated. He accepted Planck's hypothesis that a quantum of radiation carries an energy $h\nu$. Thus, if an electron is bound in a metal with an energy Φ, a quantum of energy $h\nu_0 = \Phi$ will be sufficient to disloge it. And any excess energy $h(\nu - \nu_0)$ will appear as kinetic energy of the ejected electron. Einstein believed that the radiation field actually did consist of quantized particles, which were later named *photons*. Although Planck himself never believed that quanta were real, Einstein's success with the photoelectric effect greatly advanced the concept of energy quantization.

Albert Einstein

1.6 Line spectra

Most of what is known about atomic (and molecular) structure and mechanics has been deduced from spectroscopy. Fig. 1.8 shows two different types of spectra. A continuous spectrum can be produced by an incandescent solid or gas at high pressure. Blackbody radiation, for example, gives a continuum. An emission spectrum can be produced by a gas at low pressure excited by heat or by collisions with electrons. An absorption spectrum results when light from a continuous source passes through a cooler gas, consisting of a series of dark lines characteristic of the composition of the gas. Frauenhofer between 1814 and 1823 discovered nearly 600 dark lines in the solar spectrum viewed at high resolution. It is now understood that these lines are caused by absorption by the outer layers of the Sun.

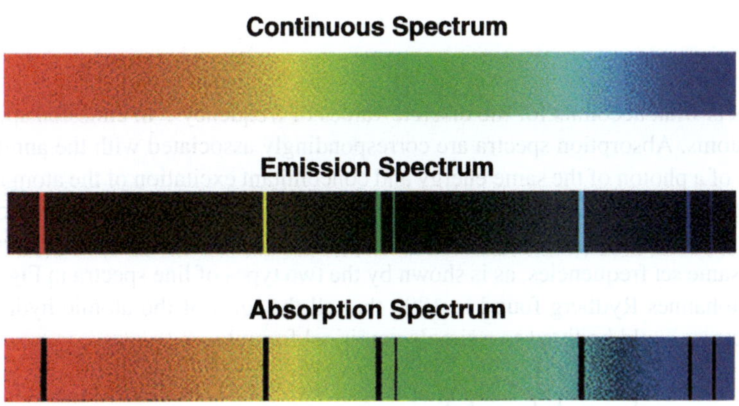

FIGURE 1.8 Continuous spectrum and two types of line spectra.

Gases heated to incandescence were found by Bunsen, Kirkhoff and others to emit light with a series of sharp wavelengths. The emitted light analyzed by a spectrometer (or even a simple prism) appears as a multitude of narrow bands of color. These so called *line spectra* are characteristic of the atomic composition of the gas. The line spectra of several elements are shown in Fig. 1.9. It is consistent with classical electromagnetic theory that motions of electrical charges within atoms can be associated with the absorption and emission of radiation. What is completely mysterious is how such radiation can occur for discrete frequencies, rather than as a continuum. The breakthrough that explained line spectra is credited to Neils Bohr in 1913. Building on the ideas of Planck and Einstein, Bohr postulated that the energy levels of atoms belong to a discrete set of values E_n, rather than a continuum as in classical mechanics. When an atom makes a downward energy transition from a higher energy level E_m to a lower energy level E_n, it caused the emission of a photon of energy

FIGURE 1.9 Emission spectra of several elements.

$$h\nu = E_m - E_n. \tag{1.5}$$

This is what accounts for the discrete values of frequency ν in emission spectra of atoms. Absorption spectra are correspondingly associated with the annihilation of a photon of the same energy and concomitant excitation of the atom from E_n to E_m. Fig. 1.10 is a schematic representation of the processes of absorption and emission of photons by atoms. Absorption and emission processes occur at the same set frequencies, as is shown by the two types of line spectra in Fig. 1.8.

Johannes Rydberg found in 1890 that all the lines of the atomic hydrogen spectrum could be fitted to a simple empirical formula

$$\frac{1}{\lambda} = \mathcal{R}\left(\frac{1}{n_1^2} - \frac{1}{n_2^2}\right), \qquad n_1 = 1, 2, 3 \ldots, \quad n_2 > n_1, \tag{1.6}$$

Neils Bohr

Johannes Rydberg

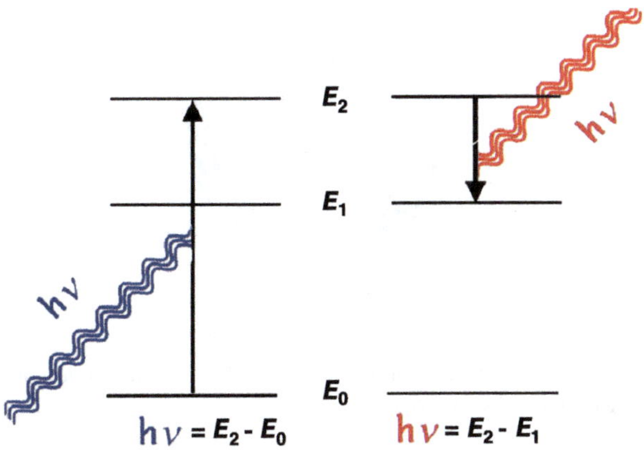

FIGURE 1.10 Origin of line spectra. Absorption of the photon shown in blue causes atomic transition from E_0 to E_2. Transition from E_2 to E_1 causes emission of the photon shown in red.

where $\mathcal{R}$, known as the Rydberg constant, has the value 109,677 cm^{-1}. This formula was found to be valid for hydrogen spectral lines in the infrared and ultraviolet regions, in addition to the four lines in the visible region. No simple formula has been found for any atom other than hydrogen. Bohr proposed a model for the energy levels of a hydrogen atom which agreed with Rydberg's formula for radiative transition frequencies. Inspired by Rutherford's nuclear atom, Bohr suggested a planetary model for the hydrogen atom in which the electron goes around the proton in one of the allowed circular orbits, as shown in Fig. 1.11. A more fundamental understanding of the discrete nature of orbits and energy levels had to await the discoveries of 1925-26, but Bohr's model provided an invaluable stepping-stone to the development of quantum mechanics. (See Fig. 1.12.) We will consider the hydrogen atom in greater detail in Chap. 8.

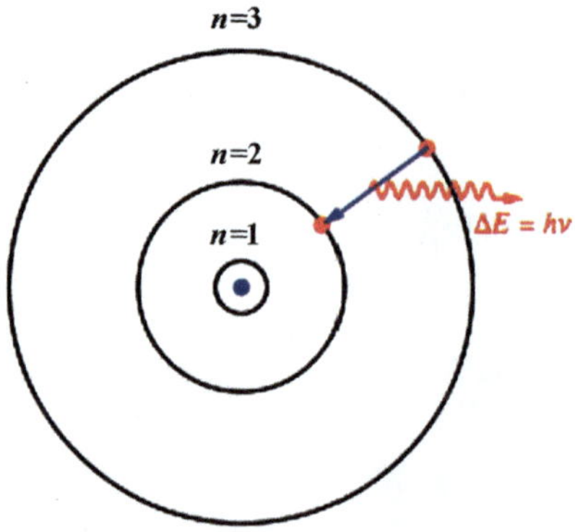

FIGURE 1.11 Bohr model of the hydrogen atom showing three lowest-energy orbits and transition $n = 3$ to $n = 2$.

FIGURE 1.12 Stylized representation of the Bohr model for a multielectron atom. From the logo of the International Atomic Energy Agency.

Supplement 1A. Maxwell's equations

These four vector relations summarize the previously discovered experimental laws describing all known electrical and magnetic phenomena. In these expressions, ρ is the electric charge density, **J**, the current density, **E**, the electric field and **B**, the magnetic induction. Maxwell's equations in free space (in the absence of dielectric or magnetic media) can be written

$$\nabla \cdot \mathbf{D} = \rho, \tag{1.7}$$

$$\nabla \cdot \mathbf{B} = 0, \tag{1.8}$$

$$\nabla \times \mathbf{E} + \frac{\partial \mathbf{B}}{\partial t} = 0, \tag{1.9}$$

$$\nabla \times \mathbf{H} = \mathbf{J} + \frac{\partial \mathbf{D}}{\partial t}. \tag{1.10}$$

The two auxiliary fields **D**, the electric displacement, and **H**, the magnetic field are defined by *constitutive relations*. In free space

$$\mathbf{D} = \epsilon_0 \mathbf{E} \quad \text{and} \quad \mathbf{B} = \mu_0 \mathbf{H}, \tag{1.11}$$

where ϵ_0 and μ_0, are the vacuum electric permittivity and magnetic permeability, respectively. Eq. (1.7) states that an electric field diverges from a distribution of electric charge. This leads to Coulomb's law. Eq. (1.8) implies the nonexistence of isolated magnetic poles–the magnetic equivalent of electric charges. The most elementary magnetic objects are *dipoles*, connected pairs of north and south poles which can *not* be isolated from one another. Eq. (1.9) is an expression of Faraday's law of electromagnetic induction, which shows how a circulating electric field can be produced by a time-varying magnetic field. Eq. (1.10) contains Ampère's law showing how a magnetic field is produced by an electric current. The second term on the right, which was added by Maxwell himself, is, in a sense, reciprocal to Faraday's law, since it implies that a circulating magnetic field can also be produced by a time-varying electric field.

In the absence of charges and currents, Maxwell equations can be transformed into three-dimensional wave equations

$$\left\{\nabla^2 - \frac{1}{c^2}\frac{\partial^2}{\partial t^2}\right\}\mathbf{E} = 0 \quad \text{and} \quad \left\{\nabla^2 - \frac{1}{c^2}\frac{\partial^2}{\partial t^2}\right\}\mathbf{B} = 0, \tag{1.12}$$

where $c = 1/\sqrt{\epsilon_0 \mu_0} = 2.9979 \times 10^8$ m/sec, representing the speed of light in vacuum. Possible solutions to Eqs. (1.12) represent synchronized transverse electric and magnetic waves propagating at the speed c, as shown in Fig. 1.3.

Even in the classical theory, electromagnetic fields can carry energy and momentum. The energy density of an electromagnetic field in free space is given by

$$\rho_{\mathrm{E}} = \tfrac{1}{2}\left(\epsilon_0 E^2 + \mu_0^{-1} B^2\right). \tag{1.13}$$

The energy flux or intensity (energy transported across unit area per unit time across unit area) is given by the Poynting vector

$$\mathbf{S} = \mathbf{E} \times \mathbf{H}. \tag{1.14}$$

It is significant that the energy density and intensity depend of the *square* of field quantities. We will exploit an analogous relationship in the interpretation of the wavefunction in quantum mechanics.

Maxwell's first equation is equivalent to Coulomb's law. In its simplest form, the force between two point charges q_1 and q_2 separated by a distance r is given by

$$F = \frac{1}{4\pi\epsilon_0}\frac{q_1 q_2}{r^2}. \tag{1.15}$$

The algebraic signs of q_1 and q_2 determine whether the force is attractive or repulsive. If q_1 and q_2 are like charges, they repel ($F > 0$), whereas opposite charges attract ($F < 0$). In our applications to atomic and molecular structure, it is clumsy and unnecessary to carry the constant $4\pi\epsilon_0$. We will instead write Coulomb's law in gaussian electromagnetic units, whereby

$$F = \frac{q_1 q_2}{r^2}. \tag{1.16}$$

The potential energy of interaction between two charges is related to the force by $F = -dV/dr$ (more generally, $\mathbf{F} = -\nabla V$). Coulomb's law therefore implies

$$V(r) = \frac{q_1 q_2}{r}. \tag{1.17}$$

In applications to the quantum theory of atoms and molecules we will also need generalizations of Eq. (1.17) for *continuous* distributions of charge. For a point charge q located at $\mathbf{r} = \mathbf{r}_0$ interacting with a charge density $\rho(\mathbf{r})$, the interaction energy is given by an integral over volume

$$V = q \int \frac{\rho(\mathbf{r})}{|\mathbf{r} - \mathbf{r}_0|} d^3\mathbf{r}. \tag{1.18}$$

For $\mathbf{r}_0 = 0$ this simplifies to

$$V = q \int \frac{\rho(\mathbf{r})}{r} d^3\mathbf{r}. \tag{1.19}$$

Analogously, the energy of interaction between two charge densities $\rho_a(\mathbf{r})$ and $\rho_b(\mathbf{r})$ is given by a double integral

$$V = \int \int \frac{\rho_a(\mathbf{r}_1)\, \rho_b(\mathbf{r}_2)}{r_{12}} d^3\mathbf{r}_1\, d^3\mathbf{r}_2, \tag{1.20}$$

where $r_{12} = |\mathbf{r}_2 - \mathbf{r}_1|$.

An electrostatic field can be represented as the gradient of a potential:

$$\mathbf{E} = -\nabla\Phi. \tag{1.21}$$

The first of Maxwell's equations, with $\mathrm{D} = \epsilon_0 \mathrm{E}$, leads to *Poisson's equation* for the electrostatic potential:

$$\nabla^2\Phi = -\rho/\epsilon_0. \tag{1.22}$$

In later work, we will apply Poisson's equation in gaussian units, which has the form

$$\nabla^2\Phi = -4\pi\rho. \tag{1.23}$$

In free space, with $\rho = 0$, this reduces to Laplace's equation

$$\nabla^2 \Phi = 0. \tag{1.24}$$

Supplement 1B. The Planck radiation law

To apply Rayleigh's idea that the radiation field can be represented as a collection of oscillators, we need to calculate the number of oscillators per unit volume for each wavelength λ. The reciprocal of the wavelength, $k = 1/\lambda$, is known as the *wavenumber*, and equals the number of wave oscillations per unit length. The wavenumber actually represents the magnitude of the *wavevector* **k**, which also determines the direction in which a wave is propagating. Now, all the vectors **k** of constant magnitude k in a 3-dimensional space can be considered to sweep out a spherical shell of radius k and infinitesimal thickness dk. The volume (in **k**-space) of this shell is equal to $4\pi k^2 dk$ and can be identified as the number of modes of oscillation per unit volume (in real space). Expressed in terms of λ, the number of modes per unit volume equals $(4\pi/\lambda^4)d\lambda$. Sir James Jeans recognized that this must be multiplied by 2 to take account of the two possible polarizations of each mode of the electromagnetic field. Assuming equipartition of energy implies that each oscillator has the energy kT, where k here is Boltzmann's constant R/N_A. Thus we obtain for the energy per unit volume per unit wavelength range

$$\rho(\lambda) = \frac{8\pi kT}{\lambda^4}, \tag{1.25}$$

which is known as the Rayleigh-Jeans law. This result gives a fairly accurate account of blackbody radiation for larger values of λ, in the infrared region and beyond. But it does suffer from the dreaded ultraviolet catastrophe, whereby $\rho(\lambda)$ increases without limit as $\lambda \to 0$.

Planck realized that the fatal flaw was equipartition, which is based on the assumption that the possible energies of each oscillator belong to a continuum ($0 \le E < \infty$). If, instead, the energy of an oscillator of wavelength λ comes in discrete bundles $h\nu = hc/\lambda$, then the possible energies are given by

$$E_{\lambda,n} = nh\nu = nhc/\lambda, \qquad \text{where } n = 0, 1, 2 \ldots. \tag{1.26}$$

By the Boltzmann distribution in statistical mechanics, the average energy of an oscillator at temperature T is given by

$$\langle E_\lambda \rangle_{\text{av}} = \frac{\sum_n E_{\lambda,n}\, e^{-E_{\lambda,n}/kT}}{\sum_n e^{-E_{\lambda,n}/kT}}. \tag{1.27}$$

Using the formula for the sum of a decreasing geometric progression

$$\sum_{n=0}^{\infty} e^{-nhc/\lambda kT} = \frac{1}{1 - e^{-hc/\lambda kT}}, \tag{1.28}$$

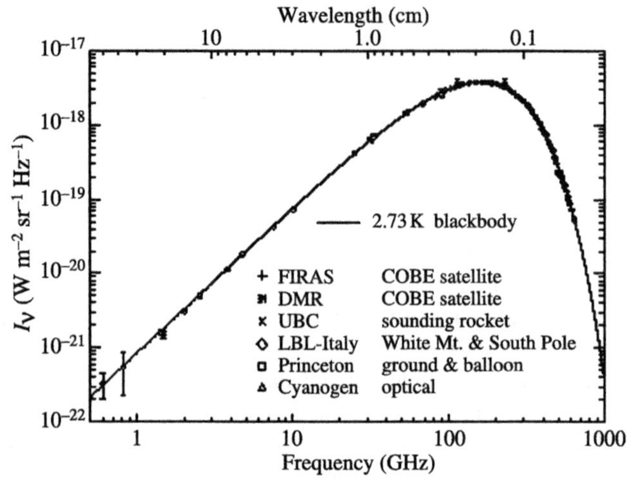

FIGURE 1.13 Cosmic Microwave Background. From G. F. Smoot and D. Scott, http://pdg.lbl.gov/2001/microwaverpp.pdf.

we obtain

$$\langle E_\lambda \rangle_{\text{av}} = \frac{hc/\lambda}{e^{hc/\lambda kT} - 1}. \tag{1.29}$$

This implies that the higher-energy modes are less populated than what is implied by the equipartition principle. Substituting this value, rather than kT, into the Rayleigh-Jeans formula (1.25), we obtain the Planck distribution law

$$\rho(\lambda) = \frac{8\pi hc}{\lambda^5} \frac{1}{e^{hc/\lambda kT} - 1}. \tag{1.30}$$

Note that, for large values of λ and/or T, the average energy (1.29) is approximated by $\langle E_\lambda \rangle_{\text{av}} \approx kT$ and the Planck formula reduces to the Rayleigh-Jeans approximation. The Planck distribution law accurately accounts for the experimental data on thermal radiation shown in Fig. 1.5. Remarkably, measurements by the Cosmic Microwave Background Explorer satellite (COBE) give a perfect fit for a blackbody distribution at temperature 2.73 K, as shown in Fig. 1.13. The cosmic microwave background radiation, which was discovered by Penzias and Wilson in 1965, is a relic of the Big Bang 13.8 billion years ago.

From the Planck distribution law one can calculate the wavelength at which $\rho(\lambda)$ is a maximum at a given T. The result agrees with the Wien displacement law with

$$\lambda_{\max} T = \frac{ch}{4.965\, k}. \tag{1.31}$$

By integration of Eq. (1.26) over all wavelengths λ, we obtain the total radiation energy density per unit volume

$$\mathcal{E} = \int_0^\infty \rho(\lambda)\, d\lambda = \frac{8\pi^5 k^4}{15c^3h^3} T^4, \tag{1.32}$$

in accord with the Stefan-Boltzmann law.

Chapter 2

Waves and particles

Quantum mechanics is the theoretical framework which describes the behavior of matter on the atomic scale. It is the most successful quantitative theory in the history of science, having withstood thousands of experimental tests without a single verifiable exception. It has correctly predicted or explained phenomena in fields as diverse as chemistry, elementary-particle physics, solid-state electronics, molecular biology and cosmology. A host of modern technological marvels, including transistors, lasers, computers and nuclear reactors are offspring of the quantum theory. Possibly 30% of the US gross national product involves technology which is based on quantum mechanics. For all its relevance, the quantum world differs quite dramatically from the world of everyday experience. To understand the modern theory of matter, challenging conceptual hurdles of both psychological and mathematical variety must be overcome.

A paradox which stimulated the early development of the quantum theory concerned the indeterminate nature of light. Light usually behaves as a wave phenomenon but occasionally it betrays a particle-like aspect, a schizoid tendency known as the wave-particle duality. We consider first the wave aspect of light.

2.1 The double-slit experiment

Fig. 2.1 shows a modernized version of the famous double-slit diffraction experiment first carried out by Thomas Young in 1801. Light from a monochromatic (single wavelength) source is passed through two narrow slits and projected onto a screen. Each slit by itself would allow just a narrow band of light to illuminate the screen. But with both slits open, a beautiful interference pattern of alternating light and dark bands appears, with maximum intensity in the center. To understand what is happening, we review some key results about electromagnetic waves.

Maxwell's theory of electromagnetism was an elegant unification of the diverse phenomena of electricity, magnetism and radiation, including light. Electromagnetic radiation is carried by transverse waves of electric and magnetic fields, propagating in vacuum at a speed $c \approx 3 \times 10^8$ m/sec, known as the "speed of light." As shown in Fig. 1.11, the **E** and **B** fields oscillate sinusoidally, in synchrony with one another. The magnitudes of **E** and **B** are proportional ($B = E/c$ in SI units). The distance between successive maxima (or minima) at a given instant of time is called the wavelength λ. At every point in space, the fields

Introduction to Quantum Mechanics. https://doi.org/10.1016/B978-0-12-822310-9.00010-0

Thomas Young

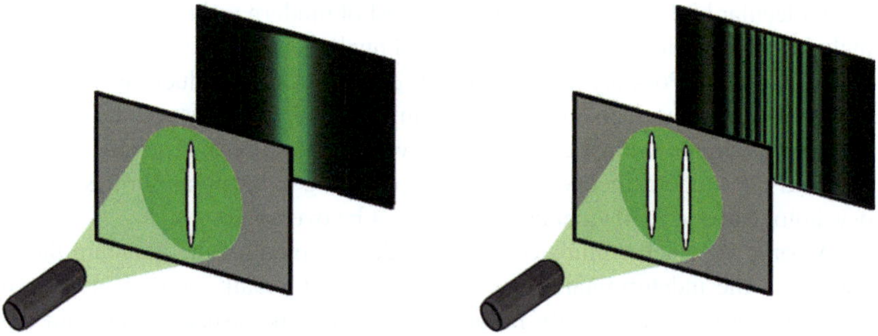

FIGURE 2.1 Modern version of Young's interference experiment using a laser gun. Single slit (left) produces an intense band of light. Double slit (right) gives a diffraction pattern.

also oscillate sinusoidally as functions of time. The number of oscillations per unit time is called the frequency ν. Since the field moves one wavelength in the time λ/c, the wavelength, frequency and propagation velocity for any wave phenomenon are related by

$$\lambda\,\nu = c. \tag{2.1}$$

In electromagnetic theory, the *intensity* of radiation—the energy flux incident on a unit area per unit time—is determined by the Poynting vector

$$\mathbf{S} = \mu_0^{-1}\,\mathbf{E} \times \mathbf{B}. \tag{2.2}$$

The energy density contained in an electromagnetic field, even a static one, is given by

$$\rho = \tfrac{1}{2}\left(\epsilon_0 E^2 + \mu_0^{-1} B^2\right). \tag{2.3}$$

Note that both of the above energy quantities depends *quadratically* on the fields **E** and **B**. To discuss the diffraction experiments described above, it is useful to define the *amplitude* of an electromagnetic wave at each point in space and time $\mathbf{r}, t$ by the function

$$\Psi(\mathbf{r}, t) = \sqrt{\epsilon_0}\, E(\mathbf{r}, t) = \frac{B(\mathbf{r}, t)}{\sqrt{\mu_0}}, \tag{2.4}$$

such that the intensity is given by

$$\rho(\mathbf{r}, t) = [\Psi(\mathbf{r}, t)]^2. \tag{2.5}$$

The function $\Psi(\mathbf{r}, t)$ will, in some later applications, have complex values. In such cases we generalize the definition of intensity to

$$\rho(\mathbf{r}, t) = |\Psi(\mathbf{r}, t)|^2 = \Psi^*(\mathbf{r}, t)\, \Psi(\mathbf{r}, t), \tag{2.6}$$

where $\Psi^*(\mathbf{r}, t)$ represents the complex conjugate of $\Psi(\mathbf{r}, t)$. In quantum-mechanical applications, the function Ψ is known as the *wavefunction*.

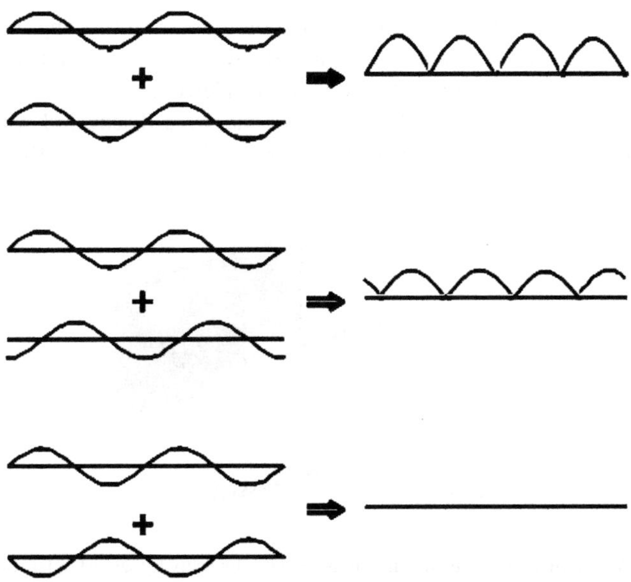

FIGURE 2.2 Interference of two equal sinusoidal waves. Top: constructive interference. Bottom: destructive interference. Center: intermediate case. The resulting intensities $\rho = \Psi^2$ is shown on the right.

The electric and magnetic fields, hence the amplitude Ψ, can have either positive and negative values at different points in space. In fact constructive and destructive interference arises from the superposition of waves, as illustrated in Fig. 2.2. By Eq. (2.5), the intensity $\rho \geq 0$ everywhere. The light and dark bands on the screen correspond to constructive and destructive interference, respectively. The wavelike nature of light is convincingly demonstrated by the fact that the intensity with both slits open is *not* the sum of the individual intensities, i.e., $\rho \neq \rho_1 + \rho_2$. Rather it is the wave amplitudes which add:

$$\Psi = \Psi_1 + \Psi_2, \tag{2.7}$$

with the intensity given by the square of the amplitude:

$$\rho = \Psi^2 = \Psi_1^2 + \Psi_2^2 + 2\Psi_1\Psi_2. \tag{2.8}$$

The cross term $2\Psi_1\Psi_2$ is responsible for the constructive and destructive interference. Where Ψ_1 and Ψ_2 have the same sign, constructive interference makes the total intensity greater than the sum of ρ_1 and ρ_2. Where Ψ_1 and Ψ_2 have opposite signs, there is destructive interference. If, in fact, $\Psi_1 = -\Psi_2$ then the two waves cancel exactly, giving a dark fringe on the screen.

2.2 Wave-particle duality

The interference phenomena demonstrated by the work of Young, Fresnel and others in the early 19th Century, apparently settled the matter that light was a wave phenomenon, contrary to the views of Newton a century earlier—case closed! But nearly a century later, phenomena were discovered which could *not* be satisfactorily accounted for by the wave theory, specifically blackbody radiation and the photoelectric effect.

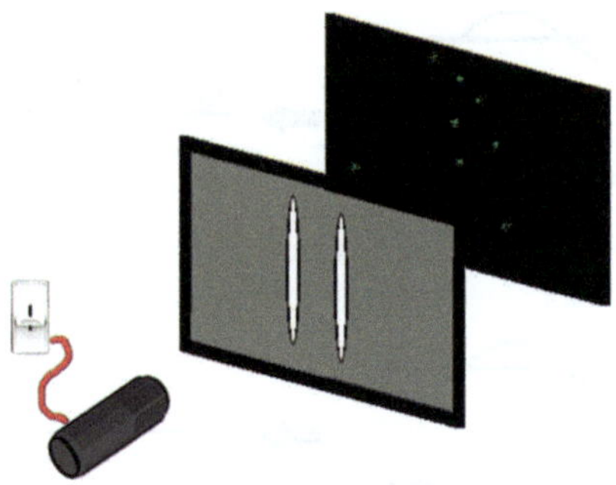

FIGURE 2.3 Scintillations observed after dimming laser intensity by several orders of magnitude. These are evidently caused by individual photons!.

Deviating from the historical development, we will illustrate these effects by a modification of the double slit experiment. Let us equip the laser source with a dimmer switch capable of reducing the light intensity by several orders of magnitude, as shown in Fig. 2.3. With each successive filter the diffraction pattern becomes dimmer and dimmer. Eventually we will begin to see localized scintillations at random positions on an otherwise dark screen. It is an almost inescapable conclusion that these scintillations are caused by *photons*, the bundles of light postulated by Planck and Einstein to explain blackbody radiation and the photoelectric effect. But wonders do not cease even here. Even though the individual scintillations appear at random positions on the screen, their statistical behavior reproduces the original high-intensity diffraction pattern. This is shown very dramatically in Fig. 2.4. Evidently the statistical behavior of the photons follows a predictable pattern, even though the behavior of individual photons is unpredictable. This implies that each individual photon, even though it behaves mostly like a particle, somehow carry with it a "knowledge" of the entire wavelike diffraction pattern. In some sense, a single photon must be able to go through *both* slits at the same time. This is what is known as the *wave-particle duality* for light: under appropriate circumstances light can behave either as a wave or as a particle.

FIGURE 2.4 Diffraction pattern built up from individual photon scintillations. Experiment by R. Austin and L. Page at Princeton University.

Planck's resolution of the problem of blackbody radiation and Einstein's explanation of the photoelectric effect can be summarized by a relation between the energy of a photon to its frequency:

$$E = h\nu, \tag{2.9}$$

where $h = 6.626 \times 10^{-34}$ J sec, known as Planck's constant. Much later, the Compton effect was discovered, the phenomenon whereby an x-ray or gamma-ray photon ejects an electron from an atom, as shown in Fig. 2.5. Assuming conservation of momentum in a photon-electron collision, the photon is found to carry a momentum of magnitude p, given by

$$p = h/\lambda. \tag{2.10}$$

Eqs. (2.9) and (2.10) constitute quantitative realizations of the wave-particle duality, each relating a particle-like property—energy or momentum—to a wave-like property—frequency or wavelength.

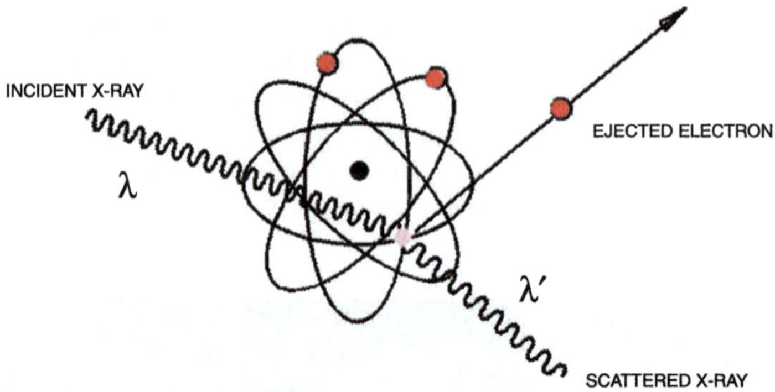

FIGURE 2.5 Compton effect. The momentum and energy carried by the incident x-ray photon are transferred to the ejected electron and the scattered photon.

According to the special theory of relativity, the last two formulas are actually different facets of the same fundamental relationship. By Einstein's famous formula, the equivalence of mass and energy is given by

$$E = mc^2. \tag{2.11}$$

The photon's rest mass is zero, but in traveling at speed c, it acquires a finite effective mass. Equating Eqs. (2.8) and (2.10) for the photon energy and taking the photon momentum to be $p = mc$, we obtain

$$p = E/c = h\nu/c = h/\lambda. \tag{2.12}$$

Thus, the wavelength-frequency relation (2.1), implies the Compton-effect formula (2.10). The best we can do is to *describe* the phenomena constituting the wave-particle duality. There is no widely accepted explanation in terms of everyday experience and common sense. Feynman referred to the "experiment with two holes" as the "central mystery of quantum mechanics." It should be mentioned that a number of models have been proposed over the years to rationalize these quantum mysteries. Bohm proposed that there might exist *hidden variables* which would make the behavior of each photon deterministic, i.e., particle-like. Everett and Wheeler proposed the "many worlds interpretation of quantum mechanics" in which each random event causes the splitting of the entire universe into disconnected parallel universes in which each possibility becomes the reality. Needless to say, not many people are willing to accept such a metaphysically unwieldy view of reality. Most scientists are content to apply the highly successful computational mechanisms of quantum theory to their work, without worrying unduly about its philosophical underpinnings. As Feynman put it, "Shut up and calculate!" Much like most of us happily using our computers without acquiring a detailed knowledge of either semiconductor technology or operating-system programming.

There was never any drawn-out controversy about whether electrons or any other constituents of matter were other than particle-like. Individual electrons produce scintillations on a phosphor screen—that is how the older TVs worked. But electrons also exhibit diffraction effects, which indicates that they too have wavelike attributes. An analog of the double-slit experiment using electrons instead of light is technically difficult but has been done. An electron gun, instead of a light source, produces a beam of electrons at a selected velocity. Then, everything that happens for photons has its analog for electrons, as shown by the diffraction pattern in Fig. 2.6.

FIGURE 2.6 Slit diffraction pattern for electrons, obtained by Claus Jönsson in Tübingen (1961). Selected as the "most beautiful" experiment in the history of physics by *Physics World* in 2002.

Diffraction experiments have been more recently carried out for particles as large as atoms and molecules, even for C_{60} fullerene molecules.

Louis de Broglie

De Broglie in 1924 first conjectured that matter might also exhibit a wave-particle duality. A wavelike aspect of the electron might, for example, be responsible for the discrete nature of Bohr orbits in the hydrogen atom (cf. Chap. 8). According to de Broglie's hypothesis, the "matter waves" associated with a particle have a wavelength given by

$$\lambda = h/p, \tag{2.13}$$

identical in form to Compton's result (2.10) (which, in fact, was discovered later). The correctness of de Broglie's conjecture was most dramatically confirmed by the observations of Davisson and Germer in 1927 of diffraction of monoenergetic beams of electrons by metal crystals, much like the diffraction of x-rays. And measurements showed that de Broglie's formula (2.13) did indeed give the correct wavelength (see Fig. 2.7).

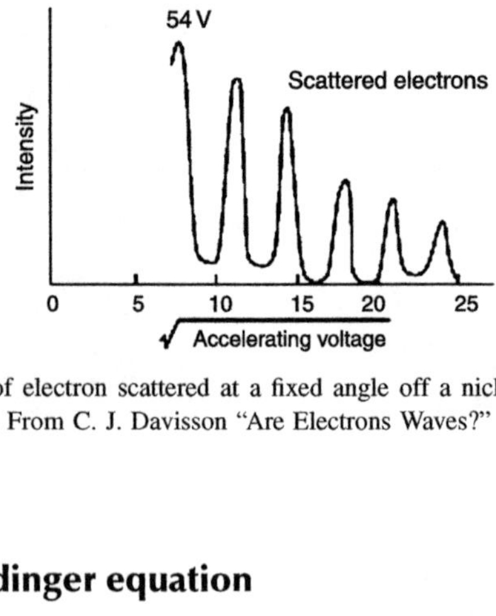

FIGURE 2.7 Intensity of electron scattered at a fixed angle off a nickel crystal, as function of incident electron energy. From C. J. Davisson "Are Electrons Waves?" *Franklin Institute Journal* **205**, 597 (1928).

2.3 The Schrödinger equation

Schrödinger in 1926 first proposed an equation for de Broglie's matter waves. This equation cannot be derived from some other principle since it constitutes a fundamental law of nature. Its correctness can be judged only by its subsequent agreement with observed phenomena (*a posteriori* proof). Nonetheless, we will attempt a heuristic argument to make the result at least plausible.

In classical electromagnetic theory, it follows from Maxwell's equations that each component of the electric and magnetic fields in vacuum is a solution of the wave equation

$$\nabla^2\Psi - \frac{1}{c^2}\frac{\partial^2\Psi}{\partial t^2} = 0, \tag{2.14}$$

where the Laplacian or "del-squared" operator is defined by

$$\nabla^2 = \frac{\partial^2}{\partial x^2} + \frac{\partial^2}{\partial y^2} + \frac{\partial^2}{\partial z^2}. \tag{2.15}$$

Erwin Schrödinger

We will attempt now to create an analogous equation for de Broglie's matter waves.

Accordingly, let us consider a very general instance of wave motion propagating in the x-direction. At a given instant of time, the form of a wave might be represented by a periodic function such as

$$\psi(x) = f(2\pi x/\lambda), \tag{2.16}$$

where $f(\theta)$ might be a sinusoidal function such as $\sin\theta$, $\cos\theta$, $e^{i\theta}$, $e^{-i\theta}$ or some linear combination of these. The most suggestive form will turn out to be the complex exponential, which is related to the sine and cosine by Euler's formula

$$e^{\pm i\theta} = \cos\theta \pm i \sin\theta. \tag{2.17}$$

Derivatives of exponentials are simpler than those of sines or cosines. Each of the above functions is periodic, with its value repeating as the argument increases by 2π. This happens whenever x increases by one wavelength λ. At a fixed point in space, the time-dependence of the wave has an analogous structure:

$$T(t) = f(2\pi \nu t), \tag{2.18}$$

where ν gives the number of cycles of the wave per unit time. Taking into account both x- and t-dependence, we consider a wavefunction of the form

$$\Psi(x,t) = \exp\left[2\pi i \left(\frac{x}{\lambda} - \nu t\right)\right], \tag{2.19}$$

representing waves traveling from left to right. Now we make use of the Planck and de Broglie formulas (2.8) and (2.12) to replace ν and λ by their particle analogs. This gives

$$\Psi(x,t) = \exp[i(px - Et)/\hbar], \tag{2.20}$$

where

$$\hbar \equiv \frac{h}{2\pi}. \tag{2.21}$$

Since Planck's constant occurs in most formulas with the denominator 2π, this symbol, pronounced "aitch-bar," was invented by Dirac.

Eq. (2.19) represents in some abstruse way the wavelike nature of a particle with energy E and momentum p. We will attempt to discover the underlying wave equation by "reverse engineering." The time derivative of (2.19) gives

$$\frac{\partial \Psi}{\partial t} = -(iE/\hbar) \times \exp[i(px - Et)/\hbar]. \tag{2.22}$$

Thus

$$i\hbar \frac{\partial \Psi}{\partial t} = E\Psi. \tag{2.23}$$

Analogously

$$-i\hbar \frac{\partial \Psi}{\partial x} = p\Psi, \tag{2.24}$$

and

$$-\hbar^2 \frac{\partial^2 \Psi}{\partial x^2} = p^2 \Psi. \tag{2.25}$$

The energy and momentum for a nonrelativistic free particle are related by

$$E = \frac{1}{2}mv^2 = \frac{p^2}{2m}. \tag{2.26}$$

This suggests that $\Psi(x,t)$ satisfies the partial differential equation

$$i\hbar \frac{\partial \Psi}{\partial t} = -\frac{\hbar^2}{2m}\frac{\partial^2 \Psi}{\partial x^2}. \tag{2.27}$$

For a particle with a potential energy $V(x)$, the analog of (2.25) is

$$E = \frac{p^2}{2m} + V(x). \tag{2.28}$$

We postulate that the equation for matter waves is then

$$i\hbar \frac{\partial \Psi}{\partial t} = \left\{ -\frac{\hbar^2}{2m}\frac{\partial^2}{\partial x^2} + V(x) \right\} \Psi. \tag{2.29}$$

For waves in three dimensions the obvious generalization is

$$i\hbar\frac{\partial}{\partial t}\Psi(\mathbf{r},t) = \left\{-\frac{\hbar^2}{2m}\nabla^2 + V(\mathbf{r})\right\}\Psi(\mathbf{r},t). \tag{2.30}$$

Here the potential energy and the wavefunction depend on the three space coordinates x, y, z, which we write for brevity as $\mathbf{r}$. We have thus arrived at the *time-dependent Schrödinger equation* for the amplitude $\Psi(\mathbf{r},t)$ of the matter waves associated with the particle. Its formulation in 1926 represents the starting point of modern quantum mechanics. (Heisenberg in 1925 proposed another version known as matrix mechanics.)

For conservative systems, in which the energy is a constant, we can separate out the time-dependent factor from (2.19) and write

$$\Psi(\mathbf{r},t) = \psi(\mathbf{r})\,e^{-iEt/\hbar}, \tag{2.31}$$

where $\psi(\mathbf{r})$ is a wavefunction dependent only on space coordinates. Putting (2.30) into (2.29) and canceling the exponential factors, we obtain the *time-independent Schrödinger equation*:

$$\left\{-\frac{\hbar^2}{2m}\nabla^2 + V(\mathbf{r})\right\}\psi(\mathbf{r}) = E\psi(\mathbf{r}). \tag{2.32}$$

Most of our applications of quantum mechanics to chemistry will be based on this equation.

2.4 Operators and eigenvalues

The bracketed object in Eq. (2.31) is called an *operator*. An operator is a generalization of the concept of a function. Whereas a function is a rule for turning one number into another, an operator is a rule for turning one function in another. The Laplacian is an example of an operator. We provisionally indicate that an object is an operator by placing a "hat" over it, e.g., $\hat{A}$. The action of an operator that turns the function f into the function g is represented by

$$\hat{A}\,f = g. \tag{2.33}$$

Eq. (2.23) implies that the operator for the x-component of momentum can be written

$$\hat{p}_x = -i\hbar\frac{\partial}{\partial x}, \tag{2.34}$$

and by analogy, we must have

$$\hat{p}_y = -i\hbar\frac{\partial}{\partial y}, \qquad \hat{p}_z = -i\hbar\frac{\partial}{\partial z}. \tag{2.35}$$

The energy, as in Eq. (2.27), expressed as a function of position and momentum is known in classical mechanics as the *Hamiltonian*. Generalizing to three dimensions,

$$H = \frac{\mathbf{p}^2}{2m} + V(\mathbf{r}) = \frac{1}{2m}(p_x^2 + p_y^2 + p_z^2) + V(x, y, z). \tag{2.36}$$

We can construct from this the corresponding quantum-mechanical operator

$$\hat{H} = -\frac{\hbar^2}{2m}\left(\frac{\partial^2}{\partial x^2} + \frac{\partial^2}{\partial y^2} + \frac{\partial^2}{\partial z^2}\right) + V(x, y, z)$$
$$= -\frac{\hbar^2}{2m}\nabla^2 + V(\mathbf{r}). \tag{2.37}$$

The time-independent Schrödinger equation (2.31) can then be written symbolically as

$$\hat{H}\,\Psi = E\,\Psi. \tag{2.38}$$

This form is applicable to *any* quantum-mechanical system, given the appropriate Hamiltonian and wavefunction. Most applications to chemistry involve systems containing several particles—the electrons and nuclei in atoms and molecules.

An operator equation of the form

$$\hat{A}\,\psi = \text{const}\,\psi \tag{2.39}$$

is called an *eigenvalue equation*. Recall that, in general, an operator acting on a function gives another function (see Eq. (2.32)). The special case (2.38) occurs when the second function is a multiple of the first. In this case, ψ is known as an *eigenfunction* and the constant is called an *eigenvalue*. (These terms are hybrids with German, the purely English equivalents being "characteristic function" and "characteristic value.") To every dynamical variable A in quantum mechanics, there corresponds an eigenvalue equation, usually written

$$\hat{A}\,\psi = a\,\psi. \tag{2.40}$$

The eigenvalues a represent the possible measured values of the variable A. The Schrödinger equation (2.37) is the best known instance of an eigenvalue equation, with its eigenvalues corresponding to the allowed energy levels of the quantum system.

2.5 The wavefunction

For a single-particle system, the wavefunction $\Psi(\mathbf{r}, t)$, or $\psi(\mathbf{r})$ for the time-independent case, represents the amplitude of the still vaguely defined matter waves. The relationship between amplitude and intensity of electromagnetic waves developed for Eq. (2.6) can be extended to matter waves. The most commonly accepted interpretation of the wavefunction is due to Max Born (1926), according to which $\rho(\mathbf{r})$, the square of the absolute value of $\psi(\mathbf{r})$ is proportional

Max Born

to the probability density (probability per unit volume) that the particle will be found at the position $\mathbf{r}$. Probability density is the three-dimensional analog of the diffraction pattern that appears on the two-dimensional screen in the double-slit diffraction experiment for electrons, as in Fig. 2.6 for example. In the latter case we have the relative probability a scintillation will appear at a given point on the screen. The function $\rho(\mathbf{r})$ becomes equal, rather than just proportional to, the probability density when the wavefunction is *normalized*, that is,

$$\int |\psi(\mathbf{r})|^2 \, d\tau = 1. \tag{2.41}$$

This simply accounts for the fact that the total probability of finding the particle *somewhere* adds up to unity. The integration in (2.40) extends over all space and the symbol $d\tau$ designates the appropriate volume element. For example, in cartesian coordinates, $d\tau = dx\,dy\,dz$; in spherical polar coordinates, $d\tau = r^2 \sin\theta \, dr\, d\theta\, d\phi$.

The physical significance of the wavefunctions makes certain demands on its mathematical behavior. The wavefunction must be a single-valued function of all its coordinates, since the probability density ought to be uniquely determined at each point in space. Moreover, the wavefunction should be finite and continuous everywhere, since a physically-meaningful probability density must have the same attributes. The conditions that the wavefunction be single-valued, finite and continuous—in short, "well-behaved"—lead to restrictions on solutions of the Schrödinger equation such that only certain values of the energy and other dynamical variables are allowed. This is called *quantization*—the feature that gives *quantum* mechanics its name.

Chapter 2. Exercises

2.1. In the theory of relativity, space and time variables can be combined into a 4-dimensional vector with $x_1 = x$, $x_2 = y$, $x_3 = z$, $x_4 = ict$. The momentum and energy analogously combine to a 4-vector with $p_1 = p_x$, $p_2 = p_y$, $p_3 = p_z$, $p_4 = iE/c$. By a suitable generalization of the quantization prescription for momentum components, deduce the time-dependent Schrödinger equation:

$$\left\{-\frac{\hbar^2}{2m}\nabla^2 + V(\mathbf{r})\right\}\Psi(\mathbf{r},t) = i\hbar\frac{\partial\Psi(\mathbf{r},t)}{\partial t}.$$

2.2. Estimate the number of photons emitted per second by a 100-watt lightbulb. Assume a wavelength of 550 nm (yellow light).

2.3. Electron diffraction makes use of 40 keV (40,000 eV) electrons. Calculate their de Broglie wavelength.

2.4. Show that the wavefunction $\Psi(x,t) = e^{i(px-Et)/\hbar}$ is a solution of the one-dimensional time-dependent Schrödinger equation.

2.5. Show that $\Psi(\mathbf{r},t) = e^{i(\mathbf{p}\cdot\mathbf{r}-Et)/\hbar}$ is a solution of the three-dimensional time-dependent Schrödinger equation.

2.6. A certain one-dimensional quantum system in $0 \le x \le \infty$ is described by the Hamiltonian:

$$\hat{H} = -\frac{\hbar^2}{2m}\frac{d^2}{dx^2} - \frac{q^2}{x}, \qquad (q = \text{constant}).$$

One of the eigenfunctions is known to be

$$\psi(x) = A\,x\,e^{-\alpha x}, \qquad \alpha \equiv mq^2/\hbar^2, \qquad A = \text{constant}$$

(i) Write down the Schrödinger equation and carry out the differentiation.
(ii) Find the corresponding energy eigenvalue (in terms of $\hbar$, m and q).
(iii) Find the value of A which normalizes the wavefunction according to

$$\int_0^\infty |\psi(x)|^2\,dx = 1.$$

You may require the definite integrals

$$\int_0^\infty x^n e^{-ax}\,dx = n!/a^{n+1}.$$

Chapter 3

Quantum mechanics of some simple system

3.1 The free particle

The simplest system in quantum mechanics has the potential energy V equal to zero everywhere. This is called a *free particle* since it has no forces acting on it. We consider the one-dimensional case, with motion only in the x-direction, represented by the Schrödinger equation

$$-\frac{\hbar^2}{2m}\frac{d^2\psi(x)}{dx^2} = E\psi(x). \tag{3.1}$$

Total derivatives can be used since there is but one independent variable. The equation simplifies to

$$\psi''(x) + k^2\,\psi(x) = 0, \tag{3.2}$$

with the definition

$$k^2 \equiv 2mE/\hbar^2. \tag{3.3}$$

Possible solutions of Eq. (3.2) are

$$\psi(x) = \text{const}\begin{cases}\sin kx\\ \cos kx\\ e^{\pm ikx}.\end{cases} \tag{3.4}$$

There is no restriction on the value of k. Thus a free particle, even in quantum mechanics, can have any non-negative value of the energy

$$E = \frac{\hbar^2k^2}{2m} \geq 0. \tag{3.5}$$

The energy levels in this case are *not* quantized and correspond to the same continuum of kinetic energy shown by a classical particle. Even for an atom or a molecule, which is most notable for its quantized energy levels, there will exist a continuum at sufficiently high energies, associated with the onset of ionization or dissociation.

Introduction to Quantum Mechanics. https://doi.org/10.1016/B978-0-12-822310-9.00011-2

It is of interest also to consider the x-component of linear momentum for the free-particle solutions (3.4). According to Eqs. (2.32) and (2.33), the eigenvalue equation for momentum should read

$$\hat{p}_x \psi(x) = -i\hbar \frac{d\psi(x)}{dx} = p\,\psi(x), \tag{3.6}$$

where we have denoted the momentum eigenvalue as p. It is easily shown that neither of the functions $\sin kx$ or $\cos kx$ from (3.4) is an eigenfunction of $\hat{p}_x$. But the $e^{\pm ikx}$ are both eigenfunctions with eigenvalues $p = \pm\hbar k$, respectively. Evidently the momentum p can take on any real value between $-\infty$ and $+\infty$. The kinetic energy, equal to $E = p^2/2m$, can correspondingly have any value between 0 and $+\infty$.

The functions $\sin kx$ and $\cos kx$, while not eigenfunctions of $\hat{p}_x$, are each *superpositions* of the two eigenfunctions $e^{\pm ikx}$, by virtue of the trigonometric identities

$$\cos kx = \frac{1}{2}(e^{ikx} + e^{-ikx}) \text{ and } \sin kx = \frac{1}{2i}(e^{ikx} - e^{-ikx}). \tag{3.7}$$

The eigenfunction e^{ikx} for $k > 0$ represents the particle moving from left to right on the x-axis, with momentum $p > 0$. Correspondingly, e^{-ikx} represents motion from right to left with $p < 0$. The functions $\sin kx$ and $\cos kx$ represent *standing waves*, obtained by superposition of opposing wave motions. Although these latter two are not eigenfunctions of $\hat{p}_x$ they *are* eigenfunctions of $\hat{p}_x^2$, hence of the Hamiltonian $\hat{H}$.

Orthogonality and normalization of the free-particle eigenfunctions involves the deltafunction. This will be discussed in Sect. 5.3, Eqs. (5.35) and (5.36).

3.2 Particle in a box

Also known as a particle in an infinite square well, this is the simplest nontrivial application of the Schrödinger equation, but one which illustrates many of the fundamental concepts of quantum mechanics. For a particle moving in one dimension (again take the x-axis), the Schrödinger equation can be written

$$-\frac{\hbar^2}{2m}\psi''(x) + V(x)\psi(x) = E\,\psi(x). \tag{3.8}$$

Assume that the particle can move freely between two endpoints $x = 0$ and $x = a$, but cannot penetrate past either end. This can be represented by a potential energy function

$$V(x) = \begin{cases} 0 & 0 \le x \le a \\ \infty & x < 0 \text{ and } x > a. \end{cases} \tag{3.9}$$

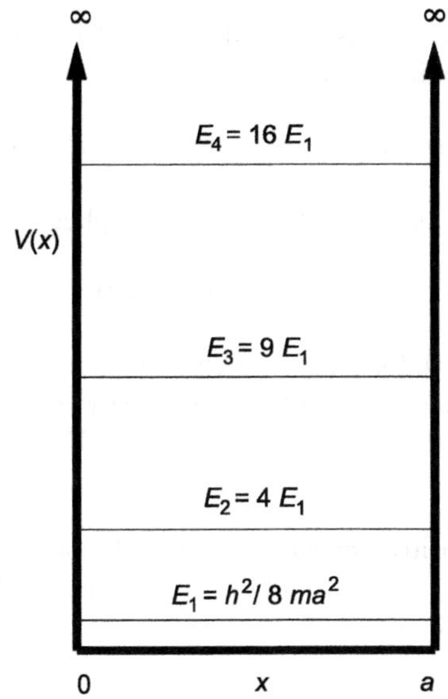

FIGURE 3.1 Potential well and lowest energy levels for particle in a box.

This potential is represented by the dark lines in Fig. 3.1. Infinite potential energy constitute an impenetrable barrier. The particle is thus bound to a *potential well*, sometimes called a *square well*. Since the particle cannot penetrate beyond the endpoints $x = 0$ or $x = a$ we must have

$$\psi(x) = 0 \quad \text{for} \quad x < 0 \quad \text{and} \quad x > a. \tag{3.10}$$

By the requirement that the wavefunction be continuous, it must be true as well that

$$\psi(0) = 0 \quad \text{and} \quad \psi(a) = 0, \tag{3.11}$$

which constitutes a pair of boundary conditions on the wavefunction *within* the box. Inside the box, $V(x) = 0$, so the Schrödinger equation reduces to the free-particle form (3.1)

$$-\frac{\hbar^2}{2m}\psi''(x) = E\,\psi(x), \qquad 0 \le x \le a. \tag{3.12}$$

We again have the differential equation

$$\psi''(x) + k^2\,\psi(x) = 0 \qquad \text{with} \qquad k^2 = 2mE/\hbar^2. \tag{3.13}$$

The general solution can be written

$$\psi(x) = A\sin kx + B\cos kx, \tag{3.14}$$

where A and B are constants to be determined by the boundary conditions (3.11). By the first condition, we find

$$\psi(0) = A \sin 0 + B \cos 0 = B = 0. \tag{3.15}$$

The second boundary condition at $x = a$ then implies

$$\psi(a) = A \sin ka = 0. \tag{3.16}$$

It is assumed that $A \neq 0$, for otherwise $\psi(x)$ would be zero everywhere and the particle would disappear. The condition that $\sin kx = 0$ implies that

$$ka = n\pi, \tag{3.17}$$

where n is a integer, positive, negative or zero. The case $n = 0$ must be excluded, for then $k = 0$ and again $\psi(x)$ would vanish everywhere. Eliminating k between (3.13) and (3.17), we obtain

$$E_n = \frac{\hbar^2 \pi^2}{2ma^2} n^2 = \frac{h^2}{8ma^2} n^2 \qquad n = 1, 2, 3 \ldots. \tag{3.18}$$

These are the only values of the energy which allow solution of the Schrödinger equation (3.12) consistent with the boundary conditions (3.11). The integer n, called a *quantum number*, is appended as a subscript on E to label the allowed energy levels. Negative values of n add nothing new because $\sin(-kx) = -\sin kx$, which represents the same quantum state. Fig. 3.1 also shows part of the energy-level diagram for the particle in a box. Classical mechanics would predict a continuum for all values $E \geq 0$. The occurrence of discrete or *quantized* energy levels is characteristic of a bound system in quantum mechanics, that is, one confined to a finite region in space. For the free particle, the absence of confinement allows an energy continuum. Note that, in both cases, the number of energy levels is infinite—denumerably infinite for the particle in a box but nondenumerably infinite for the free particle.

The particle in a box assumes its lowest possible energy when $n = 1$, namely

$$E_1 = \frac{h^2}{8ma^2}. \tag{3.19}$$

The state of lowest energy for a quantum system is termed its *ground state*. Higher energies are called *excited states*. An interesting point is that the ground-state energy $E_1 > 0$, whereas the corresponding classical system would have a minimum energy of zero. This is a recurrent phenomenon in quantum mechanics. The residual energy of the ground state, that is, the energy in excess of the classical minimum, is known as *zero point energy*. In effect, the kinetic energy, hence the momentum, of a bound particle cannot be reduced to zero. The minimum value of momentum is found by equating E_1 to $p^2/2m$, giving

$p_{\min} = \pm h/2a$. This can be expressed as an *uncertainty* in momentum approximated by $\Delta p \approx h/a$. Coupling this with the uncertainty in position, $\Delta x \approx a$, the size of the box, we can write

$$\Delta x \, \Delta p \approx h. \tag{3.20}$$

This is in accord with the *Heisenberg uncertainty principle*, which we will discuss in greater detail later.

The particle-in-a-box eigenfunctions are given by Eq. (3.14), with $B = 0$ and $k = n\pi/a$, in accordance with (3.17):

$$\psi_n(x) = A \sin \frac{n\pi x}{a}, \qquad n = 1, 2, 3 \ldots. \tag{3.21}$$

These, like the energies, can be labeled by the quantum number n. The constant A, thus far arbitrary, can be adjusted so that $\psi_n(x)$ is normalized. The normalization condition (2.40) becomes, in this case,

$$\int_0^a [\psi_n(x)]^2 \, dx = 1 \tag{3.22}$$

the integration running over the domain of the particle, $0 \leq x \leq a$. Substituting (3.21) into (3.22),

$$A^2 \int_0^a \sin^2 \frac{n\pi x}{a} \, dx = A^2 \frac{a}{n\pi} \int_0^{n\pi} \sin^2 \theta \, d\theta = A^2 \frac{a}{2} = 1. \tag{3.23}$$

We have made the substitution $\theta = n\pi x/a$ and used the fact that the average value of $\sin^2 \theta$ over an integral number of half wavelengths equals 1/2. (Alternatively, one could refer to standard integral tables.) From (3.23) we can identify the normalization constant $A = (2/a)^{1/2}$ for all values of n. Thus we obtain the normalized eigenfunctions:

$$\psi_n(x) = \left(\frac{2}{a}\right)^{1/2} \sin \frac{n\pi x}{a}, \qquad n = 1, 2, 3 \ldots. \tag{3.24}$$

The first few eigenfunctions and the corresponding probability distributions are plotted in Fig. 3.2. There is a close analogy between the states of this quantum system and the modes of vibration of a string. The patterns of standing waves on the string are, in fact, identical in form with the wavefunctions (3.24).

A notable feature of the particle-in-a-box quantum states is the occurrence of *nodes*. These are points, other than the two end points (which are fixed by the boundary conditions), at which the wavefunction vanishes. At a node there is exactly zero probability of finding the particle. The nth quantum state has, in fact, $n - 1$ nodes. It is generally true that the number of nodes increases with the energy of a quantum state, which can be rationalized by the following qualitative argument. As the number of nodes increases, so does the number and steepness

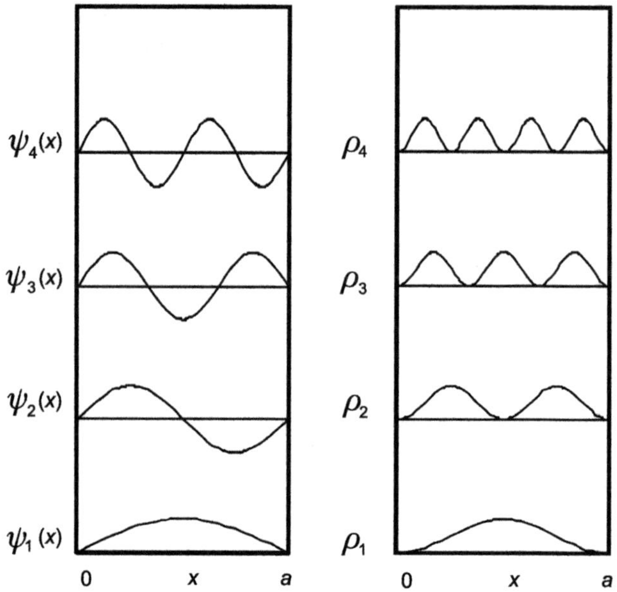

FIGURE 3.2 Eigenfunctions and probability densities for particle in a box.

of the 'wiggles' in the wavefunction. It's like skiing down a slalom course. Accordingly, the average curvature, given by the second derivative, must increase. But the second derivative is proportional to the kinetic energy operator. Therefore, the more nodes, the higher the energy. This will prove to be an invaluable guide in more complex quantum systems.

Another important property of the eigenfunctions (3.24) applies to the integral over a product of two *different* eigenfunctions. The following relationship is easy to see from Fig. 3.3:

$$\int_0^a \psi_2(x)\,\psi_1(x)\,dx = 0. \tag{3.25}$$

To prove this result in general, use the trigonometric identity

$$\sin\alpha\,\sin\beta = \frac{1}{2}[\cos(\alpha-\beta) - \cos(\alpha+\beta)] \tag{3.26}$$

to show that

$$\int_0^a \psi_m(x)\,\psi_n(x)\,dx = 0 \quad \text{if} \quad m \neq n. \tag{3.27}$$

This property is called *orthogonality*. We will show in the Chap. 4 that this is a general result for quantum-mechanical eigenfunctions. The normalization (3.25)

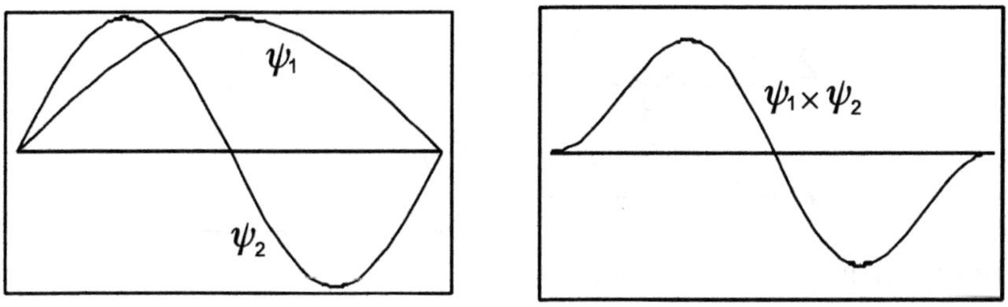

FIGURE 3.3 Product of n=1 and n=2 eigenfunctions.

together with the orthogonality (3.27) can be combined into a single relationship

$$\int_0^a \psi_m(x)\,\psi_n(x)\,dx = \delta_{mn}, \tag{3.28}$$

in terms of the *Kronecker delta*

$$\delta_{mn} \equiv \begin{cases} 1 \text{ if } m = n \\ 0 \text{ if } m \neq n. \end{cases} \tag{3.29}$$

A set of functions $\{\psi_n\}$ which obeys (3.28) is called *orthonormal.*

3.3 Free-electron model

The simple quantum-mechanical problem we have just solved can provide an instructive application to chemistry: the *free-electron model* (FEM) for delocalized π-electrons. The simplest case is the 1,3-butadiene molecule

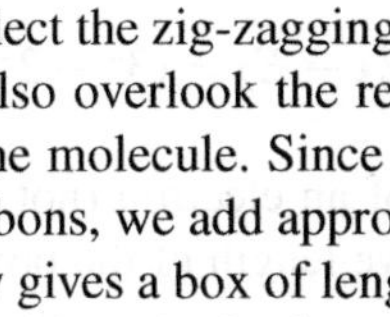

The four π-electrons are assumed to move freely over the four-carbon framework of single bonds. We neglect the zig-zagging of the C–C bonds and assume a one-dimensional box. We also overlook the reality that π-electrons actually have a node in the plane of the molecule. Since the electron wavefunction extends beyond the terminal carbons, we add approximately one-half bond length at each end. This conveniently gives a box of length equal to the number of carbon atoms times the C–C bond length, for butadiene, approximately 4×1.40 Å (recalling 1 Å=100 pm=10^{-10} m). Now, in the lowest energy state of butadiene, the 4 delocalized electrons will fill the two lowest FEM "molecular orbitals." The total π-electron density will be given (as shown in Fig. 3.4) by

$$\rho = 2\psi_1^2 + 2\psi_2^2. \tag{3.30}$$

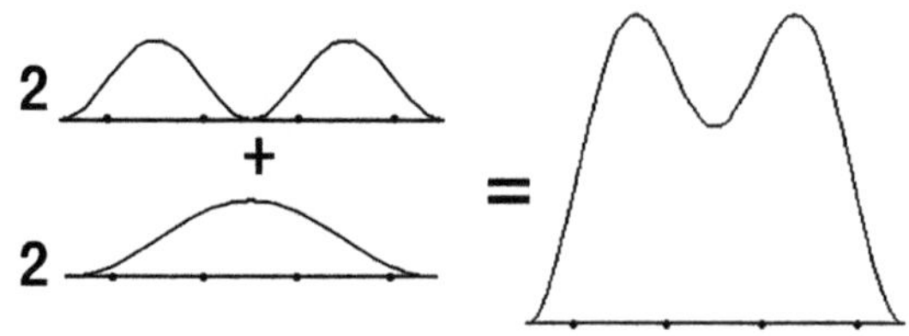

FIGURE 3.4 Pi-electron density in butadiene.

For a chemical interpretation of this picture, note that the π-electron density is concentrated between carbon atoms 1 and 2, and between 3 and 4. Thus the predominant structure of butadiene has double bonds between these two pairs of atoms. Each double bond consists of a π-bond, in addition to the underlying σ-bond. However, this is not the complete story, because we must also take account of the residual π-electron density between carbons 2 and 3 and beyond the terminal carbons. In the terminology of valence-bond theory, butadiene would be described as a *resonance hybrid* with the predominant structure $CH_2{=}CH{-}CH{=}CH_2$ but with a secondary contribution from $^{\circ}CH_2{-}CH{=}CH{-}CH_2{}^{\circ}$. The reality of the latter structure is suggested by the ability of butadiene to undergo 1,4-addition reactions.

The free-electron model can also be applied to the electronic spectrum of butadiene and other linear polyenes. The lowest unoccupied molecular orbital (LUMO) in butadiene corresponds to the $n = 3$ particle-in-a-box state. Neglecting electron-electron interaction, the longest-wavelength (lowest-energy) electronic transition should occur from $n = 2$, the highest occupied molecular orbital (HOMO), as shown below:

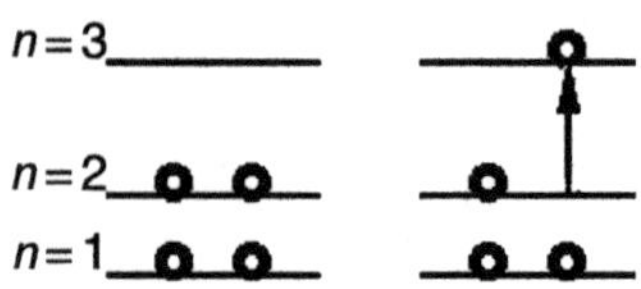

The energy difference is given by

$$\Delta E = E_3 - E_2 = (3^2 - 2^2)\frac{h^2}{8mL^2}. \tag{3.31}$$

Here m represents the mass of an electron (not a butadiene molecule!), 9.1×10^{-31} kg, and L is the effective length of the box, $4 \times 1.40 \times 10^{-10}$ m. By the Bohr frequency condition

$$\Delta E = h\nu = \frac{hc}{\lambda}. \tag{3.32}$$

The wavelength is predicted to be 207 nm. This compares well with the experimental maximum of the first electronic absorption band, $\lambda_{max} \approx 210$ nm, in the ultraviolet region.

We might therefore be emboldened to apply the model to predict absorption spectra in higher polyenes $CH_2{=}(CH{-}CH{=})_{n-1}CH_2$. For the molecule with $2n$ carbon atoms (n double bonds), the HOMO → LUMO transition corresponds to $n \rightarrow n+1$, thus

$$\frac{hc}{\lambda} \approx \left[(n+1)^2 - n^2\right] \frac{h^2}{8m(2nL_{CC})^2}. \tag{3.33}$$

A useful constant in this computation is the Compton wavelength $h/mc = 2.426 \times 10^{-12}$ m. For $n = 3$, hexatriene, the predicted wavelength is 332 nm, while experiment gives $\lambda_{max} \approx 250$ nm. For $n = 4$, octatetraene, FEM predicts 460 nm, while $\lambda_{max} \approx 300$ nm. Clearly the model has been pushed beyond it range of quantitative validity, although the trend of increasing absorption band wavelength with increasing n is correctly predicted. A compound should be colored if its absorption includes any part of the visible range 400–700 nm. Retinol (vitamin A), which contains a polyene chain with $n = 5$, has a pale yellow color. This is its structure:

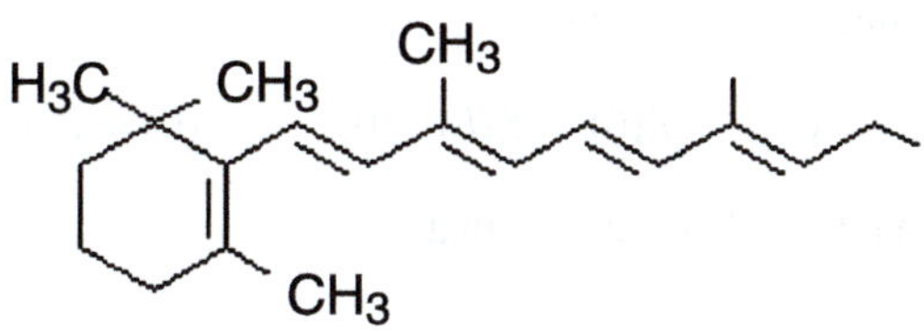

3.4 Particle in a three-dimensional box

A *real* box has three dimensions. Consider a particle which can move freely with in rectangular box of dimensions $a \times b \times c$ with impenetrable walls, as shown in Fig. 3.5.

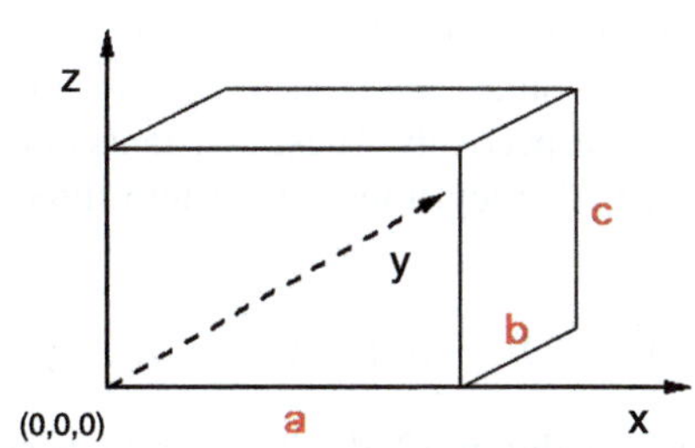

FIGURE 3.5 Coordinate system for particle in a box.

In terms of potential energy, we can write

$$V(x,y,z) = \begin{cases} 0 & \text{inside box} \\ \infty & \text{outside box.} \end{cases} \tag{3.34}$$

Again, the wavefunction vanishes everywhere outside the box. By the continuity requirement, the wavefunction must also vanish in the six surfaces of the box. Orienting the box so its edges are parallel to the cartesian axes, with one corner at (0, 0, 0), the following boundary conditions must be satisfied:

$$\psi(x,y,z)=0 \text{ when } x=0, x=a, y=0, y=b, z=0 \text{ or } z=c. \tag{3.35}$$

Inside the box, where the potential energy is everywhere zero, the Hamiltonian is simply the three-dimensional kinetic energy operator and the Schrödinger equation reads

$$-\frac{\hbar^2}{2m}\nabla^2\psi(x,y,z)=E\,\psi(x,y,z), \tag{3.36}$$

subject to the boundary conditions (3.35). This second-order partial differential equation is separable in cartesian coordinates, with a solution of the form

$$\psi(x,y,z)=X(x)\,Y(y)\,Z(z), \tag{3.37}$$

subject to the boundary conditions

$$X(0)=X(a)=0, \qquad Y(0)=Y(b)=0, \qquad Z(0)=Z(c)=0. \tag{3.38}$$

Substitute (3.37) into (3.36) and note that

$$\frac{\partial^2}{\partial x^2}X(x)\,Y(y)\,Z(z)=X''(x)\,Y(y)\,Z(z), \qquad \text{etc.} \tag{3.39}$$

Dividing through by (3.37), we obtain

$$\frac{X''(x)}{X(x)}+\frac{Y''(y)}{Y(y)}+\frac{Z''(z)}{Z(z)}+\frac{2mE}{\hbar^2}=0. \tag{3.40}$$

Each of the first three terms in (3.40) depends on one variable only, independent of the other two. This is possible only if each term separately equals a constant, say, $-\alpha^2$, $-\beta^2$ and $-\gamma^2$, respectively. These constants must be negative in order that $E>0$. Eq. (3.40) is thereby transformed into three ordinary differential equations

$$X''+\alpha^2X=0, \qquad Y''+\beta^2Y=0, \qquad Z''+\gamma^2Z=0, \tag{3.41}$$

subject to the boundary conditions (3.39). The constants are related by

$$\frac{2mE}{\hbar^2}=\alpha^2+\beta^2+\gamma^2. \tag{3.42}$$

Each of Eqs. (3.41), with its associated boundary conditions is equivalent to the one-dimensional problem (3.13) with boundary conditions (3.11). The normalized solutions $X(x)$, $Y(y)$, $Z(z)$ can therefore be written down in complete analogy with (3.24):

$$X_{n_1}(x) = \left(\frac{2}{a}\right)^{1/2} \sin\frac{n_1\pi x}{a}, \qquad n_1 = 1, 2\ldots$$
$$Y_{n_2}(y) = \left(\frac{2}{b}\right)^{1/2} \sin\frac{n_2\pi y}{b}, \qquad n_2 = 1, 2\ldots$$
$$Z_{n_3}(x) = \left(\frac{2}{c}\right)^{1/2} \sin\frac{n_3\pi z}{c}, \qquad n_3 = 1, 2\ldots \tag{3.43}$$

The constants in Eq. (3.41) are given by

$$\alpha = \frac{n_1\pi}{a}, \qquad \beta = \frac{n_2\pi}{b}, \qquad \gamma = \frac{n_3\pi}{c}, \tag{3.44}$$

so that the allowed energy levels are

$$E_{n_1,n_2,n_3} = \frac{h^2}{8m}\left(\frac{n_1^2}{a^2} + \frac{n_2^2}{b^2} + \frac{n_3^2}{c^2}\right), \qquad n_1, n_2, n_3 = 1, 2\ldots. \tag{3.45}$$

Three quantum numbers are required to specify the state of this three-dimensional system. The corresponding eigenfunctions are

$$\psi_{n_1,n_2,n_3}(x, y, z) = \left(\frac{8}{V}\right)^{1/2} \sin\frac{n_1\pi x}{a} \sin\frac{n_2\pi y}{b} \sin\frac{n_3\pi z}{c}, \tag{3.46}$$

where $V = abc$, the volume of the box. These eigenfunctions form an orthonormal set [cf. Eq. (3.28)] such that

$$\int_0^a \int_0^b \int_0^c \psi_{n_1',n_2',n_3'}(x, y, z)\, \psi_{n_1,n_2,n_3}(x, y, z)\, dx\, dy\, dz = \delta_{n_1',n_1}\, \delta_{n_2',n_2}\, \delta_{n_3',n_3}. \tag{3.47}$$

Note that two eigenfunctions will be orthogonal unless *all three* quantum numbers match. The three-dimensional matter waves represented by (3.46) are comparable with the modes of vibration of a solid block. The nodal surfaces are planes parallel to the sides, as shown in Fig. 3.6.

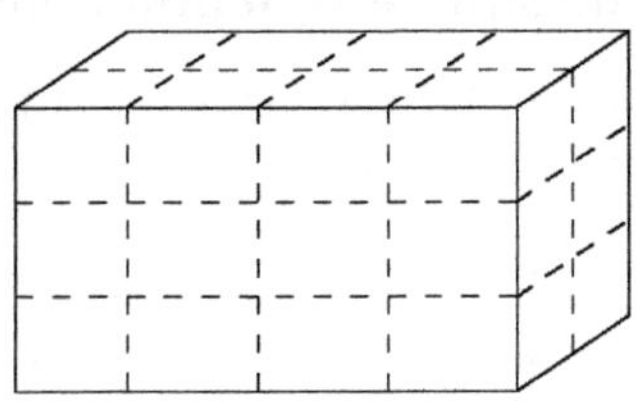

FIGURE 3.6 Nodal planes for particle in a box, for $n_1 = 4, n_2 = 2, n_3 = 3$.

When a box has the symmetry of a cube, with $a = b = c$, the energy formula (3.45) simplifies to

$$E_{n_1,n_2,n_3} = \frac{h^2}{8ma^2}(n_1^2 + n_2^2 + n_3^2), \quad n_1, n_2, n_3 = 1, 2 \ldots. \tag{3.48}$$

Quantum systems with symmetry generally exhibit *degeneracy* in their energy levels. This means that there can exist distinct eigenfunctions which share the same eigenvalue. An eigenvalue which corresponds to a unique eigenfunction is termed *nondegenerate* while one which belongs to n different eigenfunctions is termed *n-fold degenerate*. As an example, we enumerate the first few levels for a cubic box, with E_{n_1,n_2,n_3} expressed in units of $h^2/8ma^2$:

$E_{1,1,1} = 3$ (nondegenerate),

$E_{1,1,2} = E_{1,2,1} = E_{2,1,1} = 6$ (3-fold degenerate),

$E_{1,2,2} = E_{2,1,2} = E_{2,2,1} = 9$ (3-fold degenerate),

$E_{1,1,3} = E_{1,3,1} = E_{3,1,1} = 11$ (3-fold degenerate),

$E_{2,2,2} = 12$ (nondegenerate),

$E_{1,2,3} = E_{1,3,2} = E_{2,1,3} = E_{2,3,1} = E_{3,1,2} = E_{3,2,1} = 14$ (6-fold degenerate).

The particle in a box is applied in statistical thermodynamics to model the perfect gas. Each molecule is assumed to move freely within the box *without* interacting with the other molecules. The total energy of N molecules, in any distribution among the energy levels (3.48), is proportional to $1/a^2$, thus

$$E = \text{const}\, V^{-2/3}. \tag{3.49}$$

From the differential of work $dw = -p\,dV$, we can identify

$$p = -\frac{dE}{dV} = \frac{2}{3}\frac{E}{V}. \tag{3.50}$$

But the energy of a perfect monatomic gas is known to equal $\frac{3}{2}nRT$, which leads to the perfect gas law

$$pV = nRT. \tag{3.51}$$

Supplement 3A. Finite square-well potential

A more challenging variant of the particle-in-a-box problem is the case when the potential well is finite. Consider a finite square-well potential

$$V(x) = \begin{cases} -V_0 & -a/2 \le x \le a/2 \\ 0 & x < -a/2 \text{ and } x > a/2, \end{cases} \tag{3.52}$$

shown in Fig. 3.7. We consider first the negative energy case $-V_0 < E < 0$.

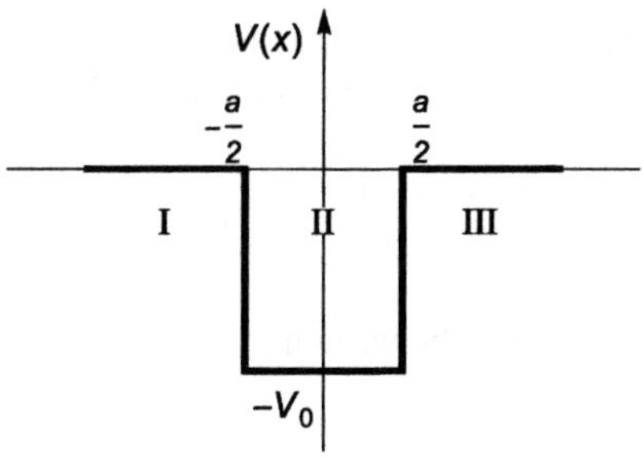

FIGURE 3.7 Finite potential well of depth V_0.

There are three regions to consider: I ($x < -\frac{a}{2}$), II ($-\frac{a}{2} \leq x \leq \frac{a}{2}$) and III ($\frac{a}{2} < x$). Since the potential is an even function of x, $V(-x) = V(x)$, the wavefunction must be either an even or an odd function of x, designated as symmetric or antisymmetric. For the symmetric case

$$\psi_{\mathrm{I}} = Ae^{\kappa x}, \qquad \psi_{\mathrm{II}} = B\cos kx, \qquad \psi_{\mathrm{III}} = Ae^{-\kappa x}, \tag{3.53}$$

while for odd parity,

$$\psi_{\mathrm{I}} = Ae^{\kappa x}, \qquad \psi_{\mathrm{II}} = B\sin kx, \qquad \psi_{\mathrm{III}} = -Ae^{-\kappa x}, \tag{3.54}$$

where

$$\kappa = \sqrt{-\frac{2mE}{\hbar^2}}, \qquad k = \sqrt{\frac{2m(E+V_0)}{\hbar^2}}. \tag{3.55}$$

We have taken into account the limits $\psi_{\mathrm{I}} \to 0$ as $x \to -\infty$ and $\psi_{\mathrm{III}} \to 0$ as $x \to \infty$. The energy eigenvalues are determined by the condition that the wavefunction and its first derivative must be continuous at the discontinuities of the potential. It is sufficient to consider just the matching conditions at $x = -a/2$, namely $\psi_{\mathrm{I}}(-a/2) = \psi_{\mathrm{II}}(-a/2)$ and $\psi_{\mathrm{I}}'(-a/2) = \psi_{\mathrm{II}}'(-a/2)$. For the even parity solutions, we find

$$Ae^{-\kappa a/2} = B\cos(ka/2) \quad \text{and} \quad -\frac{\kappa}{2}Ae^{-\kappa a/2} = -\frac{k}{2}B\sin(ka/2). \tag{3.56}$$

Dividing the second equation by the first, we obtain

$$\kappa = k\tan(ka/2) \quad \text{(symmetric)}. \tag{3.57}$$

Analogously, for antisymmetric wavefunctions

$$\kappa = -k\cot(ka/2) \quad \text{(antisymmetric)}. \tag{3.58}$$

By virtue of (3.55), these constitute transcendental equations for the energies of a particle in a finite potential well. Let us define the dimensionless variables

$$\xi = \frac{ka}{2} = \sqrt{\frac{ma^2(-E)}{2\hbar^2}} \quad \text{and} \quad \xi_0 = \sqrt{\frac{ma^2 V_0}{2\hbar^2}}. \tag{3.59}$$

Eqs. (3.57) and (3.58) can then be written

$$\sqrt{\xi_0^2 - \xi^2} = \xi \tan \xi \quad \text{and} \quad \sqrt{\xi_0^2 - \xi^2} = -\xi \cot \xi. \tag{3.60}$$

Graphical solutions for the energies can be obtained by plots of the functions $f(\xi) = \sqrt{\xi_0^2 - \xi^2}$ and either $\xi \tan \xi$ or $-\xi \cot \xi$, as shown in Fig. 3.8. The energies are determined by the values of ξ at the points of intersection. For example,

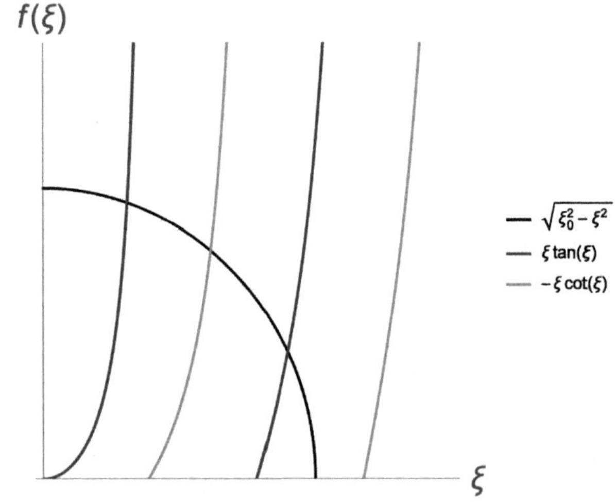

FIGURE 3.8 Graphical determination of energies for finite square well.

with $\xi_0 = 4$, we obtain three solutions $\xi = 1.252$, 2.475 and 3.595. The corresponding energies are then determined using (3.59).

For energies $E > 0$, the solutions of the Schrödinger equation in regions I, II and III have the form

$$\psi(x) = a\, e^{ikx} + b\, e^{-ikx}. \tag{3.61}$$

We will not work out the detailed values of a and b but note simply that solutions exist for all values of $k \geq 0$, corresponding to an energy continuum for $E \geq 0$, similar to the case of a free particle. The finite square well problem thus has a feature in common with atomic and molecular systems: a finite set of negative-energy bound states plus a continuum for $E \geq 0$, corresponding to ionization or dissociation of the system. (A fine point: the hydrogen atom and most positive ions have an infinite number of bound states.)

Chapter 3. Exercises

3.1. Which of the following is *not* a solution $y(x)$ of the differential equation $y''(x) + k^2 y(x) = 0$ (k = constant): (i) $\sin(kx)$ (ii) $\cos(kx)$ (iii) e^{ikx} (iv) e^{-kx} (v) $\sin(kx + \alpha)$ (α = constant).

3.2. For a particle in a 1-dimensional box, calculate the probability that the particle will be found in the middle third of the box: $L/3 \leq x \leq 2L/3$. From the general formula for arbitrary n, find the limiting value as $n \to \infty$.

3.3. Redo the solution of the 1-dimensional particle-in-a-box, when the particle is confined to the region $-a/2 \leq x \leq a/2$.

3.4. Predict the wavelength (in nm) of the lowest-energy electronic transition in the following polymethine ion:

$$(CH_3)_2N^+ = CH - CH = CH - CH = CH - N(CH_3)_2$$

Assume that all the C–C and C–N bond lengths equal 1.40 Å. Note that N^+ and N contribute 1 and 2 π-electrons, respectively.

3.5. In this calculation you will determine the order of magnitude of nuclear energies. Assume that a nucleus can be represented as a cubic box of side 10^{-14} m. The particles in this box are the nucleons (protons and neutrons). Calculate the lowest allowed energy of a nucleon. Express your result in MeV (1 MeV = 10^6 eV = 1.602×10^{-13} J).

3.5. Consider the hypothetical reaction of two "cube-atoms" to form a "molybox":

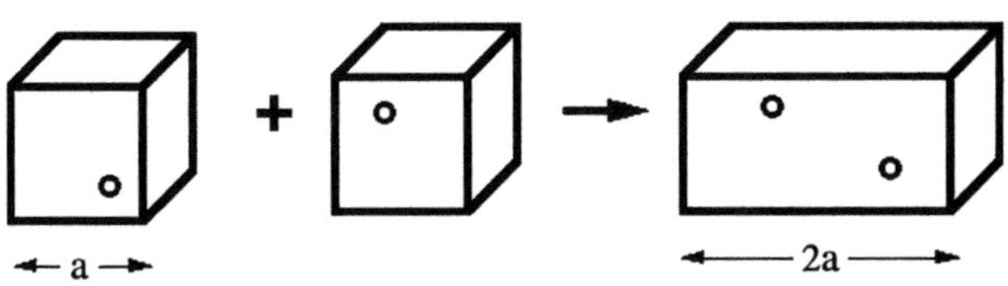

Each cube-atom contains one electron. The interaction between electrons can be neglected. Determine the energy change in the above reaction.

3.7. Consider the two-dimensional particle-in-a-box—a particle free to move on a square plate of side a. Solve the Schrödinger equation to obtain the eigenvalues and eigenfunctions. You should be able to do this entirely by analogy with solutions we have already obtained. Discuss the degeneracies of the lowest few energy levels.

3.8. As a variant on the free-electron model applied to benzene, assume that the six π electrons are delocalized within a square plate of side a. Calculate the value of a that would account for the 268 nm ultraviolet absorption in benzene.

Chapter 4

Principles of quantum mechanics

In this Chapter we will continue to develop the mathematical formalism of quantum mechanics, using heuristic arguments as necessary. This will lead to a system of postulates which will be the basis of our subsequent applications of quantum mechanics.

4.1 Hermitian operators

An important property of operators is suggested by considering the Hamiltonian for the particle in a box:

$$\hat{H} = -\frac{\hbar^2}{2m}\frac{d^2}{dx^2}. \tag{4.1}$$

Let $f(x)$ and $g(x)$ be arbitrary functions which obey the same boundary values as the eigenfunctions of $\hat{H}$, namely that they vanish at $x = 0$ and $x = a$. Consider the integral

$$\int_0^a f(x)\,\hat{H}\,g(x)\,dx = -\frac{\hbar^2}{2m}\int_0^a f(x)\,g''(x)\,dx. \tag{4.2}$$

Now, using integration by parts,

$$\int_0^a f(x)\,g''(x)\,dx = -\int_0^a f'(x)\,g'(x)\,dx + \Big[f(x)\,g'(x)\Big]_0^a. \tag{4.3}$$

The boundary terms vanish by the assumed conditions on f and g. A second integration by parts transforms (4.3) to

$$+\int_0^a f''(x)\,g(x)\,dx - \Big[f'(x)\,g(x)\Big]_0^a.$$

It follows therefore that

$$\int_0^a f(x)\,\hat{H}\,g(x)\,dx = \int_0^a g(x)\,\hat{H}\,f(x)\,dx. \tag{4.4}$$

Introduction to Quantum Mechanics. https://doi.org/10.1016/B978-0-12-822310-9.00012-4

An obvious generalization for complex functions will read

$$\int_0^a f^*(x)\,\hat{H}\,g(x)\,dx = \left(\int_0^a g^*(x)\,\hat{H}\,f(x)\,dx\right)^*. \tag{4.5}$$

In mathematical terminology, an operator $\hat{A}$ for which

$$\int f^*\,\hat{A}\,g\,d\tau = \left(\int g^*\,\hat{A}\,f\,d\tau\right)^* \tag{4.6}$$

for all functions f and g which obey specified boundary conditions is denoted as *hermitian* or *self-adjoint*. Evidently, the Hamiltonian is a hermitian operator. It is postulated that *all* quantum-mechanical operators that represent dynamical variables are hermitian.

4.2 Eigenvalues and eigenfunctions

The sets of energies and wavefunctions obtained by solving any quantum-mechanical problem can be summarized symbolically as solutions of the eigenvalue equation

$$\hat{H}\,\psi_n = E_n\,\psi_n. \tag{4.7}$$

For another value of the quantum number, we can write

$$\hat{H}\,\psi_m = E_m\,\psi_m. \tag{4.8}$$

Let us multiply (4.7) by ψ_m^* and the complex conjugate of (4.8) by ψ_n. Then we subtract the two expressions and integrate over $d\tau$. The result is

$$\int \psi_m^*\,\hat{H}\,\psi_n\,d\tau - \left(\int \psi_n^*\,\hat{H}\,\psi_m\,d\tau\right)^* = (E_n - E_m^*)\int \psi_m^*\,\psi_n\,d\tau. \tag{4.9}$$

But by the hermitian property (4.6), the left-hand side of (4.9) equals zero. Thus

$$(E_n - E_m^*)\int \psi_m^*\,\psi_n\,d\tau = 0. \tag{4.10}$$

Consider first the case $m = n$. The second factor in (4.10) then becomes the normalization integral $\int \psi_n^*\,\psi_n\,d\tau$, which equals 1 (or at least a nonzero constant). Therefore the first factor in (4.10) must equal zero, meaning that

$$E_n^* = E_n. \tag{4.11}$$

This implies that the energy eigenvalues must be real numbers, which is quite reasonable from a physical point of view since eigenvalues represent possible results of measurement. Consider next the case when $E_m \neq E_n$. Then it is the second factor in (4.10) that must vanish and

$$\int \psi_m^* \psi_n \, d\tau = 0 \qquad \text{when} \quad E_m \neq E_n. \tag{4.12}$$

Thus eigenfunctions belonging to different eigenvalues are orthogonal. In the case that ψ_m and ψ_n are degenerate eigenfunctions, so $m \neq n$ but $E_m = E_n$, the above proof of orthogonality does not apply. But it is always possible to construct degenerate functions that are mutually orthogonal. A general result is therefore the orthonormalization condition

$$\int \psi_m^* \psi_n \, d\tau = \delta_{mn}. \tag{4.13}$$

It is easy to prove that a linear combination of degenerate eigenfunctions is itself an eigenfunction of the same energy. Let

$$\hat{H} \psi_{nk} = E_n \psi_{nk}, \qquad k = 1, 2 \ldots d, \tag{4.14}$$

where the ψ_{nk} represent a d-fold degenerate set of eigenfunctions with the same eigenvalue E_n. Consider now the linear combination

$$\psi = c_1 \psi_{n,1} + c_2 \psi_{n,2} + \cdots + c_d \psi_{n,d}. \tag{4.15}$$

Operating on ψ with the Hamiltonian and using (4.14), we find

$$\begin{aligned} \hat{H}\psi = c_1 \hat{H} \psi_{n,1} + c_2 \hat{H} \psi_{n,2} + \cdots = \\ E_n(c_1\psi_{n,1} + c_2\psi_{n,2} + \ldots) = E_n\psi, \end{aligned} \tag{4.16}$$

which shows that the linear combination ψ is also an eigenfunction of the same energy. There is evidently a limitless number of possible eigenfunctions for a degenerate eigenvalue. However, only d of these will be linearly independent.

4.3 Expectation values

One of the extraordinary features of quantum mechanics is the possibility for superposition of states. The state of a quantum system can sometimes exist as a linear combination of other states, such that

$$\psi = c_1 \psi_1 + c_2 \psi_2. \tag{4.17}$$

For example, the electronic ground state of the butadiene can (to an approximation) be considered a superposition of the valence-bond structures CH_2=CH-CH=CH_2 and °CH_2-CH=CH-CH_2°. Assuming that all three functions in (4.17) are normalized and that ψ_1 and ψ_2 are orthogonal, we find

$$\int \psi^* \psi \, d\tau = \int (c_1^* \psi_1^* + c_2^* \psi_2^*)(c_1 \psi_1 + c_2 \psi_2) \, d\tau =$$

$$c_1^* c_1 \int \psi_1^* \psi_1 \, d\tau + c_2^* c_2 \int \psi_2^* \psi_2 \, d\tau + c_1^* c_2 \int \psi_2^* \psi_1 \, d\tau + c_2^* c_1 \int \psi_2^* \psi_1 \, d\tau = \\ c_1^* c_1 \times 1 + c_2^* c_2 \times 1 + c_1^* c_2 \times 0 + c_2^* c_1 \times 0 = |c_1|^2 + |c_2|^2 = 1. \tag{4.18}$$

We can interpret $|c_1|^2$ and $|c_2|^2$ as the probabilities that a system in a state described by ψ can have the attributes of the states ψ_1 and ψ_2, respectively. (In the butadiene example above, $|c_1|^2$ and $|c_2|^2$ might approximate the fraction of molecules which undergo 1,3-addition and 1,4-addition, respectively, in the reaction with Cl_2.)

Suppose ψ_1 and ψ_2 represent eigenstates of an observable A, satisfying the respective eigenvalue equations

$$\hat{A}\psi_1 = a_1 \psi_1 \qquad \text{and} \qquad \hat{A}\psi_2 = a_2 \psi_2. \tag{4.19}$$

Then a large number of measurements of the variable A in the state ψ will register the value a_1 with a probability $|c_1|^2$ and the value a_2 with a probability $|c_2|^2$. The average value or *expectation value* of A will be given by

$$\langle A \rangle = |c_1|^2 a_1 + |c_2|^2 a_2. \tag{4.20}$$

This can be obtained directly from ψ by the "sandwich construction"

$$\langle A \rangle = \int \psi^* \hat{A} \psi \, d\tau, \tag{4.21}$$

or, if ψ is not normalized,

$$\langle A \rangle = \frac{\int \psi^* \hat{A} \psi \, d\tau}{\int \psi^* \psi \, d\tau}. \tag{4.22}$$

Note that the expectation value need not itself be a possible result of a single measurement (like the centroid of a donut, which is located in the hole!). When the operator $\hat{A}$ is a simple function, not containing differential operators or the like, then (4.21) reduces to the classical formula for an average value:

$$\langle A \rangle = \int A \, \rho \, d\tau. \tag{4.23}$$

4.4 More on operators

An operator represents a prescription for turning one function into another—in symbols, $\hat{A}\psi = \phi$. From a physical point of view, the action of an operator on a wavefunction can be pictured as the process of measuring the observable A on the state ψ. The transformed wavefunction ϕ then represents the state of the system *after* the measurement is performed. In general ϕ is different from ψ, consistent with the fact that the process of measurement on a quantum system can produce an irreducible perturbation of its state. Only in the special case that ψ is an eigenstate of A, does a measurement preserve the original state. The function ϕ is then equal to an eigenvalue a times ψ.

The product of two operators, say $\hat{A}\,\hat{B}$, represents the successive action of the operators, reading from *right to left*—i.e., first $\hat{B}$ then $\hat{A}$. In general, the action of two operators in the reversed order, say $\hat{B}\,\hat{A}$, gives a different result, which can be written $\hat{A}\,\hat{B} \neq \hat{B}\,\hat{A}$. We say that the operators do not *commute*. This can be attributed to the perturbing effect one measurement on a quantum system can have on subsequent measurements. Here's an example of non-commuting operators from everyday life: in our usual routine each morning, we shower and we get dressed. But the result of carrying out these operations in reversed order will be dramatically different!

The *commutator* of two operators is defined by

$$[\hat{A},\,\hat{B}] \equiv \hat{A}\,\hat{B} - \hat{B}\,\hat{A}. \tag{4.24}$$

When $[\hat{A},\,\hat{B}] = 0$, the two operators are said to *commute*. This means their combined effect will be the same whatever order they are applied (like brushing your teeth and showering).

Werner Heisenberg

The uncertainty principle for simultaneous measurement of two observables A and B is determined by their commutator. The uncertainty Δa in the observable A is defined in terms of the mean square deviation from the average:

$$(\Delta a)^2 = \langle(\hat{A} - \langle A\rangle)^2\rangle = \langle A^2\rangle - \langle A\rangle^2. \tag{4.25}$$

This corresponds to the *standard deviation* σ in statistics. The following inequality can be proven for the product of two uncertainties:

$$\Delta a\,\Delta b \geq \frac{1}{2}|\langle[\hat{A},\,\hat{B}]\rangle|. \tag{4.26}$$

The best known application of (4.26) is to the position and momentum operators, say $\hat{x}$ and $\hat{p}_x$. Their commutator is given by

$$[\hat{x},\ \hat{p}_x] = i\hbar, \tag{4.27}$$

so that

$$\Delta x\,\Delta p \geq \hbar/2, \tag{4.28}$$

which is known as the *Heisenberg uncertainty principle*. This fundamental consequence of quantum theory implies that the position and momentum of a particle cannot be determined with arbitrary precision—the more accurately one is known, the more uncertain is the other. For example, if the momentum is known exactly, as in a momentum eigenstate, then the position is completely undetermined.

The uncertainty relation for energy and time can be deduced by using the "energy operator" $\hat{E} = i\hbar\partial/\partial t$ in Eq. (4.26). The result is

$$\Delta E\,\Delta t \geq \hbar/2. \tag{4.29}$$

This aspect of the uncertainty principle, first proposed by Bohr, refers to a measurement of the energy carried out during a time interval Δt. It implies that a short-lived excited state of an atom or molecule, for which Δt is small, will be associated with a relatively large uncertainty in energy ΔE. This is one factor contributing to the broadening of spectral lines.

If two operators commute, there is no restriction on the accuracy of their simultaneous measurement. For example, the x and y coordinates of a particle can be known at the same time. An important theorem states that two commuting observables can have simultaneous eigenfunctions. To prove this, write the eigenvalue equation for an operator $\hat{A}$

$$\hat{A}\,\psi_n = a_n\,\psi_n, \tag{4.30}$$

then operate with $\hat{B}$ and use the commutativity of $\hat{A}$ and $\hat{B}$ to obtain

$$\hat{B}\,\hat{A}\,\psi_n = \hat{A}\,\hat{B}\,\psi_n = a_n\,\hat{B}\,\psi_n. \tag{4.31}$$

This shows that $\hat{B}\,\psi_n$ is also an eigenfunction of $\hat{A}$ with the same eigenvalue a_n. This implies that

$$\hat{B}\,\psi_n = \text{const}\,\psi_n \equiv b_n\,\psi_n, \tag{4.32}$$

so that ψ_n is a simultaneous eigenfunction of $\hat{A}$ and $\hat{B}$ with eigenvalues a_n and b_n, respectively. The derivation becomes slightly more complicated in the case of degenerate eigenfunctions, but the same conclusion follows.

After the Hamiltonian, the operators for angular momenta are probably the most important in quantum mechanics. The definition of angular momentum in classical mechanics is $\mathbf{L} = \mathbf{r} \times \mathbf{p}$. In terms of its cartesian components,

$$L_x = yp_z - zp_y, \qquad L_y = zp_x - xp_z, \qquad L_z = xp_y - yp_x. \tag{4.33}$$

In future, we will write such sets of equation as "$L_x = yp_z - zp_y$, *et cyc*," meaning that we add to one explicitly stated relation, the versions formed by successive cyclic permutation $x \to y \to z \to x$. The general prescription for turning a classical dynamical variable into a quantum-mechanical operator was developed in Chap. 2. The key relations were the momentum components

$$\hat{p}_x = -i\hbar\frac{\partial}{\partial x}, \quad \hat{p}_y = -i\hbar\frac{\partial}{\partial y}, \quad \hat{p}_z = -i\hbar\frac{\partial}{\partial z}, \tag{4.34}$$

with the coordinates x, y, z simply carried over into multiplicative operators. Applying (4.34) to (4.33), we construct the three angular momentum operators

$$\hat{L}_x = -i\hbar\left(y\frac{\partial}{\partial z} - z\frac{\partial}{\partial y}\right) \quad \textit{et cyc}, \tag{4.35}$$

while the total angular momentum is given by

$$\hat{L}^2 = \hat{L}_x^2 + \hat{L}_y^2 + \hat{L}_z^2. \tag{4.36}$$

The angular momentum operators obey the following commutation relations:

$$[\hat{L}_x, \hat{L}_y] = i\hbar\hat{L}_z \quad \textit{et cyc}, \tag{4.37}$$

but

$$[\hat{L}^2, \hat{L}_z] = 0 \tag{4.38}$$

and analogously for $\hat{L}_x$ and $\hat{L}_y$. This is consistent with the existence of simultaneous eigenfunctions of $\hat{L}^2$ and any one component, conventionally designated $\hat{L}_z$. But then these states *cannot* be eigenfunctions of either $\hat{L}_x$ or $\hat{L}_y$.

We will eventually drop the "hat" notation for operators and write simply A in place of $\hat{A}$.

4.5 Postulates of quantum mechanics

Our development of quantum mechanics is now sufficiently complete that we can reduce the theory to a set of postulates.

Postulate 1. The state of a quantum-mechanical system is completely specified by a wavefunction Ψ that depends on the coordinates and time. The square of this function $\Psi^*\Psi$ gives the probability density for finding the system with a specified set of coordinate values.

The wavefunction must fulfill certain mathematical requirements because of its physical interpretation. It must be single-valued, finite and continuous. It must also satisfy a normalization condition

$$\int \Psi^* \Psi \, d\tau = 1. \tag{4.39}$$

Postulate 2. Every observable in quantum mechanics is represented by a linear, hermitian operator.

The hermitian property was defined in Eq. (4.6). A linear operator is one which satisfies the identity

$$\hat{A}(c_1 \psi_1 + c_2 \psi_2) = c_1 \hat{A} \psi_1 + c_2 \hat{A} \psi_2, \tag{4.40}$$

which is necessary for explaining superpositions of quantum states. The form of an operator which has an analog in classical mechanics is derived by the prescriptions

$$\hat{\mathbf{r}} = \mathbf{r}, \qquad \hat{\mathbf{p}} = -i\hbar\nabla, \tag{4.41}$$

which we have previously expressed in terms of cartesian components (cf. Eq. (4.34)).

Postulate 3. In any measurement of an observable A, associated with an operator $\hat{A}$, the only possible results are the eigenvalues a_n, which satisfy an eigenvalue equation

$$\hat{A} \psi_n = a_n \, \psi_n. \tag{4.42}$$

This postulate captures the essence of quantum mechanics—the quantization of measured dynamical variables. A continuum of eigenvalues is not excluded, however, as in the case of an unbound particle.

Every measurement of A invariably gives one of its eigenvalues. For an arbitrary state (not an eigenstate of A), these measurements will be individually unpredictable—they can introduce "noise" into a system—but they follow a definite statistical law, according to the fourth postulate:

Postulate 4. For a system in a state described by a normalized wavefunction Ψ, the average or expectation value of the observable corresponding to A is given by

$$\langle A \rangle = \int \Psi^* \, \hat{A} \, \Psi \, d\tau. \tag{4.43}$$

Finally, we state

Postulate 5. The wavefunction of a system evolves in time in accordance with the time-dependent Schrödinger equation

$$i\hbar\frac{\partial\Psi}{\partial t}=\hat{H}\,\Psi. \tag{4.44}$$

For time-independent problems this reduces to the time-independent Schrödinger equation

$$\hat{H}\,\psi=E\,\psi, \tag{4.45}$$

which is equivalent to the eigenvalue equation for the Hamiltonian operator.

4.6 Dirac bra-ket notation

P. A. M. Dirac

The term *orthogonal* has been used to refer to both perpendicular vectors and to functions whose product integrates to zero. This actually connotes a deep connection between vectors and functions. Consider two orthogonal vectors **a** and **b**. Then, in terms of their x, y, z components, labeled by 1, 2, 3, respectively, the scalar product can be written

$$\mathbf{a}\cdot\mathbf{b}=a_1b_1+a_2b_2+a_3b_3=0. \tag{4.46}$$

Suppose now that we consider an analogous relationship involving vectors in n-dimensional space (which you need not visualize!). We could then write

$$a\cdot b=\sum_{k=1}^{n}a_kb_k=0. \tag{4.47}$$

Finally let the dimension of the space become nondenumerably infinite, turning into a continuum. The sum (4.47) would then be replaced by an integral such as

$$\int a(x)\,b(x)\,dx = 0. \tag{4.48}$$

But this is just the relation for orthogonal functions. A function can therefore be regarded as an abstract vector in a higher-dimensional continuum, known as *Hilbert space*. This is true for eigenfunctions as well. Dirac denoted the vector in Hilbert space corresponding to the eigenfunction ψ_n by the symbol $|n\rangle$. Correspondingly, the complex conjugate ψ_m^* is denoted by $\langle m|$. The integral over the product of the two functions is then analogous to a scalar product of the abstract vectors, written

$$\int \psi_m^*\,\psi_n\,d\tau = \langle m|\cdot|n\rangle \equiv \langle m|n\rangle. \tag{4.49}$$

The last quantity is known as a *bracket*, which led Dirac to designate the vectors $\langle m|$ and $|n\rangle$ as a "bra" and a "ket," respectively. The orthonormality conditions (4.13) can be written

$$\langle m|n\rangle = \delta_{mn}. \tag{4.50}$$

A matrix element, an integral of a "sandwich" containing an operator $\hat{A}$, can be written very compactly in the form

$$\int \psi_m^*\,\hat{A}\,\psi_n\,d\tau = \langle m|A|n\rangle. \tag{4.51}$$

The hermitian condition on $\hat{A}$ [cf. Eq. (4.6)] is therefore expressed as

$$\langle m|A|n\rangle = \langle n|A|m\rangle^*. \tag{4.52}$$

For any operator $\hat{A}$ the *adjoint* operator $\hat{A}^\dagger$ is defined by

$$\langle m|A^\dagger|n\rangle = \langle n|A|m\rangle^*. \tag{4.53}$$

A hermitian operator is thus *self-adjoint* since $\hat{A}^\dagger = \hat{A}$.

In matrix terminology, a ket $|n\rangle$ is analogous to a *column vector*, a bra $\langle m|$ to a *row vector* and an operator A to a square matrix. Thus the expressions $\langle m|n\rangle$ and $\langle m|A|n\rangle$ represent compatible combinations for matrix products. A bra can be considered to be the adjoint of a ket

$$\langle n| = |n\rangle^\dagger, \tag{4.54}$$

with the elements of a column vector being rearrayed as a row of corresponding complex conjugates. Note that

$$\langle n|A^\dagger = (A|n\rangle)^\dagger, \tag{4.55}$$

so that

$$\langle n|A^\dagger A|n\rangle = \int \psi_n^*\hat{A}^\dagger\hat{A}\psi_n\,d\tau = \int |\hat{A}\psi_n|^2\,d\tau \geq 0. \tag{4.56}$$

4.7 The variational principle

Except for a small number of intensively-studied examples, the Schrödinger equation for most problems of chemical interest *cannot* be solved exactly. The variational principle provides a guide for constructing the best possible approximate solutions of a specified functional form. Suppose that we seek an approximate solution for the ground state of a quantum system described by a Hamiltonian $\hat{H}$. We presume that the Schrödinger equation

$$\hat{H}\,\psi_0 = E_0\,\psi_0 \tag{4.57}$$

is too difficult to solve exactly. Suppose, however, that we have a function $\tilde{\psi}$ which we think is an approximation to the true ground-state wavefunction. According to the variational principle (or variational theorem), the following formula provides an *upper bound* to the exact ground-state energy E_0:

$$\tilde{E} \equiv \frac{\int \tilde{\psi}^*\,\hat{H}\,\tilde{\psi}\,d\tau}{\int \tilde{\psi}^*\,\tilde{\psi}\,d\tau} \geq E_0. \tag{4.58}$$

Note that this ratio of integrals has the same form as the expectation value $\langle H \rangle$ defined by (4.22). The better the approximation $\tilde{\psi}$, the lower will be the computed energy $\tilde{E}$, though it will still be greater than the exact value. To prove Eq. (4.58), we suppose that the approximate function can, in concept, be represented as a superposition of the actual eigenstates of the Hamiltonian, analogous to (4.17),

$$\tilde{\psi} = c_0\psi_0 + c_1\psi_1 + \ldots. \tag{4.59}$$

This means that $\tilde{\psi}$, the approximate ground state, might be close to the actual ground state ψ_0 but is "contaminated" by contributions from excited states $\psi_1, \psi_2, \ldots$. Of course, none of the states or coefficients on the right-hand side is actually known, otherwise there would no need to worry about approximate computations. By Eq. (4.20), the expectation value of the Hamiltonian in the state (4.59) is given by

$$\tilde{E} = |c_0|^2 E_0 + |c_1|^2 E_1 + \cdots. \tag{4.60}$$

Since all the excited states have *higher* energy than the ground state, $E_1, E_2 \cdots \geq E_0$, we find

$$\tilde{E} \geq \left(|c_0|^2 + |c_1|^2 + \cdots\right) E_0 = E_0, \tag{4.61}$$

assuming $\tilde{\psi}$ has been normalized. Thus $\tilde{E}$ must be greater than the true ground-state energy E_0, as implied by (4.58).

As a very simple, although artificial, illustration of the variational principle, consider the ground state of the particle in a box. Suppose we had never studied trigonometry and knew nothing about sines or cosines. Then a reasonable approximation to the ground state might be an inverted parabola such as the normalized function

$$\tilde{\psi}(x) = \left(\frac{30}{a^5}\right)^{1/2} x\,(a-x). \tag{4.62}$$

Fig. 4.1 shows this function along with the exact ground-state eigenfunction

$$\psi_1(x) = \left(\frac{2}{a}\right)^{1/2} \sin\left(\frac{\pi x}{a}\right). \tag{4.63}$$

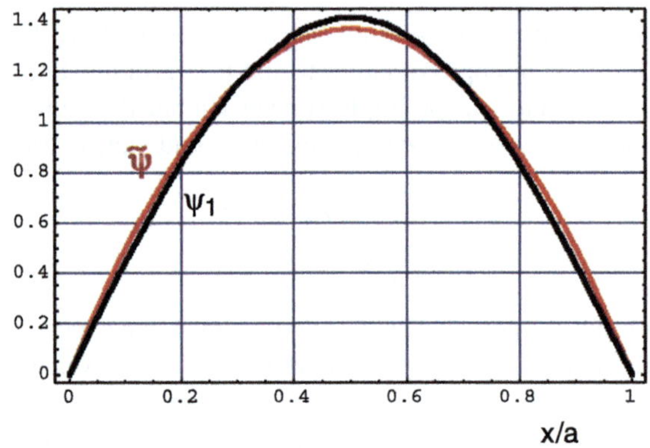

FIGURE 4.1 Variational approximation for particle in a box.

A variational calculation gives

$$\tilde{E} = \int_0^a \tilde{\psi}(x)\left(-\frac{\hbar^2}{2m}\right)\tilde{\psi}''(x)\,dx =$$

$$\frac{5}{4\pi^2}\frac{h^2}{ma^2} = \frac{10}{\pi^2}E_1 = 1.01321E_1, \tag{4.64}$$

in terms of the exact ground state energy $E_1 = h^2/8ma^2$. In accord with the variational theorem, $\tilde{E} > E_1$. The computation is in error by about 1%.

4.8 Spectroscopic transitions

Interactions of atoms or molecules with electromagnetic radiation can induce transitions between quantum states. To deal with transitions, we need to consider the full *time-dependent* Schrödinger equation (TDSE)

$$i\hbar\frac{d\Psi}{\partial t} = \hat{H}\Psi. \tag{4.65}$$

We consider a Hamiltonian of the form

$$\hat{H}(t) = \hat{H}_0 + \hat{V}(t), \tag{4.66}$$

where $\hat{H}_0$, the *unperturbed Hamiltonian*, is independent of time and $\hat{V}(t)$, a *perturbation*, is much smaller than $\hat{H}_0$ but contains dependence on time. The unperturbed Hamiltonian possesses a set of eigenstates satisfying

$$\hat{H}_0\psi_n = E_n\psi_n. \tag{4.67}$$

The perturbation $\hat{V}(t)$ can induce *transitions* among eigenstates ψ_n and ψ_m of $\hat{H}_0$ if the integral of $\hat{V}(t)$ connecting the two states, the *matrix element*

$$V_{mn}(t) = \int \psi_m^* \, \hat{V}(t) \, \psi_n \, d\tau \tag{4.68}$$

is *not* equal to zero. This is consistent with our discussion of operators in Sect. 2.4. The action of $\hat{V}(t)$ on ψ_n transforms the system into a state $\hat{V}(t)\psi_n$. Then the overlap (4.68) of this function with ψ_m gives the probability amplitude that ψ_m is the resulting state. We will show in Supplement 4A that the rate of transitions $m \leftarrow n$ is proportional to $|V_{mn}(t)|^2$.

The interaction of an atom or molecule with an radiation field is, to a first approximation,

$$V(t) = -\boldsymbol{\mu} \cdot \mathbf{E}(t), \tag{4.69}$$

where $\boldsymbol{\mu}$ is the instantaneous electric dipole moment of the atomic or molecular charge distribution and $\mathbf{E}$, the electric field of the radiation. The contribution to the dipole for each electron is given by $\boldsymbol{\mu} = -e\mathbf{r}$, where $\mathbf{r}$ is measured from the centroid of positive charge. There is, in addition, a contribution to the interaction energy from the magnetic dipole, with the analogous form $-\boldsymbol{\mu}_{\text{mag}} \cdot \mathbf{B}$, but this is generally small compared to the electric dipole energy. The magnetic dipole becomes dominant however in magnetic resonance phenomena, to be considered in Chap. 18. The key quantities for electric dipole transitions between quantum states n and m are the matrix elements

$$x_{mn} = \int \psi_m^* \, x \, \psi_n \, d\tau \tag{4.70}$$

and analogous ones for y_{mn} and z_{mn}. These determine the probability of a transition for radiation polarized in the corresponding directions x, y, z. If x_{mn}, y_{mn} and z_{mn} are all equal to zero for two states m and n, then the transition between these states is *electric dipole forbidden*. The conditions on the quantum numbers m and n for allowed transitions to occur are called *selection rules*.

Transitions among states of the particle in a box are determined by the integrals

$$x_{n'n} = \int_0^a \psi_{n'}(x) \, x \, \psi_n(x) \, dx. \tag{4.71}$$

It can be deduced that the selection rule for transitions is $\Delta n = \pm 1, \pm 3, \pm 5 \ldots$, where $n' = n + \Delta n$. Transitions with $\Delta n = \pm 2, \pm 4 \ldots$ are forbidden.

Supplement 4A. Perturbation theory

The number of problems in quantum mechanics which can be solved exactly is relatively small. The great majority of problems, including those involving atoms or molecules more complicated than the hydrogen atom, must be approached using approximate methods. The two most important approximation techniques are the variational method, discussed in Sect. 4.7, and perturbation theory, which we will now describe.

Assume that the Hamiltonian for a problem of interest can be represented by a relatively small addition to the Hamiltonian for an exactly solvable problem. We write

$$H = H_0 + \lambda V, \tag{4.72}$$

where the Schrödinger equation for *unperturbed Hamiltonian* H_0 is assumed to be solvable for a complete set of eigenvalues and eigenfunctions

$$H_0 \psi_n^{(0)} = E_n^{(0)} \psi_n^{(0)}. \tag{4.73}$$

The *perturbation operator* V is scaled by an auxiliary parameter λ, which will serve to track the different orders of approximation; $\lambda = 0$ corresponds to the unperturbed problem. The equation we wish to solve is

$$(H_0 + \lambda V)\psi_n = E_n \psi_n. \tag{4.74}$$

Assuming that the effect of the perturbation on the eigenfunctions ψ_n and eigenvalues E_n is relatively small, we can expand them in power series in the parameter λ:

$$\psi_n = \psi_n^{(0)} + \lambda \psi_n^{(1)} + \lambda^2 \psi_n^{(2)} + \ldots, \tag{4.75}$$
$$E_n = E_n^{(0)} + \lambda E_n^{(1)} + \lambda^2 E_n^{(2)} + \ldots. \tag{4.76}$$

Substituting these expansions into (4.74) we find

$$\begin{aligned} H_0 \psi_n^{(0)} + \lambda(H_0 \psi_n^{(1)} + V \psi_n^{(0)}) + \lambda^2 (H_0 \psi_n^{(2)} + V \psi_n^{(1)}) + \cdots = \\ E_n^{(0)} \psi_n^{(0)} + \lambda(E_n^{(0)} \psi_n^{(1)} + E_n^{(1)} \psi_n^{(0)}) + \\ \lambda^2 (E_n^{(0)} \psi_n^{(2)} + E_n^{(1)} \psi_n^{(1)} + E_n^{(2)} \psi_n^{(0)}) + \ldots. \end{aligned} \tag{4.77}$$

Since the parameter λ is arbitrary, the coefficients of like powers of λ can be equated. This leads to a series of equations:

$$H_0 \psi_n^{(0)} = E_n^{(0)} \psi_n^{(0)}, \tag{4.78}$$
$$H_0 \psi_n^{(1)} + V \psi_n^{(0)} = E_n^{(0)} \psi_n^{(1)} + E_n^{(1)} \psi_n^{(0)}, \tag{4.79}$$
$$H_0 \psi_n^{(2)} + V \psi_n^{(1)} = E_n^{(0)} \psi_n^{(2)} + E_n^{(1)} \psi_n^{(1)} + E_n^{(2)} \psi_n^{(0)}. \tag{4.80}$$

The first of these, is just the unperturbed result (4.73). The second equation can be solved to give $E_n^{(1)}$ and $\psi_n^{(1)}$, the third to give $E_n^{(2)}$ and $\psi_n^{(2)}$ and so on. For compactness, we will now omit the superscripts (0) pertaining to the unperturbed eigenvalues and eigenfunctions.

To find the solutions, it is assumed that the function $\psi_n^{(1)}$ can be expanded in the complete orthonormal set of unperturbed eigenfunctions

$$\psi_n^{(1)} = c_{n0}\psi_0 + c_{n1}\psi_1 + c_{n2}\psi_2 + \cdots = \sum_k c_{nk}\psi_k. \tag{4.81}$$

If we multiply Eq. (4.79) by ψ_n^* and integrate, we find

$$\int \psi_n^* H_0 \psi_n^{(1)}\, d\tau + \int \psi_n^* V \psi_n\, d\tau = E_n \int \psi_n^* \psi_n^{(1)}\, d\tau + E_n^{(1)} \int \psi_n^* \psi_n\, d\tau. \tag{4.82}$$

Noting that

$$\int \psi_n^* H_0 \psi_n^{(1)}\, d\tau = \int \psi_n^{(1)*} H_0 \psi_n\, d\tau = E_n \int \psi_n^* \psi_n^{(1)}\, d\tau \tag{4.83}$$

and $\int \psi_n^* \psi_n\, d\tau = 1$, we determine the first-order perturbation energy

$$E_n^{(1)} = \int \psi_n^* V \psi_n\, d\tau = V_{nn}. \tag{4.84}$$

Note that the first-order energy depends only on the unperturbed eigenfunction. We introduce the matrix notation

$$V_{mn} = \int \psi_m^* V \psi_n\, d\tau, \tag{4.85}$$

where the indices refer to the unperturbed eigenfunctions.

If we multiply Eq. (4.79) instead by ψ_m^*, where $m \neq n$ and integrate, we find

$$\int \psi_m^* H_0 \psi_n^{(1)}\, d\tau + \int \psi_m^* V \psi_n\, d\tau = E_n \int \psi_m^* \psi_n^{(1)}\, d\tau, \tag{4.86}$$

noting that $\int \psi_m^* \psi_n\, d\tau = 0$ by orthogonality of the unperturbed eigenfunctions. Now, inserting the expansion (4.81) for $\psi_n^{(1)}$ and using matrix notation,

$$\sum_k E_m c_{nk} \delta_{mk} + V_{mn} = \sum_k E_n c_{nk} \delta_{kn}, \tag{4.87}$$

where we have noted that $\int \psi_k^* \psi_n\, d\tau = \delta_{kn}$ and $\int \psi_k^* H_0 \psi_n\, d\tau = E_n \delta_{kn}$. Noting that the Kronecker deltas reduce each sum to a single term, we determine the expansion coefficients

$$c_{nm} = \frac{V_{mn}}{E_n - E_m}. \tag{4.88}$$

The first-order eigenfunction $\psi_n^{(1)}$ can be assumed to be orthogonal to the unperturbed ψ_n, so that

$$c_{nn} = 0. \tag{4.89}$$

Thus the first order function is given by

$$\psi_n^{(1)} = \sum_m{}' \frac{V_{mn}\psi_m}{E_n - E_m}, \tag{4.90}$$

where the prime on the summation indicates that the term $m = n$ is omitted.

To find the second-order energy we multiply Eq. (4.80) by ψ_n^* and integrate. Noting that

$$\int \psi_n^* H_0 \psi_n^{(2)}\, d\tau = \int \psi_n^{(2)*} H_0 \psi_n\, d\tau = E_n \int \psi_n^* \psi_n^{(2)}\, d\tau \tag{4.91}$$

and

$$\int \psi_n^* \psi_n^{(1)}\, d\tau = 0. \tag{4.92}$$

The result is

$$E_n^{(2)} = \int \psi_n V \psi_n^{(1)}\, d\tau. \tag{4.93}$$

Using (4.90), this gives the well-known formula for the second-order energy

$$E_n^{(2)} = \sum_m{}' \frac{V_{nm}V_{mn}}{E_n - E_m} = \sum_m{}' \frac{|V_{nm}|^2}{E_n - E_m}. \tag{4.94}$$

Supplement 4B. Time-dependent perturbation theory for radiative transitions

If a Hamiltonian H is independent of time, it will possess a set of eigenstates satisfying

$$H\psi_n = E_n\psi_n. \tag{4.95}$$

The functions $\Psi_n = \psi_n e^{-iE_nt/\hbar}$ will be also solutions of the time-dependent Schrödinger equation

$$i\hbar\frac{d\Psi}{\partial t} = H\Psi. \tag{4.96}$$

Consider now a superposition of functions Ψ_n

$$\Psi = \sum_n c_n\, e^{-iE_n t/\hbar}\, \psi_n = c_0\, e^{-iE_0 t/\hbar}\, \psi_0 + c_1\, e^{-iE_1 t/\hbar}\, \psi_1 + \dots, \tag{4.97}$$

where $c_0, c_1 \dots$ are constants. It is readily shown that Ψ is a general solution of the time-dependent Schrödinger equation. This represents a *nonstationary state*, one which does *not* have a definite energy and is *not* a solution of the time-independent Schrödinger equation (4.72).

For a Hamiltonian $H(t)$ which has explicit dependence on time, solutions of the TISE do not generally exist, but it is still possible to solve the TDSE (4.73). Again consider a Hamiltonian of the form

$$H(t) = H_0 + V(t), \tag{4.98}$$

where H_0 is independent of time and $V(t)$ is much smaller than H_0 but contains dependence on time. Assume that we can determine the eigenstates of the unperturbed Hamiltonian, such that

$$H_0 \psi_n = E_n \psi_n. \tag{4.99}$$

Let the perturbation $V(t)$ now be "turned on." The wavefunction of the perturbed system Ψ will now be determined by the TDSE

$$i\hbar \frac{d\Psi}{\partial t} = (H_0 + V)\Psi. \tag{4.100}$$

The solution can be represented by a generalization of (4.74)

$$\Psi = c_0(t)\, e^{-iE_0 t/\hbar}\, \psi_0 + c_1(t)\, e^{-iE_1 t/\hbar}\, \psi_1 + \dots, \tag{4.101}$$

in which the coefficients $c_0, c_1 \dots$ are now functions of t, to reflect the time-dependence of the new Hamiltonian. We will suppose that the perturbation has been operative for only a short time, so that the system deviates only slightly from its unperturbed state Ψ_0. With this approximation,

$$|c_0(t)| \approx 1 \quad \text{and} \quad |c_1(t)|, |c_2(t)| \cdots << 1. \tag{4.102}$$

Substituting (4.78) into (4.77) with the approximations (4.79), we obtain

$$\begin{aligned} &E_0 e^{-iE_0 t/\hbar} \psi_0 + i\hbar \frac{dc_1}{dt} e^{-iE_1 t/\hbar} \psi_1 + c_1 E_1 e^{-iE_1 t/\hbar} \psi_1 + \dots \\ &\quad \approx E_0 e^{-iE_0 t/\hbar} \psi_0 + V e^{-iE_0 t/\hbar} \psi_0 + c_1 E_1 e^{-iE_1 t/\hbar} \psi_1 + \dots, \end{aligned} \tag{4.103}$$

having neglected terms of second order containing the product of two "small" quantities V and c_n. After cancelation of terms by virtue of the eigenvalue equation (4.76), this simplifies to

$$i\hbar \frac{dc_1}{dt} e^{-iE_1 t/\hbar} \psi_1 + \dots \approx V e^{-iE_0 t/\hbar} \psi_0. \tag{4.104}$$

Multiplying by one of the ψ_n^* (with $n \neq 0$) and integrating, we obtain

$$i\hbar \frac{dc_n}{dt} \approx V_{n0}(t) e^{i\,\omega_{n0} t}, \tag{4.105}$$

where

$$V_{n0}(t) \equiv \int \psi_n \, V(t) \, \psi_0 \, d\tau \tag{4.106}$$

and

$$\omega_{n0} \equiv \frac{E_n - E_0}{\hbar}. \tag{4.107}$$

We have made use of the orthonormality of the eigenfunctions ψ_n. Assuming that $c_n = 0$ for all $n \neq 0$ at time $t = 0$, Eq. (4.82) can be integrated to give

$$c_n(t) \approx -\frac{i}{\hbar} \int_0^t V_{n0}(t') e^{i\,\omega_{n0} t'} \, dt'. \tag{4.108}$$

The probability of a transition from the ground state ψ_0 to the excited state ψ_n, to this first-order approximation, is given by

$$P_{n \leftarrow 0} = |c_n(t)|^2 = c_n(t)^* \, c_n(t). \tag{4.109}$$

An electromagnetic field interacting with an atom or molecule can be represented by perturbation with an oscillatory time dependence of the form

$$V(t) = V \cos \omega t = \frac{1}{2} V (e^{i\omega t} + e^{-i\omega t}). \tag{4.110}$$

Evaluating the matrix element V_{n0} and using (4.85) and (4.86), we obtain

$$P_{n \leftarrow 0} \approx \frac{|V_{n0}|^2}{\hbar^2} \frac{4 \sin^2[(\omega - \omega_{n0})t/2]}{(\omega - \omega_{n0})^2}. \tag{4.111}$$

If the perturbation frequency ω is approximately in resonance with ω_{n0}, thus satisfying the Bohr frequency condition $\hbar\omega = E_n - E_0$, then (4.88) reduces to

$$P_{n \leftarrow 0} \approx \frac{|V_{n0}|^2}{\hbar^2} 2\pi t, \tag{4.112}$$

implying a transition probability per unit time given by

$$W_{n \leftarrow 0} \approx \frac{2\pi}{\hbar^2} \, |V_{n0}|^2. \tag{4.113}$$

This result is valid for times t sufficiently short that the ground state ψ_0 is not significantly depopulated yet sufficiently long so that $\omega t >> 2\pi$. The dependence on the transition rate on the square of the perturbation matrix element, as assumed in Sect. 4.8, has thus been proven.

Chapter 4. Exercises

4.1. If ψ happens to be an eigenfunction of an operator $\hat{A}$ with the eigenvalue a, evaluate the expectation value $\langle A \rangle$.

4.2. Discuss why the noncommutativity of observables is not generally significant in everyday life. For example, why can we simultaneously measure the instantaneous position and momentum of a pitched baseball with confidence?

4.3. Evaluate the commutator $[x, p_x]$ used to derive the Heisenberg uncertainty principle. Hint: First compute the quantity $x\hat{p}_x f(x) - \hat{p}_x x f(x)$, where $f(x)$ is a arbitrary function.

4.4. Convince yourself of the correctness of the commutation relation

$$[L_x, L_y] = i\hbar L_z.$$

4.5. Can you measure simultaneously a particle's y-position coordinate and x-component of momentum?

4.6. Can you measure simultaneously a particle's z-component of linear momentum and z-component of angular momentum? Give proof.

4.7. The one-dimensional harmonic oscillator involves a particle of mass m subject to a potential energy $V(x) = \frac{1}{2}kx^2$. Write down the Lagrangian and Hamiltonian for this system and show that Newton's second law, Lagrange's equation and Hamilton's equations all lead to the same solution.

Chapter 5

Special functions

Solutions for the simple quantum systems considered thus far have involved only *elementary functions*, the term applied to algebraic, exponential, logarithmic and trigonometric functions. In subsequent Chapters, we will take up quantum mechanical problems including the harmonic oscillator, angular momentum and the hydrogen atom which require more complex functions. These are known collectively as *special functions* and are usually named in honor of mathematicians who studied them a couple of centuries ago, including Hermite, Legendre, Laguerre, Bessel and several others. Hopefully, previous familiarity with this mathematics will expedite our treatment of more complex quantum systems as we encounter them.

A comprehensive reference on special functions is M. Abramowitz and I. A. Stegun, eds., *Handbook of Mathematical Functions with Formulas, Graphs and Mathematical Tables* (National Bureau of Standards Applied Mathematics Series, Vol 55, Washington, DC, 1964). The National Institute of Standards and Technology (NIST) has published a Digital Library of Mathematical Functions (DLMF) available online at http://dlmf.nist.gov. This features numerous hypertext links and extensive graphics.

In the first few Sections, we will also introduce some useful mathematics, in addition to the above named special functions. This Chapter can be skipped by the reader, without loss of continuity. The relevant Sections can then be referred to as the need arises.

5.1 Gaussian functions

A gaussian function (Fig. 5.1) is given by

$$g(x) = \frac{1}{\sigma\sqrt{2\pi}} e^{-x^2/2\sigma^2}. \tag{5.1}$$

The function is normalized, with

$$\int_{-\infty}^{\infty} g(x)\, dx = 1. \tag{5.2}$$

Laplace (1778) proved that

$$\int_{-\infty}^{\infty} e^{-x^2}\, dx = \sqrt{\pi}. \tag{5.3}$$

Introduction to Quantum Mechanics. https://doi.org/10.1016/B978-0-12-822310-9.00013-6

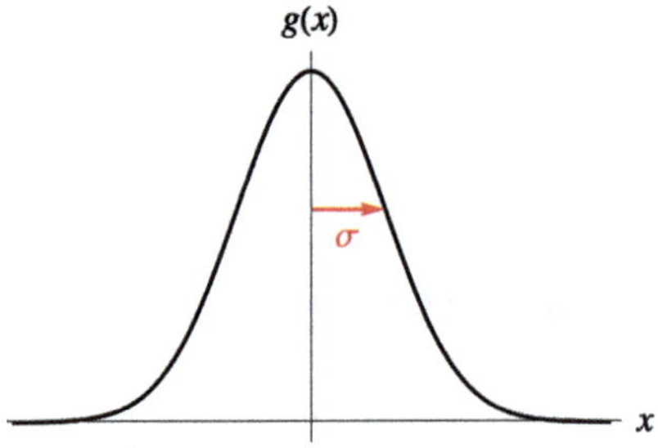

FIGURE 5.1 Normalized gaussian function.

This remarkable result can be obtained as follows. Denoting the integral by I, we can write

$$I^2 = \left(\int_{-\infty}^{\infty} e^{-x^2}\, dx\right)^2 = \int_{-\infty}^{\infty} e^{-x^2}\, dx \int_{-\infty}^{\infty} e^{-y^2}\, dy, \tag{5.4}$$

where the dummy variable y has been substituted for x in the last integral. The product of two integrals can be expressed as a double integral:

$$I^2 = \int_{-\infty}^{\infty} \int_{-\infty}^{\infty} e^{-(x^2+y^2)}\, dx\, dy. \tag{5.5}$$

The differential $dx\, dy$ represents an element of area in cartesian coordinates, with the domain of integration extending over the entire xy-plane. An alternative representation of the last integral can be expressed in plane polar coordinates r, θ. The two coordinate systems are related by

$$x = r\cos\theta, \qquad y = r\sin\theta, \tag{5.6}$$

so that

$$r^2 = x^2 + y^2. \tag{5.7}$$

The element of area in polar coordinates is given by $r\, dr\, d\theta$, so that the double integral becomes

$$I^2 = \int_0^{\infty} \int_0^{2\pi} e^{-r^2}\, r\, dr\, d\theta. \tag{5.8}$$

Integration over θ gives a factor 2π. The integral over r can be done after the substitution $u = r^2$, $du = 2r\, dr$:

$$\int_0^{\infty} e^{-r^2}\, r\, dr = \tfrac{1}{2} \int_0^{\infty} e^{-u}\, du = \tfrac{1}{2}. \tag{5.9}$$

Therefore $I^2 = 2\pi \times \frac{1}{2}$ and Laplace's result (5.68) is proven. A slightly more general result is

$$\int_{-\infty}^{\infty} e^{-\alpha x^2}\, dx = \left(\frac{\pi}{\alpha}\right)^{1/2}, \tag{5.10}$$

obtained by scaling the variable x to $\sqrt{\alpha}\, x$.

We also require definite integrals of the type

$$\int_{-\infty}^{\infty} x^n\, e^{-\alpha x^2}\, dx, \qquad n = 1, 2, 3\ldots, \tag{5.11}$$

for computations involving harmonic oscillator wavefunctions. For odd n, these integrals all equal zero since the contributions from $\{-\infty, 0\}$ exactly cancel those from $\{0, \infty\}$. The following stratagem produces successive integrals for even n. Differentiate each side of (5.75) wrt the parameter α and cancel minus signs to obtain

$$\int_{-\infty}^{\infty} x^2\, e^{-\alpha x^2}\, dx = \frac{\pi^{1/2}}{2\alpha^{3/2}}. \tag{5.12}$$

Differentiation under an integral sign is valid provided that the integrand is a continuous function. Differentiating again, we obtain

$$\int_{-\infty}^{\infty} x^4\, e^{-\alpha x^2}\, dx = \frac{3\pi^{1/2}}{4\alpha^{5/2}}. \tag{5.13}$$

The general result is

$$\int_{-\infty}^{\infty} x^n\, e^{-\alpha x^2}\, dx = \frac{1 \cdot 3 \cdot 5 \cdots |n-1|\, \pi^{1/2}}{2^{n/2} \alpha^{(n+1)/2}}, \qquad n = 0, 2, 4\ldots. \tag{5.14}$$

Finite-interval integrations over a gaussian defines the *error function*

$$\operatorname{erf}(x) = \frac{2}{\sqrt{\pi}} \int_0^x e^{-t^2}\, dt \tag{5.15}$$

and the related *complementary error function*

$$\operatorname{erfc}(x) = 1 - \operatorname{erf}(x) = \frac{2}{\sqrt{\pi}} \int_x^{\infty} e^{-t^2}\, dt. \tag{5.16}$$

5.2 The gamma function

The gamma function is also most conveniently introduced by a definite integral. The following integral can be evaluated exactly,

$$\int_0^{\infty} e^{-\alpha x}\, dx = \frac{1}{\alpha}. \tag{5.17}$$

The derivative of this integral with respect to α gives

$$\frac{d}{d\alpha}\int_0^\infty e^{-\alpha x}\,dx = \int_0^\infty (-x)e^{-\alpha x}\,dx = \frac{d}{d\alpha}\left(\frac{1}{\alpha}\right) = -\frac{1}{\alpha^2}. \tag{5.18}$$

We have therefore obtained a new definite integral:

$$\int_0^\infty xe^{-\alpha x}\,dx = \frac{1}{\alpha^2}. \tag{5.19}$$

Taking $d/d\alpha$ again we find

$$\int_0^\infty x^2e^{-\alpha x}\,dx = \frac{2}{\alpha^3}. \tag{5.20}$$

Repeating the process n times

$$\int_0^\infty x^n e^{-\alpha x}\,dx = \frac{2\cdot 3\cdot 4\cdots n}{\alpha^{n+1}}. \tag{5.21}$$

Setting $\alpha = 1$, now that its job is done, we wind up a neat integral formula for $n!$

$$\int_0^\infty x^n e^{-x}\,dx = n!. \tag{5.22}$$

This is certainly not the most convenient way to evaluate $n!$, but suppose we replace n by a *noninteger* ν. In conventional notation, this defines the *gamma function*:

$$\Gamma(\nu) \equiv \int_0^\infty x^{\nu-1}e^{-x}\,dx. \tag{5.23}$$

When ν is an integer, this reduces to the factorial by the relation

$$\Gamma(n+1) = n! \qquad n = 0, 1, 2, 3\ldots. \tag{5.24}$$

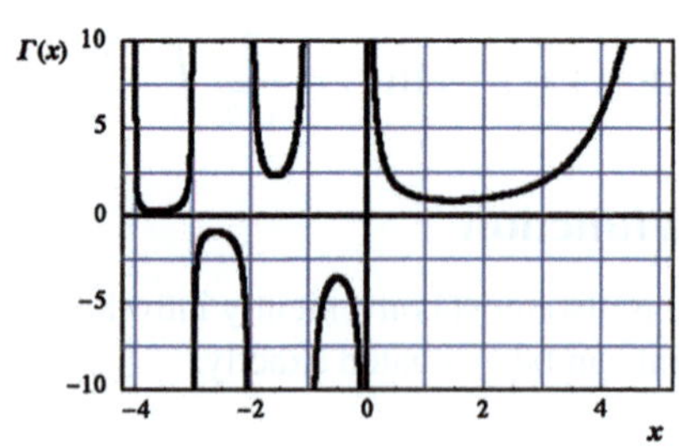

FIGURE 5.2 The gamma function on the real axis.

For the case $\nu = 1/2$,

$$\Gamma\left(\frac{1}{2}\right) = \int_0^\infty x^{-1/2} e^{-x}\, dx. \tag{5.25}$$

The integral can be evaluated with a change of variables $x = y^2$, $dx = 2y\, dy$ giving

$$\Gamma\left(\frac{1}{2}\right) = 2\int_0^\infty e^{-y^2}\, dy = \sqrt{\pi}, \tag{5.26}$$

where we have recalled Laplace's famous result from Eq. (5.3).

Fig. 5.2 shows a plot of the gamma function. For $x > 0$, the function is a smooth interpolation between integer factorials. $\Gamma(x)$ becomes infinite for $x = 0, -1, -2\cdots$.

5.3 The Dirac deltafunction

The Dirac deltafunction can intuitively be understood as the limit of the density of a finite distribution of mass, say $\rho(\mathbf{r})$, as the object is shrunken to a point mass at $\mathbf{r} = 0$, in which case we write $\rho(\mathbf{r}) = \delta(\mathbf{r})$. In one dimension, the deltafunction is the analog of the Kronecker delta

$$\delta_{mn} \equiv \begin{cases} 1 \text{ if } m = n \\ 0 \text{ if } m \neq n, \end{cases} \tag{5.27}$$

when the discrete variables n and m are replaced by continuous variables. A summation over Kronecker deltas reduces to a single term, such that

$$\sum_{n=0}^{\infty} f_n\, \delta_{nm} = f_m. \tag{5.28}$$

The analog of this relation for continuous variables, with $n \to x$ and $m \to x_0$, introduces the deltafunction $\delta(x - x_0)$. We can take as the defining property

$$\int_{-\infty}^{\infty} f(x)\, \delta(x - x_0)\, dx = f(x_0). \tag{5.29}$$

This also implies the normalization condition

$$\int_{-\infty}^{\infty} \delta(x - x_0)\, dx = 1. \tag{5.30}$$

Evidently

$$\delta(x - x_0) \equiv \begin{cases} 0 & \text{if } x \neq x_0 \\ \infty & \text{if } x = x_0. \end{cases} \tag{5.31}$$

The approach to ∞ is sufficiently tame, however, that the integral has a finite value.

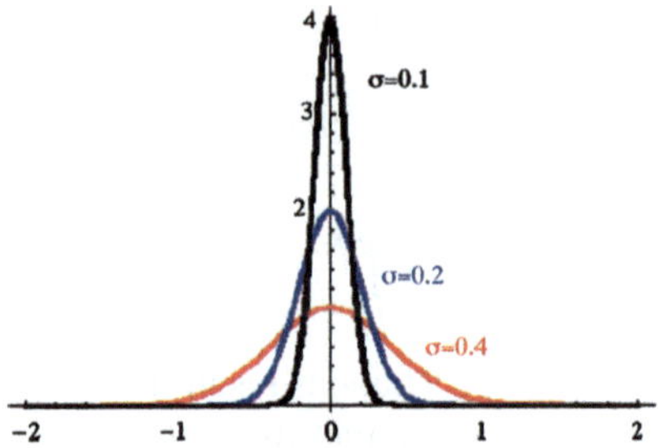

FIGURE 5.3 Dirac deltafunction with $x_0 = 0$, as limit of normalized gaussian.

A simple representation for the deltafunction is the limit of a normalized gaussian as the standard deviation approaches zero:

$$\delta(x - x_0) = \lim_{\sigma \to 0} \frac{1}{\sqrt{2\pi}\,\sigma} e^{-(x-x_0)^2/2\sigma^2}. \tag{5.32}$$

This is shown pictorially in Fig. 5.3. The deltafunction is the limit of a function which becomes larger and larger in an interval which becomes narrower and narrower.

An integral which reduces to a deltafunction is the following:

$$\int_{-\infty}^{\infty} e^{i(k-k')x}\,dx = 2\pi\,\delta(k - k'). \tag{5.33}$$

To prove this, consider the related integral

$$\int_{-\infty}^{\infty} e^{-\sigma x^2/2} e^{i(k-k')x}\,dx = \frac{\sqrt{2\pi}}{\sigma} e^{-\frac{(k-k')^2}{2\sigma^2}}. \tag{5.34}$$

The right hand side can be written $2\pi\,g(k - k')$, in terms of the normalized gaussian (5.1). By (5.32) the limit as $\sigma \to 0$ gives the deltafunction, thus verifying (5.33).

The free-particle eigenfunctions, considered in Sect. 3.1, can now be deltafunction orthonormalized. Defining

$$\psi_k(x) = \frac{1}{\sqrt{2\pi}} e^{ikx}, \tag{5.35}$$

Eq. (5.33) determines the continuum orthonormalization relation

$$\int_{-\infty}^{\infty} \psi_k^*(x)\psi_k'(x)\,dx = \delta(k - k'). \tag{5.36}$$

Differentiation of a function at a finite discontinuity produces a deltafunction. Consider, for example, the *Heaviside unit step function*:

$$H(x-x_0) \equiv \begin{cases} 0 & \text{if } x < x_0 \\ 1 & \text{if } x \geq x_0. \end{cases} \tag{5.37}$$

Sometimes $H(0)$ (for $x = x_0$) is defined as $\frac{1}{2}$. The derivative of the Heaviside function $H'(x-x_0)$ is clearly equal to zero when $x \neq x_0$. In addition

$$\int_{-\infty}^{\infty} H'(x-x_0)\,dx = H(\infty) - H(-\infty) = 1 - 0. \tag{5.38}$$

Thus, we find

$$H'(x-x_0) = \delta(x-x_0). \tag{5.39}$$

There is related result for the sign function $\text{sgn}(x) = \pm 1$, for $x > 0$ and $x < 0$, respectively. We find

$$\text{sgn}'(x-x_0) = 2\delta(x-x_0). \tag{5.40}$$

The deltafunction can be generalized to multiple dimensions. In three dimensions, the defining relation for a deltafunction can be expressed

$$\int f(\mathbf{r})\,\delta(\mathbf{r}-\mathbf{r}_0)\,d^3\mathbf{r} = f(\mathbf{r}_0). \tag{5.41}$$

For example, the limit of a continuous distribution of electrical charge $\rho(\mathbf{r})$ shrunken to a point charge q at $\mathbf{r}_0$ can be represented by

$$\rho(\mathbf{r}) = q\,\delta(\mathbf{r}-\mathbf{r}_0). \tag{5.42}$$

The potential energy of interaction between two continuous charge distributions is given by

$$V = \int\!\!\int \frac{\rho_1(\mathbf{r}_1)\rho_2(\mathbf{r}_2)}{|\mathbf{r}_2-\mathbf{r}_1|}\,d^3\mathbf{r}_1\,d^3\mathbf{r}_2. \tag{5.43}$$

If the distribution $\rho_2(\mathbf{r}_2)$ is reduced to a point charge q_2 at $\mathbf{r}_2$, this reduces to

$$V = q_2 \int \frac{\rho_1(\mathbf{r}_1)}{|\mathbf{r}_2-\mathbf{r}_1|}\,d^3\mathbf{r}_1. \tag{5.44}$$

If the analogous thing then happens to $\rho_1(\mathbf{r}_1)$, the formula reduces to the Coulomb potential energy between two point charges

$$V = \frac{q_1 q_2}{r_{12}}, \qquad \text{where} \qquad r_{12} = |\mathbf{r}_2-\mathbf{r}_1|. \tag{5.45}$$

A note on the deltafunction involving spherically symmetrical functions in spherical polar coordinates. Applying (5.41) to the case when $\mathbf{r}_0 = 0$ and $f(\mathbf{r}) = f(r)$, we can write

$$\int f(r)\,\delta(\mathbf{r})\,d^3\mathbf{r} = \int_0^\infty f(r)\delta(\mathbf{r})\,4\pi r^2\,dr = f(0). \tag{5.46}$$

But a 1-dimensional deltafunction should also produce the result

$$\int_0^\infty f(r)\delta(r)\,dr = \frac{1}{2}f(0), \tag{5.47}$$

with the factor $\frac{1}{2}$ reflecting the fact that the deltafunction is located at one of the limits, so that, in effect, only half of it is within the range of integration. Comparing the last two integrals we can surmise the relationship between 1- and 3-dimensional deltafunctions in spherical coordinates:

$$\delta(\mathbf{r}) = \frac{\delta(r)}{2\pi r^2}. \tag{5.48}$$

An interesting identity involving the deltafunction in spherical polar coordinates is

$$\nabla^2\frac{1}{r} = -4\pi\,\delta(\mathbf{r}). \tag{5.49}$$

To derive this result, we note that the variable r in spherical coordinates can have only non-negative values. Thus r might harmlessly be replaced by $|r|$. Accordingly, we can write

$$\nabla^2\frac{1}{r} = \frac{1}{r^2}\frac{d}{dr}\left(r^2\frac{d}{dr}\frac{1}{|r|}\right). \tag{5.50}$$

Working through the steps in the above formula, we find

$$r^2\frac{d}{dr}\frac{1}{|r|} = -r^2r^{-2}\frac{d}{dr}|r| = -\mathrm{sgn}(r). \tag{5.51}$$

Finally,

$$-\frac{1}{r^2}\frac{d}{dr}\mathrm{sgn}(r) = -\frac{2}{r^2}\delta(r) = -4\pi\,\delta(\mathbf{r}). \tag{5.52}$$

Poisson's equation (1.23) applied to the potential of a point charge gives

$$\nabla^2\Phi(r) = \nabla^2\frac{q}{r} = -4\pi q\,\delta(\mathbf{r}), \tag{5.53}$$

showing the density of a point charge $\rho(\mathbf{r}) = q\delta(\mathbf{r})$.

5.4 Leibniz's formula

In some of the following work, we will make use of a formula for the nth derivative of the product of two functions. This can be derived stepwise as follows:

$$\frac{d}{dx}\Big[f(x)g(x)\Big] = f'(x)g(x) + f(x)g'(x),$$
$$\frac{d^2}{dx^2}\Big[f(x)g(x)\Big] = f''(x)g(x) + 2f'(x)g'(x) + f(x)g''(x),$$
$$\frac{d^3}{dx^3}\Big[f(x)g(x)\Big] = f'''(x)g(x) + 3f''(x)g'(x) + 3f'(x)g''(x) + f(x)g'''(x),$$
$$\ldots. \tag{5.54}$$

Clearly, we are generating a series containing the binomial coefficients

$$\binom{n}{m} = \frac{n!}{m!(n-m)!} \qquad m = 0, 1 \ldots n \tag{5.55}$$

and the general result is Leibniz's formula

$$\frac{d^n}{dx^n}\Big[f(x)g(x)\Big] = \sum_{m=0}^{n} \binom{n}{m} f^{(n-m)}(x)g^{(m)}(x) =$$
$$f^{(n)}(x)g(x) + nf^{(n-1)}(x)g'(x) + \frac{n(n-1)}{2} f^{(n-2)}(x)g''(x) + \ldots. \tag{5.56}$$

5.5 Hermite polynomials

The Schrödinger equation for the quantum-mechanical harmonic oscillator can be reduced to a second-order differential equation with nonconstant coefficients:

$$\psi''(x) + (\lambda - x^2)\psi(x) = 0. \tag{5.57}$$

A useful first step is to determine the asymptotic solution to this equation, giving the form of $\psi(x)$ as $x \to \pm\infty$. For sufficiently large values of $|x|$, $x^2 >> \lambda$, so that the differential equation can be approximated by

$$\psi''(x) - x^2\psi(x) \approx 0. \tag{5.58}$$

This suggests the following manipulation:

$$\left(\frac{d^2}{dx^2} - x^2\right)\psi(x) \approx \left(\frac{d}{dx} - x\right)\left(\frac{d}{dx} + x\right)\psi(x) \approx 0. \tag{5.59}$$

Now, the first-order differential equation

$$\psi'(x) + x\psi(x) \approx 0 \tag{5.60}$$

can be solved exactly to give

$$\psi(x) \approx \text{const}\, e^{-x^2/2} \quad \text{for} \quad |x| \to \infty. \tag{5.61}$$

To build in this asymptotic behavior, let

$$\psi(x) = H(x)\, e^{-x^2/2}. \tag{5.62}$$

This reduces Eq. (5.57) to a differential equation for $H(x)$:

$$H''(x) - 2xH'(x) + (\lambda - 1)H(x) = 0. \tag{5.63}$$

To construct a solution to Eq. (5.63), we begin with the function

$$u(x) = e^{-x^2}, \tag{5.64}$$

which is clearly the solution of the *first-order* differential equation

$$u'(x) + 2x\, u(x) = 0. \tag{5.65}$$

Differentiating this equation $(n+1)$ times using Leibniz's formula (5.56), we obtain

$$w''(x) + 2x\, w'(x) + 2(n-1)w(x) = 0, \tag{5.66}$$

where

$$w(x) = \frac{d^n u}{dx^n} = \frac{d^n}{dx^n} e^{-x^2} = H(x)e^{-x^2}. \tag{5.67}$$

We find that $H(x)$ satisfies

$$H''(x) - 2x\, H'(x) + 2n\, H(x) = 0, \tag{5.68}$$

which is known as *Hermite's differential equation.* The solutions in the form

$$H_n(x) = (-)^n e^{x^2} \frac{d^n}{dx^n} e^{-x^2} \tag{5.69}$$

are known as *Hermite polynomials*, the first few of which are enumerated below:

$$\begin{gathered} H_0(x) = 1, \qquad H_1(x) = 2x, \qquad H_2(x) = 4x^2 - 2, \\ H_3(x) = 8x^3 - 12x, \qquad H_4(x) = 16x^4 - 48x^2 + 12. \end{gathered} \tag{5.70}$$

Comparing Eq. (5.68) with Eq. (5.63), we can relate the parameters

$$\lambda - 1 = 2n, \tag{5.71}$$

which, as we will see in the following Chapter, leads to a general formula for the energy eigenvalues.

A generating function for Hermite polynomials is given by

$$e^{x^2-(t-x)^2} = e^{2tx-t^2} = \sum_{n=0}^{\infty} \frac{H_n(x)}{n!} t^n. \tag{5.72}$$

Using the generating function, we can evaluate integrals over products of Hermite polynomials, such as

$$\int_{-\infty}^{\infty} H_n(x) H_{n'}(x) e^{-x^2} dx = 2^n n! \sqrt{\pi}\, \delta_{n,n'}. \tag{5.73}$$

Thus, the functions

$$\psi_n(x) = (2^n n! \sqrt{\pi})^{-1/2} H_n(x)\, e^{-x^2/2}, \qquad n = 0, 1, 2 \ldots \tag{5.74}$$

form an orthonormal set with

$$\int_{-\infty}^{\infty} \psi_n(x)\, \psi_{n'}(x)\, dx = \delta_{n,n'}. \tag{5.75}$$

5.6 Spherical polar coordinates

Most three-dimensional applications of quantum mechanics, particularly to atomic and molecular systems, make use of the spherical polar coordinate system. This is the most natural coordinate system for treating problems of spherical symmetry and for consideration of angular momentum. Following is a brief review of spherical polar coordinates, sufficient for our purposes.

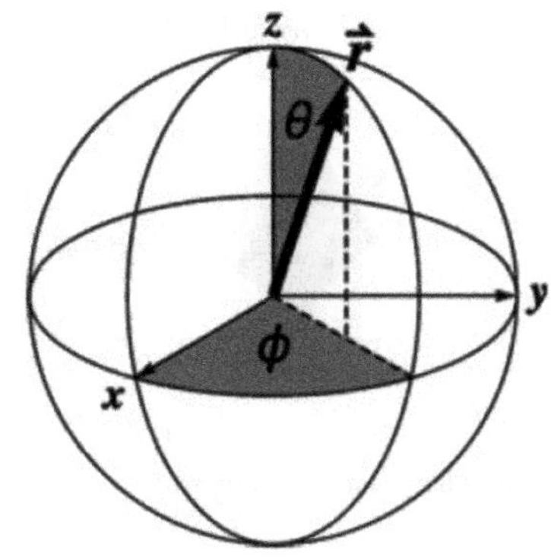

FIGURE 5.4 Spherical polar coordinates.

The position of an arbitrary point **r** in three-dimensional space is described by the coordinates r, θ, ϕ, as shown in Fig. 5.4. As can readily be deduced from the figure, these are connected to cartesian coordinates by the relations

$$x = r\sin\theta\cos\phi,$$
$$y = r\sin\theta\sin\phi,$$
$$z = r\cos\theta. \tag{5.76}$$

Spherical coordinates are closely analogous to the geographical coordinate system, which locates points by latitude, longitude and altitude. Referring to the familiar globe of the world, r represents the distance from the center of the globe, with the range $0 \le r \le \infty$. The *azimuthal angle* θ is the angle between the vector $\vec{r}$ and the z-axis or North Pole, with the range $0 \le \theta \le \pi$. Thus $\theta = 0$ points to the North Pole, $\theta = \pi$, to the South Pole and $\theta = \pi/2$ runs around the Equator. The circles of constant θ on the surface of a sphere are analogous to the parallels of latitude on the globe (although the geographic conventions are different, with the equator at 0° latitude, while the poles are at 90° N and S latitude). The *polar angle* ϕ measures the rotation of the vector $\vec{r}$ around the z-axis, with $0 \le \phi < 2\pi$, counterclockwise from the x-axis. The loci of constant ϕ on the surface of a sphere are great circles through both poles. These clearly correspond to *meridians* in the geographic specification of *longitude* (measured in degrees, 0° to 180° E and W of the Greenwich Meridian). The radial variable r represents the distance from $\mathbf{r}$ to the origin, the length of the vector $\mathbf{r}$:

$$r = \sqrt{x^2 + y^2 + z^2}. \tag{5.77}$$

The volume element in spherical polar coordinates is given by

$$d\tau = r^2 \sin\theta\, dr\, d\theta\, d\phi, \qquad (0 \le r \le \infty,\ 0 \le \theta \le \pi,\ 0 \le \phi \le 2\pi), \tag{5.78}$$

and represented graphically by the distorted cube in Fig. 5.5.

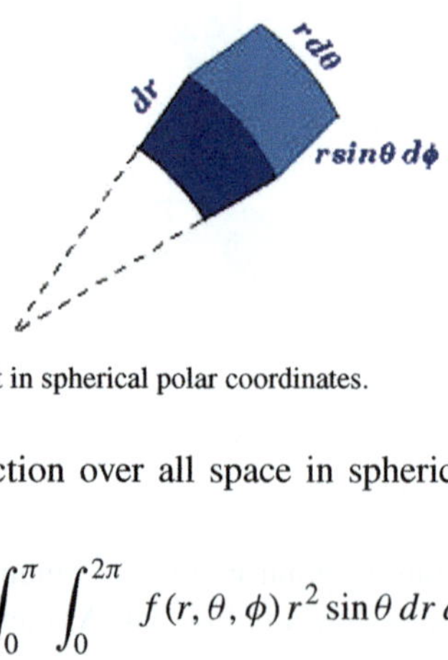

FIGURE 5.5 Volume element in spherical polar coordinates.

Integration of a function over all space in spherical coordinates takes the form:

$$\int_0^{\infty}\int_0^{\pi}\int_0^{2\pi} f(r,\theta,\phi)\, r^2 \sin\theta\, dr\, d\theta\, d\phi. \tag{5.79}$$

For a spherically-symmetric function $f(r)$, independent of θ or ϕ, this integral simplifies to

$$\int_0^\infty f(r)\, 4\pi r^2\, dr, \tag{5.80}$$

as implied by the division of space into spherical shells of area $4\pi r^2$ and thickness dr.

Laplace's equation in spherical polar coordinates can be written

$$\nabla^2\Psi(r,\theta,\phi) = \frac{1}{r^2}\frac{\partial}{\partial r} r^2 \frac{\partial \Psi}{\partial r} + \frac{1}{r^2 \sin\theta}\frac{\partial}{\partial \theta}\sin\theta\frac{\partial \Psi}{\partial \theta} + \frac{1}{r^2\sin^2\theta}\frac{\partial^2\Psi}{\partial \phi^2} = 0. \tag{5.81}$$

We consider separable solutions

$$\Psi(r,\theta,\phi) = R(r)Y(\theta,\phi), \tag{5.82}$$

where the functions $Y(\theta,\phi)$ are known as a *spherical harmonics*. Substitution of (5.82) into (5.81), followed by division by $R(r)Y(\theta,\phi)$ and multiplication by r^2, separates the radial and angular variables:

$$\frac{1}{Y(\theta,\phi)}\left[\frac{1}{\sin\theta}\frac{\partial}{\partial Y}\sin\theta\frac{\partial Y}{\partial \theta} + \frac{1}{\sin^2\theta}\frac{\partial^2 Y}{\partial \phi^2}\right] + \frac{1}{R(r)}\left[r^2 R''(r) + 2r R'(r)\right] = 0. \tag{5.83}$$

This can hold true for all values of r, θ, ϕ only if each of the two parts of this partial differential equation equals a constant. Let the first part equal $-\lambda$ and the second equal λ. The ordinary differential equation for r becomes

$$r^2 R''(r) + 2r R(r) - \lambda R(r) = 0. \tag{5.84}$$

It is easily verified that the general solution is

$$R(r) = c_1 r^\ell + c_2 r^{-\ell-1} \qquad \text{where} \qquad \lambda = \ell(\ell+1). \tag{5.85}$$

The spherical harmonics evidently satisfy a partial differential equation in two variables:

$$\frac{1}{\sin\theta}\frac{\partial}{\partial \theta}\sin\theta\frac{\partial Y}{\partial \theta} + \frac{1}{\sin^2\theta}\frac{\partial^2 Y}{\partial \phi^2} + \lambda Y = 0. \tag{5.86}$$

Again we assume a separable solution with

$$Y(\theta,\phi) = \Theta(\theta)\Phi(\phi). \tag{5.87}$$

Substituting (5.87) into (5.86), dividing by $\Theta(\theta)\Phi(\phi)$ and multiplying by $\sin^2\theta$, we achieve separation of variables:

$$\frac{\sin\theta}{\Theta(\theta)}\frac{d}{d\theta}\sin\theta\frac{d\Theta}{d\theta}+\frac{1}{\Phi(\phi)}\frac{d^2\Phi}{d\phi^2}+\lambda\sin^2\theta=0. \tag{5.88}$$

Setting the term containing ϕ to a constant $-m^2$, we obtain the familiar differential equation

$$\Phi''(\phi)+m^2\Phi(\phi)=0. \tag{5.89}$$

Solutions periodic in 2π can be chosen as

$$\Phi_m(\phi)=\sqrt{\frac{1}{2\pi}}\,e^{im\phi} \quad \text{with} \quad m=0,\pm1,\pm2\ldots. \tag{5.90}$$

Alternative solutions are $\sin m\phi$ and $\cos m\phi$.

5.7 Legendre polynomials

Separation of variables in Laplace's equation leads to an ordinary differential equation for the function $\Theta(\theta)$:

$$\left\{\frac{1}{\sin\theta}\frac{d}{d\theta}\sin\theta\frac{d}{d\theta}-\frac{m^2}{\sin^2\theta}+\lambda\right\}\Theta(\theta)=0. \tag{5.91}$$

Let us first consider the case $m=0$ and define a new independent variable

$$x=\cos\theta \quad \text{with} \quad P(x)=\Theta(\theta). \tag{5.92}$$

This transforms Eq. (5.91) to

$$(1-x^2)P''(x)-2x\,P'(x)+\lambda P(x)=0, \tag{5.93}$$

which is known as *Legendre's differential equation*. We can construct a solution to this linear second-order equation by exploiting Leibniz's formula (5.56). Begin with the function

$$u=(1-x^2)^\ell, \tag{5.94}$$

which is a solution of the first-order equation

$$(1-x^2)u'(x)+2\ell x\,u(x)=0. \tag{5.95}$$

Differentiating $(\ell+1)$ times, we obtain

$$(1-x^2)P''(x)-2x\,P'(x)+\ell(\ell+1)P(x)=0, \tag{5.96}$$

where

$$P(x)=\frac{d^\ell u}{dx^\ell}=\frac{d^\ell}{dx^\ell}(1-x^2)^\ell. \tag{5.97}$$

This is a solution of Eq. (5.93) for

$$\lambda=\ell(\ell+1) \qquad \ell=0,1,2,\ldots. \tag{5.98}$$

With a choice of constant such that $P_\ell(1)=1$, the Legendre polynomials can be defined by *Rodrigues' formula*:

$$P_\ell(x)=\frac{1}{2^\ell \ell!}\frac{d^\ell}{dx^\ell}(1-x^2)^\ell. \tag{5.99}$$

Reverting to the original variable θ, the first few Legendre polynomials are

$$P_0(\cos\theta)=1, \qquad P_1(\cos\theta)=\cos\theta, \qquad P_2(\cos\theta)=\frac{1}{2}(3\cos^2\theta-1),$$
$$P_3(\cos\theta)=\frac{1}{2}(5\cos^3\theta-3\cos\theta), \quad P_4(\cos\theta)=\frac{1}{8}(35\cos^4\theta-30\cos^2\theta+3). \tag{5.100}$$

The Legendre polynomials obey the orthonormalization relations

$$\int_{-1}^{+1} P_\ell(x)P_{\ell'}(x)\,dx=\int_0^\pi P_\ell(\cos\theta)P_{\ell'}(\cos\theta)\sin\theta\,d\theta=\frac{2}{2\ell+1}\delta_{\ell,\ell'}. \tag{5.101}$$

A generating function for Legendre polynomials is given by

$$\left(1-2tx+t^2\right)^{-1}=\sum_{\ell=0}^{\infty}P_\ell(x)\,t^\ell. \tag{5.102}$$

Returning to Eq. (5.91) for arbitrary values of m, the analog of Eq. (5.93) can be written

$$(1-x^2)P''(x)-2x\,P'(x)+\left[\ell(\ell+1)-\frac{m^2}{1-x^2}\right]P(x)=0. \tag{5.103}$$

The solutions are readily found to be

$$P_\ell^m(x)=(1-x^2)^{|m|/2}\frac{d^{|m|}}{dx^{|m|}}P_\ell(x), \tag{5.104}$$

known as *associated Legendre functions.* These reduce to the Legendre polynomials (5.99) when $m=0$. Since $P_\ell(x)$ is a polynomial of degree ℓ, $|m|$ is limited to the values $0, 1, 2 \ldots \ell$. The associated Legendre functions obey the orthonormalization relations

$$\int_{-1}^{+1} P_\ell^m(x) P_{\ell'}^m(x)\, dx = \int_0^\pi P_\ell^m(\cos\theta) P_{\ell'}^m(\cos\theta) \sin\theta\, d\theta$$
$$= \frac{2}{2\ell+1} \frac{(\ell+|m|)!}{(\ell-|m|)!} \delta_{\ell,\ell'}. \tag{5.105}$$

The orthonormalized solutions to Eq. (5.91) are thus given by

$$\Theta_{\ell m}(\theta) = \left[\frac{2\ell+1}{2} \frac{(\ell-|m|)!}{(\ell+|m|)!}\right]^{1/2} P_\ell^m(\cos\theta). \tag{5.106}$$

5.8 Spherical harmonics

Combining the above functions of θ and ϕ, we obtain the spherical harmonics (which have been used in applied mathematics long before quantum mechanics).

$$Y_{\ell m}(\theta, \phi) = \epsilon_m \Theta_{\ell m}(\theta)\, \Phi_m(\phi) = \epsilon_m \left[\frac{2\ell+1}{4\pi} \frac{(\ell-|m|)!}{(\ell+|m|)!}\right]^{1/2} P_\ell^m(\cos\theta) e^{im\phi}, \tag{5.107}$$

where

$$\epsilon_m = \begin{cases} 1 & \text{for } m \le 0 \\ (-1)^m & \text{for } m > 0. \end{cases} \tag{5.108}$$

The factor ϵ_m is appended in applications to quantum mechanics, in accordance with the Condon and Shortley phase convention. Two important special cases are

$$Y_{\ell 0}(\theta, \phi) = \sqrt{\frac{2\ell+1}{4\pi}} P_\ell(\cos\theta) \tag{5.109}$$

and

$$Y_{\ell\ell}(\theta, \phi) = \frac{(-)^\ell}{2^\ell \ell!} \sqrt{\frac{(2\ell+1)}{4\pi}} \sin^\ell\theta e^{i\ell\phi}. \tag{5.110}$$

Following is a listing of the spherical harmonics $Y_{\ell m}(\theta, \phi)$ through $\ell = 2$, which will be sufficient for our applications:

$$Y_{00} = \sqrt{\frac{1}{4\pi}}, \quad Y_{10} = \sqrt{\frac{3}{4\pi}} \cos\theta,$$
$$Y_{1\pm1} = \mp\sqrt{\frac{3}{4\pi}} \sin\theta\, e^{\pm i\phi}, \quad Y_{20} = \sqrt{\frac{5}{16\pi}}(3\cos^2\theta - 1), \tag{5.111}$$
$$Y_{2\pm1} = \mp\sqrt{\frac{15}{8\pi}} \cos\theta \sin\theta\, e^{\pm i\phi}, \quad Y_{2\pm2} = \sqrt{\frac{15}{32\pi}} \sin^2\theta\, e^{\pm 2i\phi}.$$

A graphical representation of these functions is given in Fig. 5.6. Surfaces of constant absolute value are drawn, with intermediate shadings representing differing complex values of the functions.

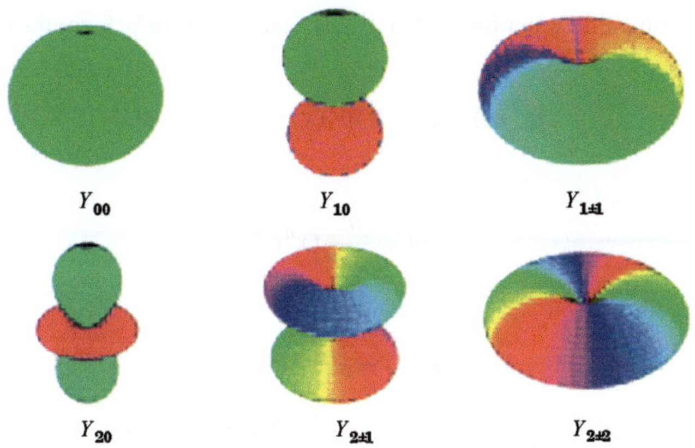

FIGURE 5.6 Contours of spherical harmonics as three-dimensional polar plots.

The spherical harmonics are an orthonormal set wrt integration over solid angle:

$$\int_0^{\pi}\int_0^{2\pi} Y^*_{\ell' m'}(\theta,\phi) Y_{\ell m}(\theta,\phi)\,\sin\theta\, d\theta\, d\phi = \delta_{\ell\ell'}\delta_{mm'}. \tag{5.112}$$

Linear combinations of $Y_{\ell m}$ and $Y_{\ell -m}$ contain the real functions:

$$P_\ell^m(\cos\theta)\begin{cases}\sin m\phi\\ \cos m\phi.\end{cases} \tag{5.113}$$

These are called *tesseral harmonics* since they divide the surface of a sphere into *tesserae*—four-sided figures bounded by nodal lines of latitude and longitude.

5.9 Laguerre polynomials

In solving the quantum-mechanical problem of the hydrogen atom, we will encounter Laguerre's differential equation:

$$xy''(x) + (1-x)y'(x) + ny(x) = 0. \tag{5.114}$$

Following the strategy used to solve the Hermite and Legendre differential equations, we begin with a function

$$u(x) = x^n e^{-x}, \tag{5.115}$$

where n is an integer. This satisfies the first-order differential equation

$$xu'(x) + (x - n)u(x) = 0. \tag{5.116}$$

Differentiating this equation $n + 1$ times using Leibniz's formula (5.56), we obtain

$$xw''(x) + (1 - x)w'(x) + nw(x) = 0, \tag{5.117}$$

where

$$w(x) = \frac{d^n}{dx^n}(x^n e^{-x}). \tag{5.118}$$

We express this as

$$w(x) = e^{-x} L(x), \tag{5.119}$$

where $L(x)$ is a *Laguerre polynomial*. Multiplying by a constant such that $L_n(0) = 0$ for all n, the Laguerre polynomials can be defined by *Rodrigues' formula*:

$$L_n(x) = \frac{e^x}{n!}\frac{d^n}{dx^n}(x^n e^{-x}). \tag{5.120}$$

A generalization of Eq. (5.114)

$$xy''(x) + (\alpha + 1 - x)y'(x) + ny(x) = 0 \tag{5.121}$$

has solutions known as *associated Laguerre polynomials*, given by the Rodrigues' formula

$$L_n^\alpha(x) = \frac{x^{-\alpha}e^x}{n!}\frac{d^n}{dx^n}(x^{n+\alpha}e^{-x}). \tag{5.122}$$

Laguerre and associated Laguerre polynomials can be determined using the following generating functions:

$$\frac{1}{(1-t)}\exp\left(-\frac{xt}{1-t}\right) = \sum_{n=0}^{\infty} L_n(x)t^n, \tag{5.123}$$

$$\frac{1}{(1-t)^{\alpha+1}}\exp\left(-\frac{xt}{1-t}\right) = \sum_{n=k}^{\infty} L_n^\alpha(x)t^n. \tag{5.124}$$

From these, one can derive the orthonormalization relations

$$\int_0^\infty e^{-x}x^\alpha L_n^\alpha(x)L_m^\alpha(x)\,dx = \frac{\Gamma(n+\alpha+1)}{n!}\delta_{nm}. \tag{5.125}$$

5.10 Series solutions of differential equations

A very general method for obtaining solutions to second-order differential equations is to expand $y(x)$ in a power series and then evaluate the coefficients term by term. In the majority of quantum mechanics textbooks, this is the method by which Hermite, Legendre and Laguerre functions are derived. Although we have made use of more direct ad hoc methods of creating the differential equations for these special functions, it is worthwhile to introduce the series solution method. We will, in fact, use it to as our approach to Bessel functions.

First we will illustrate the method with a trivial example which we have already solved, namely the equation with constant coefficients:

$$y''(x) + k^2 y(x) = 0. \tag{5.126}$$

Assume that $y(x)$ can be expanded in a power series about $x = 0$:

$$y(x) = \sum_{n=0}^{\infty} a_n x^n. \tag{5.127}$$

The first derivative is given by

$$y'(x) = \sum_{n=1}^{\infty} n a_n x^{n-1} \overset{n\to n+1}{\longrightarrow} \sum_{n=0}^{\infty} (n+1) a_{n+1} x^n. \tag{5.128}$$

We have redefined the summation index in order to retain the dependence on x^n. Analogously,

$$y''(x) = \sum_{n=2}^{\infty} n(n-1) a_n x^{n-2} \overset{n\to n+2}{\longrightarrow} \sum_{n=0}^{\infty} (n+2)(n+1) a_{n+2} x^n. \tag{5.129}$$

Eq. (5.126) then implies

$$\sum_{n=0}^{\infty} \Big[(n+2)(n+1) a_{n+2} + k^2 a_n\Big] x^n = 0. \tag{5.130}$$

Since this is true for all values of x, every quantity in square brackets must equal zero. This leads to the *recursion relation*

$$a_{n+2} = -\frac{k^2}{(n+2)(n+1)} a_n. \tag{5.131}$$

Let a_0 be treated as one constant of integration. We then find

$$a_2 = -\frac{k^2}{2\cdot 1} a_0, \quad a_4 = -\frac{k^2}{4\cdot 3} a_2 = +\frac{k^4}{4!} a_0, \quad \cdots. \tag{5.132}$$

It is convenient to rewrite the coefficient a_1 as ka_1. We find thereby

$$a_3 = -\frac{k^3}{3 \cdot 2} a_1, \quad a_5 = +\frac{k^5}{5!} a_1, \quad \cdots. \tag{5.133}$$

The general power-series solution of the differential equation is thus given by

$$y(x) = \left(1 - \frac{k^2x^2}{2!} + \frac{k^4x^4}{4!} - \cdots\right) a_0 + \left(x - \frac{k^3x^3}{3!} + \frac{k^5x^5}{5!} - \cdots\right) a_1, \tag{5.134}$$

which is recognized as the expansion for

$$y(x) = a_0 \cos kx + a_1 \sin kx. \tag{5.135}$$

We consider next the more general case of a linear homogeneous second-order differential equation with nonconstant coefficients:

$$y''(x) + p(x)y'(x) + q(x)y(x) = 0. \tag{5.136}$$

If the functions $p(x)$ and $q(x)$ are both finite at $x = x_0$, then x_0 is called a *regular point* of the differential equation. If either $p(x)$ or $q(x)$ diverges as $x \to x_0$, then x_0 is called a *singular point*. If both $(x - x_0)p(x)$ and $(x - x_0)^2 q(x)$ have finite limits as $x \to x_0$, then x_0 is called a *regular singular point* or *nonessential singularity*. If either of these limits continues to diverge, the point is an *essential singularity*.

For regular singular points, a series solution of the differential equation can be obtained by the *method of Frobenius*. This is based on the following generalization of the power series expansion:

$$y(x) = x^\alpha \sum_{k=0}^{\infty} a_k x^k = \sum_{k=0}^{\infty} a_k x^{k+\alpha}. \tag{5.137}$$

The derivatives are then given by

$$y'(x) = \sum_{k=0}^{\infty} (\alpha + k) a_k x^{k+\alpha-1} \tag{5.138}$$

and

$$y''(x) = \sum_{k=0}^{\infty} (\alpha + k)(\alpha + k - 1) a_k x^{k+\alpha-2}. \tag{5.139}$$

The possible values of α are obtained from the *indicial equation*, which is based on the presumption that a_0 is the first nonzero coefficient in the series (5.137).

5.11 Bessel functions

The solutions of *Bessel's differential equation*:

$$x^2 y''(x) + xy'(x) + (x^2 - n^2)y(x) = 0 \tag{5.140}$$

are one of the classics of mathematical physics. Bessel's equation occurs, in particular, in a number of applications involving cylindrical coordinates. Dividing the standard form (5.140) by x^2 shows that $x = 0$ is a regular singular point of Bessel's equation. The method of Frobenius is thus applicable. Substituting the power series expansion (5.137) into (5.140), we obtain

$$\sum_{k=0}^{\infty} \Big[(\alpha + k)(\alpha + k - 1)a_k + (\alpha + k)a_k + a_{k-2} - n^2 a_k\Big] x^{k+\alpha} = 0. \tag{5.141}$$

This leads to the recursion relation

$$a_{k-2} = -[(\alpha + k)(\alpha + k - 1) + (\alpha + k) - n^2]a_k. \tag{5.142}$$

Setting $k = 0$ in the recursion relation and noting that $a_{-2} = 0$ (a_0 is the first nonvanishing coefficient), we obtain the indicial equation

$$\alpha(\alpha - 1) + \alpha - n^2 = 0. \tag{5.143}$$

The roots are $\alpha = \pm n$. With the choice $\alpha = n$, the recursion relation simplifies to

$$a_k = -\frac{a_{k-2}}{k(2n + k)}. \tag{5.144}$$

Since $a_{-1} = 0$, $a_1 = 0$ (assuming $n \neq -\frac{1}{2}$). Likewise $a_3, a_5, \cdots = 0$, as do all odd a_k. For even k, we have

$$a_2 = -\frac{a_0}{2(2n + 2)}, \quad a_4 = -\frac{a_2}{4(2 + 4n)} = \frac{a_0}{2 \cdot 4(2n + 2)(2n + 4)}. \tag{5.145}$$

For $n = 0, 1, 2 \ldots$, the coefficients can be represented by

$$a_{2k} = (-)^k \frac{n! a_0}{2^{2k} k!(n + k)!}. \tag{5.146}$$

From now on, we will use the compact notation

$$(-)^k \equiv (-1)^k. \tag{5.147}$$

Setting $a_0 = 1/2^n n!$, we obtain the conventional definition of a *Bessel function of the first kind*:

$$J_n(x) = \left(\frac{x}{2}\right)^n \left[1 - \frac{(x/2)^2}{2!(n + 2)!} + \frac{(x/2)^4}{4!(n + 4)!} - \cdots\right]. \tag{5.148}$$

The first three Bessel functions are plotted in Fig. 5.7. Their general behavior can be characterized as damped oscillation. The Bessel functions are normalized in the sense that

$$\int_0^\infty J_n(x)\,dx = 1. \tag{5.149}$$

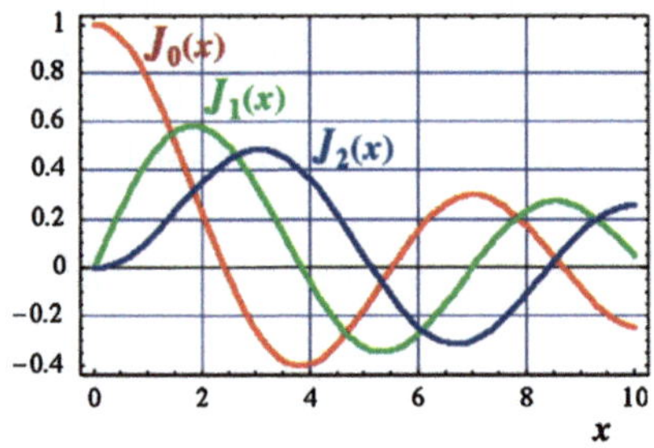

FIGURE 5.7 Bessel functions of the first kind $J_n(x)$.

Bessel functions can be generalized for noninteger index ν, as follows:

$$J_\nu(x) = \left(\frac{x}{2}\right)^\nu \left[1 - \frac{(x/2)^2}{2!\Gamma(\nu+3)} + \frac{(x/2)^4}{4!\Gamma(\nu+5)} - \cdots\right]. \tag{5.150}$$

The general solution of Bessel's differential equation for noninteger ν is given by

$$y(x) = c_1 J_\nu(x) + c_2 J_{-\nu}(x). \tag{5.151}$$

For integer n, however,

$$J_{-n}(x) = (-)^n J_n(x), \tag{5.152}$$

so that $J_{-n}(x)$ is *not* a linearly independent solution. We can construct a second solution by defining

$$Y_\nu(x) = \frac{\cos\nu\pi\, J_\nu(x) - J_{-\nu}(x)}{\sin\nu\pi}. \tag{5.153}$$

In the limit as ν approaches an integer n, we obtain

$$Y_n(x) = \frac{1}{\pi}\left[\frac{\partial}{\partial n}J_n(x) - (-1)^n\frac{\partial}{\partial n}J_{-n}(x)\right]. \tag{5.154}$$

This defines a *Bessel function of the second kind*, also known as a *Neumann function*. The computational details are horrible, but fortunately, mathematicians have worked them all out for us and these functions have been extensively tabulated. Fig. 5.8 shows the first three functions $Y_n(x)$.

The limiting behavior of the $Y_n(x)$ as $x \to 0$ is apparent from the leading term $J_n(x) \approx (x/2)^n$. Using the definition (5.154) we find

$$Y_n(x) = \frac{2}{\pi} \ln\left(\frac{x}{2}\right) J_n(x) + \text{complicated series.} \tag{5.155}$$

Fig. 5.8 shows the logarithmic singularities as $x \to 0$.

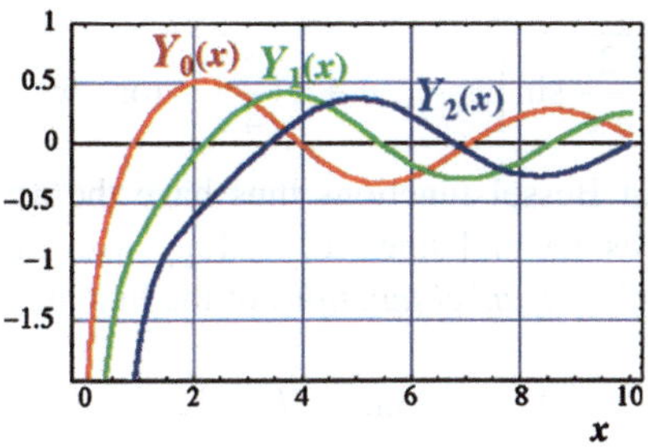

FIGURE 5.8 Bessel functions of the second kind $Y_n(x)$.

Bessel functions of integer order can be determined from expansion of a *generating function*:

$$\exp\left[\frac{x}{2}\left(t - \frac{1}{t}\right)\right] = \sum_{n=-\infty}^{\infty} J_n(x)t^n. \tag{5.156}$$

To see how this works, expand the product of the two exponentials:

$$e^{xt/2}e^{-x/2t} = \left[1 + \frac{x}{2}t + \frac{1}{2!}\left(\frac{x}{2}\right)^2 t^2 + \cdots\right]\left[1 - \frac{x}{2}t^{-1} + \frac{1}{2!}\left(\frac{x}{2}\right)^2 t^{-2} - \cdots\right]. \tag{5.157}$$

The terms which contribute to t^0 are

$$1 - \left(\frac{x}{2}\right)^2 + \frac{1}{(2!)^2}\left(\frac{x}{2}\right)^4 - \cdots, \tag{5.158}$$

which gives the expansion for $J_0(x)$. The expansion for $J_1(x)$ is found from the terms proportional to t^1 and so forth.

Asymptotic forms for the Bessel function for small and large values of the argument are given by

$$J_n(x) \approx \frac{1}{n!}\left(\frac{x}{2}\right)^n \quad \text{for} \quad x \to 0 \tag{5.159}$$

and

$$J_n(x) \approx \sqrt{\frac{2}{\pi x}} \cos\left[x - (2n+1)\frac{\pi}{4}\right] \quad \text{for} \quad x >> n. \tag{5.160}$$

Bessel functions of the second kind show the following limiting behavior:

$$Y_0(x) \approx \frac{2}{\pi} \ln x, \qquad Y_n(x) \approx -\frac{2^n (n-1)!}{\pi} x^{-n} \ (n > 0) \quad \text{for} \quad x \to 0 \tag{5.161}$$

and

$$Y_n(x) \approx \sqrt{\frac{2}{\pi x}} \sin\left[x - (2n+1)\frac{\pi}{4}\right] \quad \text{for} \quad x >> n. \tag{5.162}$$

The two kinds of Bessel functions thus have the asymptotic dependence of slowly damped cosines and sines. In analogy with Euler's formula $e^{\pm ix} = \cos x \pm i \sin x$, we define *Hankel functions* of the first and second kind:

$$H_n^{(1)}(x) = J_n(x) + iY_n(x) \quad \text{and} \quad H_n^{(2)}(x) = J_n(x) - iY_n(x). \tag{5.163}$$

The asymptotic forms of the Hankel functions is given by

$$H_n^{(1,2)}(x) \approx \sqrt{\frac{2}{\pi x}} e^{\pm i\,[x-(2n+1)\pi/4]} \quad \text{for} \quad x >> n, \tag{5.164}$$

having the character of waves propagating to the right and left, respectively.

Trigonometric functions with imaginary arguments suggested the introduction of hyperbolic functions: $\sinh x = -i \sin ix$ and $\cosh x = \cos ix$. Analogously, we can define *modified Bessel functions* (or *hyperbolic Bessel functions*) of the first and second kind as follows:

$$I_n(x) = i^{-n} J_n(ix) \tag{5.165}$$

and

$$K_n(x) = \frac{\pi}{2} i^{n+1} \Big[J_n(ix) + iY_n(ix) \Big] = \frac{\pi}{2} i^{n+1} H_n^{(1)}(ix). \tag{5.166}$$

Their asymptotic forms for large x are

$$I_n(x) \approx \frac{e^x}{\sqrt{2\pi x}} \quad \text{and} \quad K_n(x) \approx \sqrt{\frac{\pi}{2x}} e^{-x} \quad \text{for}\, x >> n. \tag{5.167}$$

We have given just a meager sampling of formulas involving Bessel functions. Many more can be found in Abramowitz & Stegun and other references. G. N. Whittaker's classic *A Treatise on the Theory of Bessel Functions* (Cambridge University Press, 1952) is a ponderous volume devoted entirely to the subject.

5.12 Spherical Bessel functions

A number of problems in spherical polar coordinates lead to a differential equation of the form

$$x^2 f''(x) + 2xf'(x) + \Big[x^2 - n(n+1)\Big] f(x) = 0. \tag{5.168}$$

With the substitution $f(x) = x^{-1/2}F(x)$, the equation reduces to

$$x^2 F''(x) + xF'(x) + \Big[x^2 - (n+\tfrac{1}{2})^2\Big] F(x) = 0, \tag{5.169}$$

which, for integer n, is recognized as Bessel's equation of odd-half order $\frac{1}{2}, \frac{3}{2}, \frac{5}{2} \ldots$. The linearly-independent solutions are $J_{n+1/2}(x)$ and $Y_{n+1/2}(x)$, the latter being proportional to $J_{-n-1/2}(x)$ since the order is not an integer. The solutions to Eq. (5.168) are known as *spherical Bessel functions* and conventionally defined by

$$j_n(x) = \sqrt{\frac{\pi}{2x}}\, J_{n+1/2}(x) \tag{5.170}$$

and

$$y_n(x) = \sqrt{\frac{\pi}{2x}}\, Y_{n+1/2}(x) = (-)^{n+1}\sqrt{\frac{\pi}{2x}}\, J_{-n-1/2}(x). \tag{5.171}$$

You can verify that Eq. (5.168) with $n = 0$ has the simple solutions $\sin x/x$ and $\cos x/x$. In fact, spherical Bessel functions have closed-form expressions in terms of trigonometric functions and powers of x, given by

$$j_n(x) = (-x)^n \left(\frac{1}{x}\frac{d}{dx}\right)^n \frac{\sin x}{x}, \quad y_n(x) = -(-x)^n \left(\frac{1}{x}\frac{d}{dx}\right)^n \frac{\cos x}{x}. \tag{5.172}$$

Explicit formulas for the first few spherical Bessel functions follow:

$$\begin{aligned} j_0(x) &= \frac{\sin x}{x}, \quad j_1(x) = \frac{\sin x}{x^2} - \frac{\cos x}{x}, \quad j_2(x) = \left(\frac{3}{x^2} - 1\right)\frac{\sin x}{x} - \frac{3\cos x}{x^2}, \\ y_0(x) &= -\frac{\cos x}{x}, \quad y_1(x) = -\frac{\cos x}{x^2} - \frac{\sin x}{x}, \quad y_2(x) = \left(-\frac{3}{x^2} + 1\right)\frac{\cos x}{x} - \frac{3\sin x}{x^2}. \end{aligned} \tag{5.173}$$

There are also spherical analogs of the Hankel functions:

$$h_n^{(1,2)}(x) = \sqrt{\frac{\pi}{2x}}\, H_{n+1/2}^{(1,2)}(x) = j_n(x) \pm i y_n(x). \tag{5.174}$$

The first few are

$$h_0^{(1,2)}(x) = \mp\frac{e^{\pm ix}}{x},\ h_1^{(1,2)}(x) = -\frac{x \pm i}{x^2}e^{\pm ix},\ h_2^{(1,2)}(x) = \pm i\frac{x^2 \pm 3ix - 3}{x^3}e^{\pm ix}. \tag{5.175}$$

Supplement 5A. Particle in a disk

With knowledge of Bessel functions, we can derive an exact solution for the two-dimensional problem of a particle confined within a disk. The Schrödinger equation, most naturally expressed in polar coordinates, has the form

$$-\frac{\hbar^2}{2M}(\nabla^2\psi(r,\theta) + V(r)\psi(r,\theta) = E\psi(r,\theta), \tag{5.176}$$

with the potential function

$$V(r) = \begin{cases} 0 & r \le r_0 \\ \infty & r > r_0. \end{cases} \tag{5.177}$$

With the definition

$$k^2 = 2ME/\hbar^2 \tag{5.178}$$

this takes the form of the Helmholtz equation for $r \le r_0$

$$(\nabla^2 + k^2)\psi(r,\theta) = 0. \tag{5.179}$$

Using the form of the Laplacian in polar coordinates, this becomes

$$\frac{1}{r}\frac{\partial}{\partial r}\left(r\frac{\partial\psi}{\partial r}\right) + \frac{1}{r^2}\frac{\partial^2\psi}{\partial\theta^2} + k^2\psi(r,\theta) = 0. \tag{5.180}$$

This is subject to the boundary condition $\psi(r_0,\theta) = 0$. Assuming a separable solution $\psi(r,\theta) = R(r)\Theta(\theta)$, Helmholtz's equation reduces to

$$R''(r)\Theta(\theta) + r^{-1}R'(r)\Theta(\theta) + r^{-2}R(r)\Theta''(\theta) + k^2R(r)\Theta(\theta) = 0 \tag{5.181}$$

or

$$R''(r) + r^{-1}R'(r) + r^{-2}R(r)\left[\frac{\Theta''(\theta)}{\Theta(\theta)}\right] + k^2R(r) = 0. \tag{5.182}$$

We complete a separation of variables by dividing by $r^{-2}R(r)$. The result implies that the function of θ in square brackets equals a constant, which we can write

$$\frac{\Theta''(\theta)}{\Theta(\theta)} = -m^2 \quad \Rightarrow \quad \Theta''(\theta) + m^2\,\Theta(\theta) = 0. \tag{5.183}$$

This is a familiar equation, with linearly-independent solutions

$$\Theta(\theta) = \begin{cases} \sin m\theta \\ \cos m\theta. \end{cases} \tag{5.184}$$

Since θ is an angular variable, the periodicity $\Theta(\theta \pm 2n\pi) = \Theta(\theta)$ is required. This restricts the parameter m to integer values. Thus we obtain an ordinary differential equation for $R(r)$:

$$R''(r) + r^{-1}R'(r) - m^2r^{-2}R(r) + k^2R(r) = 0. \tag{5.185}$$

Changing the variable to $x \equiv kr$, with $J(x) \equiv R(r)$, we can write

$$x^2J''(x) + xJ'(x) + (x^2 - m^2)J(x) = 0. \tag{5.186}$$

This is recognized as Bessel's equation (5.140). Only the Bessel functions $J_m(kr)$ are finite at $r = 0$, the Neumann functions $Y_m(kr)$ being singular there. Thus the solutions to Helmholtz's equation are

$$\psi(r,\theta) = J_m(kr)\begin{cases} \sin m\theta & m = 1, 2, 3\ldots \\ \cos m\theta & m = 0, 1, 2\ldots. \end{cases} \tag{5.187}$$

The boundary condition at $r = r_0$ requires that

$$J_m(kr_0) = 0. \tag{5.188}$$

The zeros of Bessel functions are extensively tabulated. Let x_{mn} represent the nth root of $J_m(x) = 0$. A tabulation of the first few zeros follows:

n	x_{0n}	x_{1n}	x_{2n}
1	2.4048	3.8317	5.1356
2	5.5201	7.0156	8.4172
3	8.6537	10.1735	11.6198

The corresponding values of k are given by

$$k_{mn} = x_{mn}/r_0. \tag{5.189}$$

Contour plots for the first few wavefunctions $\psi_{nm}(r,\theta)$, labeled by the quantum numbers m and n, are sketched in Fig. 5.9. The states with $m > 0$ are two-fold degenerate, corresponding to the factors $\cos m\theta$ and $\sin m\theta$. Only one of the degenerate pair is shown. The second would be obtained by a rotation of $\pi/2m$. Incidentally, an identical diagram would pertain to modes of vibration (m,n) of a circular membrane, rigidly fixed on a circular boundary of radius r_0.

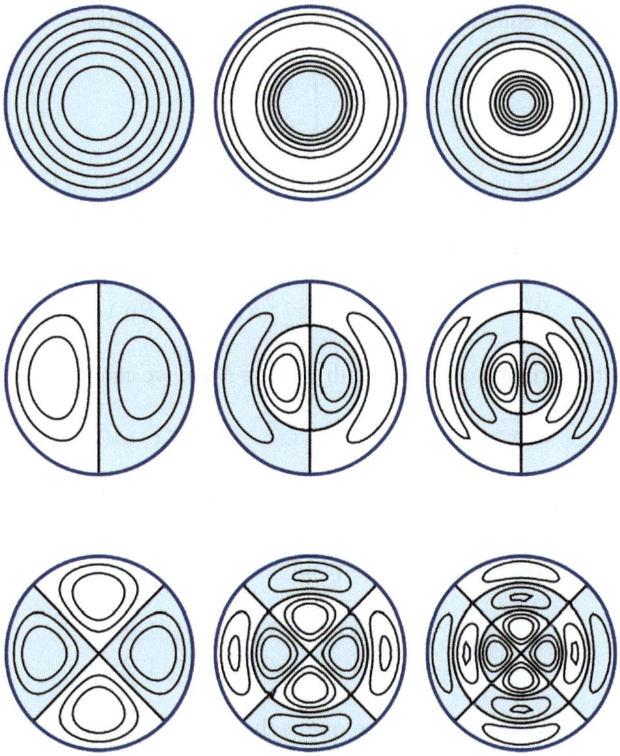

FIGURE 5.9 Several eigenfunctions $\psi_{mn}(r, \theta)$ for the particle in a disk. The functions are positive in blue areas, negative in white. Rows show $m = 0, 1, 2$, columns $n = 1, 2, 3$.

Supplement 5B. Particle in an infinite spherical well

A particle of mass in an infinite spherical potential well of radius is described by the Schrödinger equation

$$-\frac{\hbar^2}{2M}\nabla^2\psi(r,\theta,\phi) + V(r)\psi(r,\theta,\phi) = E\psi(r,\theta,\phi), \tag{5.190}$$

where

$$V(r) = \begin{cases} 0 & r \le r_0 \\ \infty & r > r_0. \end{cases} \tag{5.191}$$

The wavefunction is separable in spherical polar coordinates, such that

$$\psi(r,\theta,\phi) = R(r)Y_{\ell m}(\theta,\phi), \tag{5.192}$$

where Y is a spherical harmonic. With the definition

$$k^2 = 2mE/\hbar^2 \tag{5.193}$$

the radial equation reduces to

$$R''(r) + \frac{2}{r}R'(r) + \left(k^2 - \frac{\ell(\ell+1)}{r^2}\right)R(r) = 0, \tag{5.194}$$

subject to the boundary condition $R(r_0) = 0$. With the scaled radial variable $x = kr$, this reduces to the equation for a spherical Bessel function (5.168)

$$x^2 R''(x) + 2xR'(x) + \left[x^2 - \ell(\ell+1)\right]R(x) = 0. \tag{5.195}$$

Thus $R(r) = \text{const}\, j_\ell(kr)$. The boundary condition is fulfilled when kr_0 is the zero of the spherical Bessel function $j_\ell(x)$. Let $x_{\ell n}$ represent the nth root of $j_\ell(x) = 0$. A tabulation of the first few zeros follows:

n	x_{0n}	x_{1n}	x_{2n}
1	3.142	4.493	5.763
2	6.283	7.725	10.904
3	9.425	10.904	12.323

The corresponding values of k are $k_{\ell n} = x_{\ell n}/r_0$. The quantized energy levels are then given by

$$E_{n\ell} = \frac{\hbar^2 k_{\ell n}^2}{2Mr_0^2}. \tag{5.196}$$

Since these are independent of the quantum number m, they are $(2\ell+1)$-fold degenerate.

Fig. 5.10 shows contour plots on a cross section of the sphere for the lower-energy eigenfunctions with $n = 1, 2, 3$ and $\ell = 0, 1, 2$, with $m = 0$. The wavefunction is positive in the blue regions and negative in the white regions.

Supplement 5C. Particle in a deltafunction well

A simple one-dimensional problem is a particle in a deltafunction potential well, described by the Schrödinger equation (in atomic units):

$$-\frac{1}{2}\psi''(x) - Z\delta(x)\psi(x) = E\psi(x). \tag{5.197}$$

Since the deltafunction has the same dimensions as the function $1/|x|$, this problem might be considered a one-dimensional analog of a hydrogenlike atom. For a bound state, $E < 0$, and it is convenient to write $E = -\kappa^2/2$. For $x \neq 0$, the Schrödinger equation reduces to

$$\psi_0''(x) - \kappa^2\psi_0(x) = 0, \tag{5.198}$$

which has the solutions $\psi_0(x) = e^{\pm\kappa x}$. To get the correct behavior as $x \to \pm\infty$, we choose, for $x > 0$, $e^{-\kappa x}$, while for $x < 0$, $e^{\kappa x}$. These can be combined as

FIGURE 5.10 Spherical-well eigenfunctions $\psi_{n\ell 0}(r,\theta,\phi)$ in plane $\theta=\pi/2$. Rows show $\ell = 0, 1, 2$, columns $n = 1, 2, 3$.

a single function $\psi_0(x) = e^{-\kappa|x|}$. We next substitute this form into the original Schrödinger equation. We need to calculate the second derivative of this function. Note first that

$$\frac{d}{dx}e^{-\kappa|x|} = -\kappa\, e^{-\kappa|x|}\frac{d|x|}{dx}. \tag{5.199}$$

Clearly $\frac{d|x|}{dx} = \text{sgn}(x)$, equal to ±1 for $x > 0$ and $x < 0$, respectively. Thus

$$\frac{d}{dx}e^{-\kappa|x|} = -\kappa\, \text{sgn}(x)e^{-\kappa|x|} \tag{5.200}$$

and

$$\frac{d^2}{dx^2}e^{-\kappa|x|} = \kappa^2\, \text{sgn}(x)^2 e^{-\kappa|x|} - \kappa\, e^{-\kappa|x|}\text{sgn}'(x). \tag{5.201}$$

Using the relation $\text{sgn}'(x) = 2\delta(x)$, the Schrödinger equation (5.197) is satisfied when

$$-\frac{1}{2}\kappa^2 + \kappa\,\delta(x) - Z\delta(x) = E. \tag{5.202}$$

Thus we identify $\kappa = Z$ and $E = -\frac{1}{2}\kappa^2 = -Z^2/2$. This is the only bound solution. Remarkably, the ground-state eigenfunction e^{-Zx} is identical to a cross-section of a hydrogenic $1s$ orbital e^{-Zr}, with the same energy $E = -Z^2/2$.

Chapter 6

The harmonic oscillator

The harmonic oscillator is a model which has several important applications in both classical and quantum mechanics. It serves as a prototype in the mathematical treatment of such diverse phenomena as elasticity, acoustics, AC circuits, molecular and crystal vibrations, electromagnetic fields and optical properties of matter.

6.1 Classical oscillator

A simple realization of the harmonic oscillator in classical mechanics is a particle which is acted upon by a restoring force proportional to its displacement from its equilibrium position. Considering motion in one dimension, this means

$$F = -kx. \tag{6.1}$$

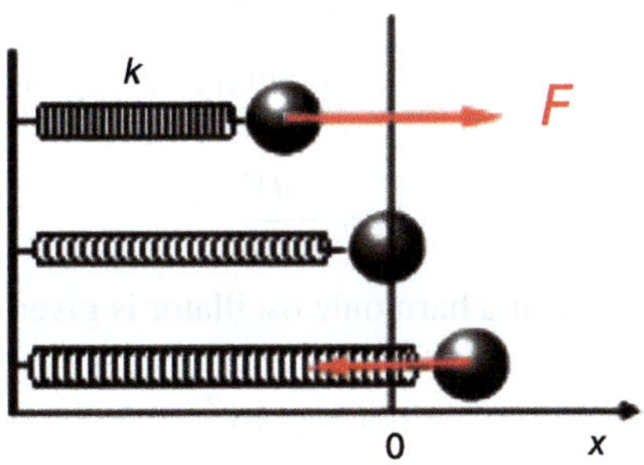

FIGURE 6.1 Spring obeying Hooke's law.

Such a force might originate from a spring which obeys Hooke's law, as shown in Fig. 6.1. According to Hooke's law, which applies to real springs for sufficiently small displacements, the restoring force is proportional to the displacement—either stretching or compression—from the equilibrium position. The *force constant* k is a measure of the stiffness of the spring. The variable x is chosen equal to zero at the equilibrium position, positive for stretching, negative for compression. The negative sign in (6.1) reflects the fact that F is a *restoring* force, always in the opposite sense to the displacement x.

Applying Newton's second law to the force from Eq. (6.1), we find

$$F = m\frac{d^2x}{dx^2} = -kx, \tag{6.2}$$

Introduction to Quantum Mechanics. https://doi.org/10.1016/B978-0-12-822310-9.00014-8

where m is the mass of the body attached to the spring, which is itself assumed massless. This leads to a differential equation of familiar form, although with different variables:

$$\ddot{x}(t) + \omega^2 x(t) = 0, \qquad \omega^2 \equiv k/m. \tag{6.3}$$

The dot notation (introduced by Newton himself) is used in place of primes when the independent variable is time. The general solution to (6.3) is

$$x(t) = A \sin \omega t + B \cos \omega t, \tag{6.4}$$

which represents periodic motion with a sinusoidal time dependence. This is known as *simple harmonic motion* and the corresponding system is known as a *harmonic oscillator*. The oscillation occurs with a constant angular frequency

$$\omega = \sqrt{\frac{k}{m}} \quad \text{radians per second.} \tag{6.5}$$

This is called the *natural frequency* of the oscillator. The corresponding circular frequency in hertz (cycles per second) is

$$\nu = \frac{\omega}{2\pi} = \frac{1}{2\pi}\sqrt{\frac{k}{m}} \quad \text{Hz.} \tag{6.6}$$

The general relation between force and potential energy in a conservative system in one dimension is

$$F = -\frac{dV}{dx}. \tag{6.7}$$

Thus the potential energy of a harmonic oscillator is given by

$$V(x) = \frac{1}{2} k x^2, \tag{6.8}$$

which has the shape of a parabola, as drawn in Fig. 6.2. A simple computation shows that the oscillator moves between positive and negative turning points $\pm x_{\max}$ where the total energy E equals the potential energy $\frac{1}{2} k x_{\max}^2$ while the kinetic energy is momentarily zero. In contrast, when the oscillator moves past $x = 0$, the kinetic energy reaches its maximum value while the potential energy equals zero.

6.2 Quantum harmonic oscillator

Given the potential energy (6.8), we can write down the Schrödinger equation for the one-dimensional harmonic oscillator:

$$-\frac{\hbar^2}{2m}\psi''(x) + \frac{1}{2}kx^2\psi(x) = E\,\psi(x). \tag{6.9}$$

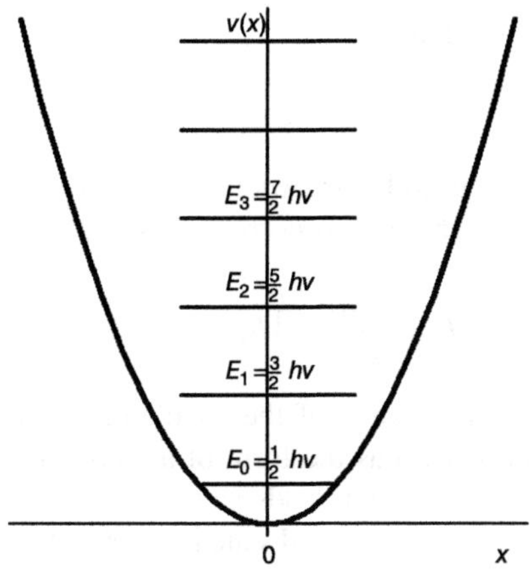

FIGURE 6.2 Potential energy function and first few energy levels for harmonic oscillator.

For the first time we encounter a differential equation with *non-constant* coefficients, which is a much greater challenge to solve. We can combine the constants in (6.9) to two parameters

$$\alpha^2 = \frac{mk}{\hbar^2} \quad \text{and} \quad \lambda = \frac{2mE}{\hbar^2 \alpha} \tag{6.10}$$

and redefine the independent variable as

$$\xi = \alpha^{1/2} x. \tag{6.11}$$

This reduces the Schrödinger equation to

$$\psi''(\xi) + (\lambda - \xi^2)\psi(\xi) = 0. \tag{6.12}$$

The range of the variable x (also ξ) must be taken from $-\infty$ to $+\infty$, there being no finite cutoff as in the case of the particle in a box.

A useful first step is to determine the asymptotic solution to (6.12), that is, the form of $\psi(\xi)$ as $\xi \to \pm\infty$. For sufficiently large values of $|\xi|$, $\xi^2 >> \lambda$ and the differential equation is approximated by

$$\psi''(\xi) - \xi^2 \psi(\xi) \approx 0. \tag{6.13}$$

This suggests the following manipulation:

$$\left(\frac{d^2}{d\xi^2} - \xi^2\right)\psi(\xi) \approx \left(\frac{d}{d\xi} - \xi\right)\left(\frac{d}{d\xi} + \xi\right)\psi(\xi) \approx 0. \tag{6.14}$$

The first-order differential equation

$$\psi'(\xi) + \xi\psi(\xi) = 0 \tag{6.15}$$

can be solved exactly to give

$$\psi(\xi) = \text{const}\, e^{-\xi^2/2}. \tag{6.16}$$

Remarkably, this turns out to be an *exact* solution of the Schrödinger equation (6.12) with $\lambda = 1$. Using (6.10), this corresponds to an energy

$$E = \frac{\hbar^2\alpha}{2m} = \tfrac{1}{2}\hbar\sqrt{\frac{k}{m}} = \tfrac{1}{2}\hbar\omega, \tag{6.17}$$

where ω is the natural frequency of the oscillator according to classical mechanics. The function (6.16) has the form of a gaussian, or bell-shaped curve. The function has no nodes, which leads us to conclude that this represents the ground state of the system. The ground state is usually designated with the quantum number $n = 0$ (the particle in a box is a exception, with $n = 1$ labeling the ground state). Reverting to the original variable x, we write

$$\psi_0(x) = \text{const}\, e^{-\alpha x^2/2}, \quad \alpha = (mk/\hbar^2)^{1/2}. \tag{6.18}$$

With help of the well-known definite integral

$$\int_{-\infty}^{\infty} e^{-\alpha x^2}\, dx = \left(\frac{\pi}{\alpha}\right)^{1/2} \tag{6.19}$$

we find the normalized eigenfunction

$$\psi_0(x) = \left(\frac{\alpha}{\pi}\right)^{1/4} e^{-\alpha x^2/2}, \tag{6.20}$$

with the corresponding eigenvalue

$$E_0 = \tfrac{1}{2}\hbar\omega. \tag{6.21}$$

6.3 Harmonic-oscillator eigenfunctions and eigenvalues

Drawing from our experience with the particle in a box, we might surmise that the first excited state of the harmonic oscillator would be a function similar to (6.20), but with a node at $x = 0$, say,

$$\psi_1(x) = \text{const}\, x\, e^{-\alpha x^2/2}. \tag{6.22}$$

This is orthogonal to $\psi_0(x)$ by symmetry and is indeed an eigenfunction with the eigenvalue

$$E_1 = \tfrac{3}{2}\hbar\omega. \tag{6.23}$$

Continuing the process, we try a function with two nodes

$$\psi_2(x) = \text{const}\, (x^2 - a)\, e^{-\alpha x^2/2}. \tag{6.24}$$

Using integrals calculated in Sect. 5.1, we determine that $a = 1/2$ makes $\psi_2(x)$ orthogonal to both $\psi_0(x)$ and $\psi_1(x)$. We verify that this is another eigenfunction, corresponding to

$$E_2 = \tfrac{5}{2}\hbar\omega. \tag{6.25}$$

The general result involves Hermite polynomials, which are discussed in Sect. 5.4. We can thereby write the normalized eigenfunctions:

$$\psi_n(x) = \left(\frac{\sqrt{\alpha}}{2^n n! \sqrt{\pi}}\right)^{1/2} H_n(\sqrt{\alpha}\, x)\, e^{-\alpha x^2/2}, \tag{6.26}$$

where $H_n(\xi)$ represents the Hermite polynomial of degree n. The first few Hermite polynomials are

$$H_0(\xi) = 1, \quad H_1(\xi) = 2\xi, \quad H_2(\xi) = 4\xi^2 - 2, \quad H_3(\xi) = 8\xi^3 - 12\xi,$$
$$H_4(\xi) = 16\xi^4 - 48\xi^2 + 12, \quad H_5(\xi) = 32\xi^5 - 160\xi^3 + 120\xi. \tag{6.27}$$

The four lowest harmonic-oscillator eigenfunctions are plotted in Fig. 6.3. Note the topological resemblance to the corresponding particle-in-a-box eigenfunctions.

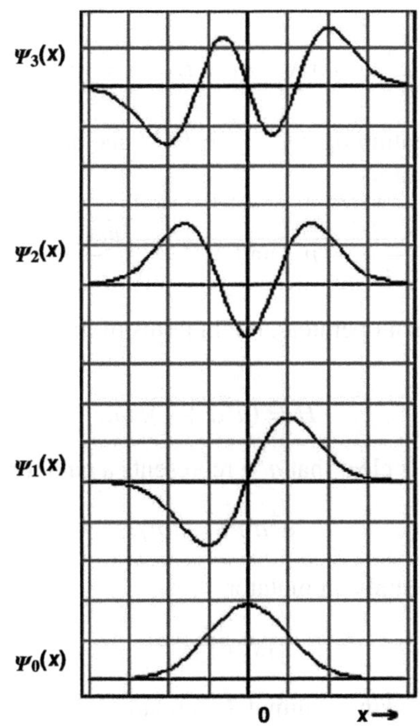

FIGURE 6.3 Harmonic oscillator eigenfunctions for n=0, 1, 2, 3.

The eigenvalues are given by the simple formula

$$E_n = (n + \tfrac{1}{2})\hbar\omega. \tag{6.28}$$

These are drawn in Fig. 6.2, on the same scale as the potential energy. The ground-state energy $E_0 = \frac{1}{2}\hbar\omega$ is greater than the classical value of zero, again a consequence of the uncertainty principle.

It is remarkable that the difference between successive energy eigenvalues has a constant value

$$\Delta E = E_{n+1} - E_n = \hbar\omega = h\nu. \tag{6.29}$$

This is reminiscent of Planck's formula for the energy of a photon. It comes as no surprise then that the quantum theory of radiation has the structure of an assembly of oscillators, with each oscillator representing a mode of electromagnetic waves of a specified frequency.

6.4 Operator formulation of harmonic oscillator

To develop the quantum theory of electromagnetic radiation, it is useful to reformulate the harmonic-oscillator problem in terms of creation and annihilation operators, following a derivation due to Dirac. The Schrödinger equation (6.9) can be written

$$H|n\rangle = \frac{1}{2m}(p^2 + m\omega^2 x^2)|n\rangle = E_n\,|n\rangle = (n + \tfrac{1}{2})\hbar\omega|n\rangle, \tag{6.30}$$

where p is the momentum operator $-i\hbar d/dx$ and $\omega = \sqrt{k/m}$. Now define the mutually adjoint operators

$$a = \sqrt{\frac{m\omega}{2\hbar}}\,x + \frac{i}{\sqrt{2m\hbar\omega}}\,p \quad \text{and} \quad a^\dagger = \sqrt{\frac{m\omega}{2\hbar}}\,x - \frac{i}{\sqrt{2m\hbar\omega}}\,p. \tag{6.31}$$

These are *not* hermitian operators, but in terms of them the Hamiltonian can be expressed

$$H = (a^\dagger a + \tfrac{1}{2})\hbar\omega. \tag{6.32}$$

Comparing (6.30), it is clear that $a^\dagger a$ represents a *number operator*, such that

$$a^\dagger a|n\rangle = n|n\rangle. \tag{6.33}$$

From the fundamental commutator

$$[x, p] = i\hbar \tag{6.34}$$

we can derive the following commutation relations:

$$[a, a^\dagger] = 1, \tag{6.35}$$

$$[a, H] = \hbar\omega a \tag{6.36}$$

and

$$[a^\dagger, H] = -\hbar\omega a^\dagger. \tag{6.37}$$

Applying (6.36) to an eigenfunction $|n\rangle$, we find

$$aH|n\rangle - Ha|n\rangle = \hbar\omega a|n\rangle. \tag{6.38}$$

Using $H|n\rangle = E_n|n\rangle$, this can be rearranged to

$$H(a|n\rangle) = (E_n - \hbar\omega)(a|n\rangle). \tag{6.39}$$

Since $E_n - \hbar\omega = E_{n-1}$ for harmonic-oscillator eigenvalues, it follows that

$$a|n\rangle = \text{const}\,|n-1\rangle, \tag{6.40}$$

meaning that $a|n\rangle$ represents an eigenfunction for quantum number $n-1$. The value of the constant in (6.40) follows from (6.33) since

$$\langle n-1|n-1\rangle = \langle n|a^\dagger a|n\rangle = n, \tag{6.41}$$

noting that $\langle n|a^\dagger = (a|n\rangle)^\dagger$ according to a general property of bras and kets. Assuming that $|n\rangle$ and $|n-1\rangle$ are both normalized, it follows that the constant in (6.40) equals $\sqrt{n}$. Thus

$$a|n\rangle = \sqrt{n}\,|n-1\rangle \tag{6.42}$$

Proceeding analogously from Eq. (6.37), we find

$$a^\dagger|n\rangle = \sqrt{n+1}\,|n+1\rangle. \tag{6.43}$$

For obvious reasons, $a^\dagger$ and a are known as *step-up* and *step-down* operators, respectively. They are also called *ladder operators* since they take us up and down the ladder of harmonic-oscillator eigenvalues. In the context of radiation theory, $a^\dagger$ and a are called *creation* and *annihilation* operators, respectively, since their action is to create or annihilate a quantum of energy.

For the harmonic-oscillator ground state $|0\rangle$, Eq. (6.42) implies that $a|0\rangle = 0$, consistent with E_0 being the lowest possible eigenvalue. Using the definition of a in (6.31), this leads to a first-order differential equation for the ground-state wavefunction:

$$\left\{\frac{d}{dx} + \frac{m\omega}{\hbar}x\right\}\psi_0(x) = 0, \tag{6.44}$$

with solution

$$\psi_0(x) = \text{const}\,e^{-m\omega x^2/2\hbar}. \tag{6.45}$$

This agrees with what we found in Eq. (6.18) as an asymptotic solution, which also proved to be exact. In fact, Eq. (6.44) is identical to Eq. (6.15).

Eqs. (6.42) and (6.43) for the ladder operators can be applied to derive matrix elements for the harmonic oscillator. Solving (6.31) for x and p we find

$$x = \sqrt{\frac{\hbar}{2m\omega}}\,(a + a^\dagger) \qquad p = -i\sqrt{\frac{m\hbar\omega}{2}}\,(a - a^\dagger). \tag{6.46}$$

Thus

$$\langle n|x|n'\rangle = \sqrt{\frac{\hbar}{2m\omega}}\langle n|(a + a^\dagger)|n'\rangle. \tag{6.47}$$

By (6.42) and (6.43), the only nonvanishing matrix elements will be for $n' = n \pm 1$, with

$$\langle n|x|n-1\rangle = \sqrt{\frac{\hbar n}{2m\omega}}, \qquad \langle n|x|n+1\rangle = \sqrt{\frac{\hbar(n+1)}{2m\omega}}. \tag{6.48}$$

It follows that the selection rule for electric-dipole transitions [cf. Eq. (4.71)] in a harmonic oscillator is $\Delta n = \pm 1$.

Matrix elements of operators x^2 and p^2 can be evaluated by taking squares of Eqs. (6.46). Some examples are given in the Exercises.

6.5 Quantum theory of radiation

Sir James Jeans' suggestion that the radiation field could be treated as an ensemble of different modes of vibration confined to an enclosure was applied to the blackbody problem in Chap. 1. The quantum theory of radiation develops this correspondence more explicitly, identifying each mode of the electromagnetic field with an abstract harmonic oscillator of frequency ω_λ. The Hamiltonian for the entire radiation field can be written

$$H = \tfrac{1}{2}\sum_\lambda (P_\lambda^2 + \omega_\lambda^2 Q_\lambda^2), \tag{6.49}$$

where λ labels the frequencies, propagation directions and polarizations of the constituent modes. We can then define, in analogy with Eq. (6.31),

$$a_\lambda = \sqrt{\frac{\omega_\lambda}{2\hbar}}\,Q_\lambda + \frac{i}{\sqrt{2\hbar\omega_\lambda}}\,P_\lambda \quad \text{and} \quad a_\lambda^\dagger = \sqrt{\frac{\omega_\lambda}{2\hbar}}\,Q_\lambda - \frac{i}{\sqrt{2\hbar\omega_\lambda}}\,P_\lambda. \tag{6.50}$$

These can be very explicitly identified as annihilation and creation operators for *photons*, the quanta of the electromagnetic field, with frequency ω_λ, propagation vector $\mathbf{k}_\lambda$ and polarization $\hat{\mathbf{e}}_\lambda$. The Hamiltonian (6.49) becomes

$$H = \sum_\lambda (a_\lambda^\dagger a_\lambda + \tfrac{1}{2})\hbar\omega_\lambda \tag{6.51}$$

with the energy of the radiation field equal to

$$E = \sum_{\lambda} (n_\lambda + \tfrac{1}{2})\hbar\omega_\lambda. \tag{6.52}$$

The n_λ represent the number of λ photons contained in a cubic box of volume L^3, as illustrated in Fig. 6.4.

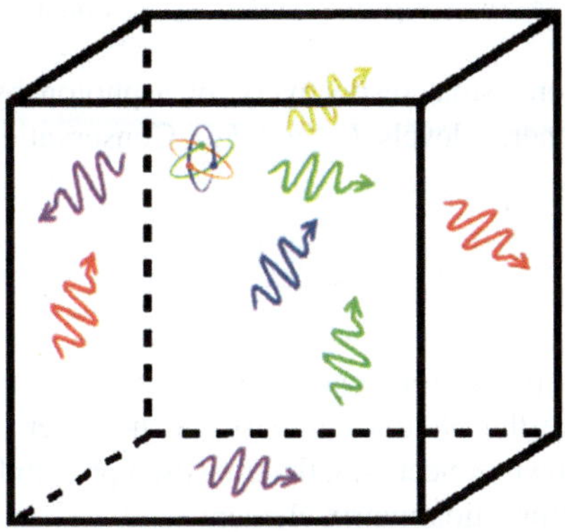

FIGURE 6.4 Schematic representation of an atom and an ensemble of photons enclosed in a cubic box.

The state of the radiation field is determined by a set of photon numbers n_λ. The *vacuum state*, designated $|0\rangle$, contains no photons. The state $|1_\lambda\rangle$ contains one photon of energy $\hbar\omega_\lambda$, propagation vector $\mathbf{k}_\lambda$ and polarization $\hat{\mathbf{e}}_\lambda$. The state $|2_\lambda\rangle$ contains two such photons, while $|1_\lambda, 1_{\lambda'}\rangle$ contains two different photons, λ and λ'. The most general state of the radiation field would be designated $|n_\lambda, n_{\lambda'}, n_{\lambda''} \ldots\rangle$. If the enclosure also contains an atom in quantum state ψ_n, the composite state is designated $|n; n_\lambda, n_{\lambda'}, n_{\lambda''} \ldots\rangle$.

The electric dipole interaction between the atom and the radiation is generally the dominant contribution, represented by a perturbation

$$V = -\boldsymbol{\mu} \cdot \mathbf{E}. \tag{6.53}$$

The energy of the radiation field within the box has the form

$$H = \tfrac{1}{2} \sum_{\lambda} \left(\epsilon_0 E_\lambda^2 + \mu_0^{-1} B_\lambda^2 \right) \times L^3, \tag{6.54}$$

where $\mathbf{E}_\lambda$ and $\mathbf{B}_\lambda$ are the electric and magnetic fields associated with the oscillator mode λ. Comparing the sums of quadratic forms (6.49) and (6.54) it is evident that the quantum-mechanical operator representing the electric field contains contributions linear in both a_λ and $a_\lambda^\dagger$. This is all we need to know.

A more detailed derivation would give the complete expression

$$\mathbf{E}(\mathbf{r}) = \frac{i}{L^{3/2}} \sum_{\lambda} \sqrt{\frac{\hbar\omega_\lambda}{2\epsilon_0}} \left(a_\lambda \hat{\mathbf{e}}_\lambda e^{i\,\mathbf{k}_\lambda \cdot \mathbf{r}} - a_\lambda^\dagger \hat{\mathbf{e}}_\lambda^* e^{-i\,\mathbf{k}_\lambda \cdot \mathbf{r}} \right). \tag{6.55}$$

A sufficient approximation for our purposes is to take

$$V_{\text{abs}} \approx \text{const} \times \boldsymbol{\mu}\, a_\lambda \quad \text{and} \quad V_{\text{em}} \approx \text{const} \times \boldsymbol{\mu}\, a_\lambda^\dagger \tag{6.56}$$

for absorption and emission, respectively, of a photon by an atom making a transition between energy levels E_n and E_m. Conservation of energy requires that the photon frequency satisfies

$$\omega_\lambda = \frac{E_m - E_n}{\hbar}. \tag{6.57}$$

The transition takes place from one of the composite states $|n; \ldots n_\lambda \ldots\rangle$ or $|m; \ldots n_\lambda \ldots\rangle$, where the photon numbers for the other (nonresonant) modes are irrelevant. As shown in Sect. 4.8, the transition probability is proportional to the square of the perturbation matrix element

$$W = \text{const} \left| \langle \text{final} | V | \text{initial} \rangle \right|^2. \tag{6.58}$$

Thus for the upward transition of the atom $(m \leftarrow n)$

$$W_{\text{abs}} = \text{const} \left| \langle m;\; n_\lambda - 1 | V_{\text{abs}} | n;\; n_\lambda \rangle \right|^2 = \text{const}\, |\boldsymbol{\mu}_{mn}|^2 n_\lambda, \tag{6.59}$$

while for the downward transition

$$W_{\text{em}} = \text{const} \left| \langle n;\; n_\lambda + 1 | V_{\text{em}} | m;\; n_\lambda \rangle \right|^2 = \text{const}\, |\boldsymbol{\mu}_{mn}|^2 (n_\lambda + 1). \tag{6.60}$$

The photon numbers come from the action of the creation and annihilation operators:

$$a_\lambda |n_\lambda\rangle = \sqrt{n_\lambda}\, |n_\lambda - 1\rangle \quad \text{and} \quad a_\lambda^\dagger |n_\lambda\rangle = \sqrt{n_\lambda + 1}\, |n_\lambda + 1\rangle \tag{6.61}$$

analogous to (6.42) and (6.43). The occurrence of different factors n_λ and $n_\lambda + 1$ is quite significant. The absorption probability is linear in n_λ, which is proportional to the radiation density in the enclosure. By contrast, the emission probability, varying as $n_\lambda + 1$, is made up of two distinct contributions. The part linear in n_λ is called *stimulated emission* while the part independent of n_λ accounts for *spontaneous emission*. Remarkably, the probability for absorption is exactly equal to that for stimulated emission. A detailed calculation gives the following transition rates:

$$W_{\text{abs}} = W_{\text{stim em}} = \frac{4\pi^2}{3\hbar^2}\,\rho(\omega)|\boldsymbol{\mu}_{mn}|^2, \tag{6.62}$$

while

$$W_{\text{spont em}} = \frac{4\omega^3}{3\hbar c^3}\,|\boldsymbol{\mu}_{mn}|^2. \tag{6.63}$$

Note that all three radiative processes depend on $|\boldsymbol{\mu}_{mn}|^2$ and thus obey the same selection rules. The dependence on ω^3 makes spontaneous emission significant only for higher-energy radiation, in practice, for optical frequencies and higher.

Einstein in 1917 showed the relation between the three radiative processes and Planck's blackbody radiation law. Suppose the radiation in a cavity is in equilibrium at a temperature T. This means that the rate of upward and downward transitions between every pair of energy levels E_n and E_m in the walls of the enclosure must exactly balance. Thus

$$N_n W_{\text{abs}} = N_m (W_{\text{stim em}} + W_{\text{spont em}}). \tag{6.64}$$

But if the populations N_n and N_m obey a Boltzmann distribution

$$\frac{N_m}{N_n} = e^{-(E_m - E_n)/kT} = e^{-\hbar\omega/kT}. \tag{6.65}$$

Putting in the transitions rates from (6.62) and (6.63) and solving for the radiation density, we obtain

$$\rho(\omega) = \frac{\hbar\omega^3/\pi^2 c^3}{e^{\hbar\omega/kT} - 1}, \tag{6.66}$$

independent of any atomic parameters such as e or m_e. This is Planck's radiation law expressed in terms of the radian frequency ω.

Supplement 6A. Anharmonic oscillator

An anharmonic oscillator is one which deviates from the exact form of the harmonic oscillator. We will consider the case of an oscillator with a *quartic anharmonicity*. We write the Hamiltonian, with conveniently scaled variables, as

$$H = -\frac{1}{2}\frac{d^2}{dx^2} + \frac{1}{2}\omega^2 x^2 + \frac{1}{4}\lambda x^4. \tag{6.67}$$

Clearly, this problem can be treated by the perturbation theory of Supplement 4A with

$$V(x) = \frac{1}{4}\lambda x^4. \tag{6.68}$$

The unperturbed eigenfunctions are given by

$$\psi_n(x) = \left(\frac{\sqrt{\omega}}{2^n n! \sqrt{\pi}}\right)^{1/2} H_n(\sqrt{\omega}\,x)\, e^{-\omega x^2/2}, \quad n = 0, 1, 2, \ldots, \tag{6.69}$$

with the corresponding eigenvalues $E_n = (n + \frac{1}{2})\omega$.

We will consider this perturbation problem for the $n = 0$ and $n = 1$ states of the oscillator. The unperturbed energies are $E_0 = \frac{1}{2}\hbar\omega$ and $E_1 = \frac{3}{2}\hbar\omega$. Referring to Supplement 4A, the first-order energies are given by

$$E_0^{(1)} = \frac{\lambda}{4}\int_{-\infty}^{\infty} \psi_0(x) x^4 \psi_0(x)\, dx = \frac{3\lambda}{16\omega^2} \tag{6.70}$$

and

$$E_1^{(1)} = \frac{\lambda}{4}\int_{-\infty}^{\infty} \psi_1(x) x^4 \psi_1(x)\, dx = \frac{15\lambda}{16\omega^2}. \tag{6.71}$$

The general result for arbitrary n is

$$E_n^{(1)} = \frac{3\lambda}{16\omega^2}(2n^2 + 2n + 1). \tag{6.72}$$

To compute the second-order perturbation energies, we require the matrix elements $(x^4)_{n,m}$. The only nonvanishing elements turn out to be those of the form $(x^4)_{n,n}$, $(x^4)_{n,n\pm 2}$ and $(x^4)_{n,n\pm 4}$. We require only the following:

$$(x^4)_{n,n+2} = \frac{1}{2\omega^2}\sqrt{(n+1)(n+2)}(2n+3), \tag{6.73}$$

$$(x^4)_{n,n+4} = \frac{1}{4\omega^2}\sqrt{(n+1)(n+2)(n+3)(n+4)}. \tag{6.74}$$

The sum representing the second-order energy for $n = 0$ and $n = 1$ thus reduces to two terms:

$$E_0^{(2)} = \frac{|V_{0,2}|^2}{E_0 - E_2} + \frac{|V_{0,4}|^2}{E_0 - E_4} = -\frac{21\lambda^2}{128\omega^5} \tag{6.75}$$

and

$$E_1^{(2)} = \frac{|V_{1,3}|^2}{E_1 - E_3} + \frac{|V_{1,5}|^2}{E_1 - E_5} = -\frac{165\lambda^2}{128\omega^5}. \tag{6.76}$$

The total energies to second order then work out to

$$E_0 = \frac{1}{2}\omega + \frac{3\lambda}{16\omega^2} - \frac{21\lambda^2}{128\omega^5}, \quad E_1 = \frac{3}{2}\omega + \frac{15\lambda}{16\omega^2} - \frac{165\lambda^2}{128\omega^5}. \tag{6.77}$$

Chapter 6. Exercises

6.1. For a classical harmonic oscillator, the particle can not go beyond the points where the total energy equals the potential energy. Identify these points for a quantum-mechanical harmonic oscillator in its ground state. Write an integral giving the probability that the particle will go beyond these classically-allowed points. (You need not evaluate the integral.)

6.2. Evaluate the average (expectation) values of potential energy and kinetic energy for the ground state of the harmonic oscillator. Comment on the relative magnitude of these two quantities.

6.3. Using ladder operators to evaluate matrix elements, calculate the average potential and kinetic energies for a harmonic oscillator in its nth quantum state.

6.4. Apply the Heisenberg uncertainty principle to the ground state of the harmonic oscillator. Applying the formula for expectation values, calculate

$$\Delta x = \sqrt{\langle x^2 \rangle - \langle x \rangle^2} \quad \text{and} \quad \Delta p = \sqrt{\langle p^2 \rangle - \langle p \rangle^2}$$

and find the product $\Delta x \Delta p$.

6.5. Using ladder operators to evaluate matrix elements, evaluate the uncertainty product $\Delta x \Delta p$ for the nth quantum state.

6.6. The probability of a radiative dipole transition between levels n and n' of a harmonic oscillator depend on the square of the matrix element

$$x_{n,n'} = \int_{-\infty}^{\infty} \psi_n(x)\, x\, \psi_{n'}(x)\, dx.$$

When $x_{n,n'} = 0$, the transition is forbidden. Determine whether each of the following harmonic-oscillator transitions is allowed or forbidden: $1 \leftarrow 0$, $2 \leftarrow 0$, $2 \leftarrow 1$. Derive the general form of the selection rule.

6.7. (optional) Verify the given results for the first- and second-order perturbation energies of a harmonic oscillator perturbed by $V = \frac{1}{4}\lambda x^4$ for the $n = 0$ and $n = 1$ states.

Chapter 7

Angular momentum

7.1 Particle in a ring

Consider a variant of the one-dimensional particle in a box problem in which the x-axis is bent into a ring of radius R. We can write the same Schrödinger equation

$$-\frac{\hbar^2}{2M}\frac{d^2\psi(x)}{dx^2} = E\psi(x). \tag{7.1}$$

There are no boundary conditions in this case since the x-axis closes upon itself. A more appropriate independent variable for this problem is the angular position on the ring given by $\phi = x/R$. The Schrödinger equation then reads

$$-\frac{\hbar^2}{2MR^2}\frac{d^2\psi(\phi)}{d\phi^2} = E\psi(\phi). \tag{7.2}$$

The kinetic energy of a body rotating in the xy-plane can be expressed as

$$E = \frac{L_z^2}{2I}, \tag{7.3}$$

where $I = MR^2$ is the moment of inertia and L_z, the z-component of angular momentum. (Since $\mathbf{L} = \mathbf{r} \times \mathbf{p}$, if $\mathbf{r}$ and $\mathbf{p}$ lie in the xy-plane, $\mathbf{L}$ points in the z-direction.) The structure of Eq. (7.2) suggests that this angular-momentum operator is given by

$$\hat{L}_z = -i\hbar\frac{\partial}{\partial\phi}. \tag{7.4}$$

This result will follow from a more general derivation in the following Section. The Schrödinger equation (7.2) can now be written more compactly as

$$\psi''(\phi) + m^2\psi(\phi) = 0, \tag{7.5}$$

where

$$m^2 \equiv 2IE/\hbar^2. \tag{7.6}$$

(We have used the symbol M for mass to avoid confusion with m.) Possible solutions to (7.5) are

$$\psi(\phi) = \text{const}\, e^{\pm im\phi}. \tag{7.7}$$

Introduction to Quantum Mechanics. https://doi.org/10.1016/B978-0-12-822310-9.00015-X

In order for this wavefunction to be physically acceptable, it must be *single-valued.* Since ϕ increased by any multiple of 2π represents the same point on the ring, we must have

$$\psi(\phi + 2\pi) = \psi(\phi) \tag{7.8}$$

and therefore

$$e^{im(\phi+2\pi)} = e^{im\phi}. \tag{7.9}$$

This requires that

$$e^{2\pi im} = 1, \tag{7.10}$$

which is true only is m is an integer:

$$m = 0, \pm 1, \pm 2 \ldots. \tag{7.11}$$

Using (7.6), this gives the quantized energy values

$$E_m = \frac{\hbar^2}{2I} m^2. \tag{7.12}$$

In contrast to the particle in a box, the eigenfunctions corresponding to $+m$ and $-m$ in Eq. (7.7) are linearly independent, so both must be included. Therefore all eigenvalues, except E_0, are two-fold (doubly) degenerate. The eigenfunctions can all be written in the form const $e^{im\phi}$, with m running over *all* integer values, as in Eq. (6.11). The normalized eigenfunctions are

$$\psi_m(\phi) = \frac{1}{\sqrt{2\pi}} e^{im\phi} \tag{7.13}$$

and can be verified to satisfy the complex generalization of the normalization condition

$$\int_0^{2\pi} \psi_m^*(\phi)\, \psi_m(\phi)\, d\phi = 1, \tag{7.14}$$

where we have noted that $\psi_m^*(\phi) = (2\pi)^{-1/2} e^{-im\phi}$. The mutual orthogonality of the functions (7.13) also follows easily, for

$$\int_0^{2\pi} \psi_{m'}^*\, \psi_m(\phi)\, d\phi = \frac{1}{2\pi} \int_0^{2\pi} e^{i(m-m')\phi}\, d\phi =$$

$$\frac{1}{2\pi} \int_0^{2\pi} [\cos(m-m')\phi + i\, \sin(m-m')\phi]\, d\phi = 0, \quad \text{for} \quad m' \neq m. \tag{7.15}$$

The solutions (7.13) are also eigenfunctions of the angular momentum operator (7.4), with

$$\hat{L}_z \psi_m(\phi) = m\hbar\, \psi_m(\phi), \quad m = 0, \pm 1, \pm 2 \ldots. \tag{7.16}$$

This is a instance of a fundamental result in quantum mechanics, that any measured component of orbital angular momentum is restricted to integral multiples of $\hbar$. The Bohr theory of the hydrogen atom, to be discussed in the next Chapter, can be derived from this assumption alone.

7.2 Free electron model for aromatic molecules

The benzene molecule contains a ring of six carbon atoms around which six delocalized π-electrons can circulate. A variant of the FEM for rings predicts the ground-state electron configuration which we can write as $1\pi^2\,2\pi^4$, as shown in Fig. 7.1:

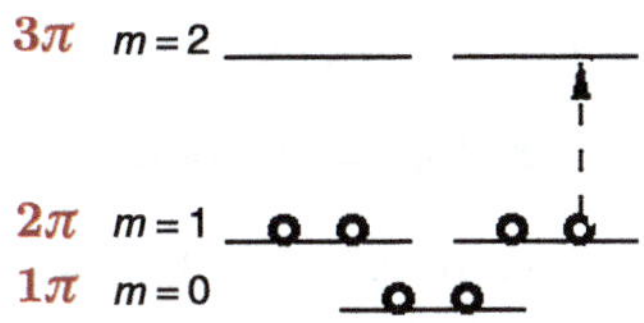

FIGURE 7.1 Free electron model for benzene. Dotted arrow shows the lowest-energy excitation.

The enhanced stability the benzene molecule can be attributed to the complete shells of π-electron orbitals, analogous to the way that noble gas electron configurations achieve their stability. Naphthalene, apart from the central C–C bond, can be modeled as a ring containing 10 electrons in the next closed-shell configuration $1\pi^2\,2\pi^4\,3\pi^4$. These molecules fulfill Hückel's "$4N + 2$ rule" for aromatic stability. The molecules cyclobutadiene ($1\pi^2\,2\pi^2$) and cyclooctatetraene ($1\pi^2\,2\pi^4\,3\pi^2$), even though they consist of rings with alternating single and double bonds, do *not* exhibit aromatic stability since they contain partially-filled orbitals.

The longest wavelength absorption in the benzene spectrum can be estimated according to this model as

$$\frac{hc}{\lambda} = E_2 - E_1 = \frac{\hbar^2}{2mR^2}(2^2 - 1^2). \tag{7.17}$$

The ring radius R can be approximated by the C–C distance in benzene, 1.39 Å. We predict $\lambda \approx 210$ nm, whereas the experimental absorption has $\lambda_{\max} \approx 268$ nm.

7.3 Rotation in three dimensions

A particle of mass M, free to move on the surface of a sphere of radius R, can be located by the two angular variables θ, ϕ. The Schrödinger equation therefore has the form

$$-\frac{\hbar^2}{2M}\nabla^2 Y(\theta,\phi) = E\,Y(\theta,\phi), \tag{7.18}$$

with the wavefunction conventionally written as $Y(\theta,\phi)$. These functions are the *spherical harmonics*, discussed in Sect. 5.7. Since $r = R$, a constant, the first term in the Laplacian does not contribute. The Schrödinger equation reduces to

$$\left\{\frac{1}{\sin\theta}\frac{\partial}{\partial\theta}\sin\theta\frac{\partial}{\partial\theta}+\frac{1}{\sin^2\theta}\frac{\partial^2}{\partial\phi^2}+\lambda\right\}Y(\theta,\phi)=0, \tag{7.19}$$

where

$$\lambda=\frac{2MR^2E}{\hbar^2}=\frac{2IE}{\hbar^2}, \tag{7.20}$$

again introducing the moment of inertia $I = MR^2$. The variables θ and ϕ can be separated in Eq. (7.19) after multiplying through by $\sin^2\theta$. If we write

$$Y(\theta,\phi)=\Theta(\theta)\Phi(\phi) \tag{7.21}$$

and follow the procedure used for the three-dimensional box, we find that dependence on ϕ alone occurs in the term

$$\frac{\Phi''(\phi)}{\Phi(\phi)}=\text{const.} \tag{7.22}$$

This is identical in form to Eq. (7.5), with the constant equal to $-m^2$, and we can write down the analogous solutions

$$\Phi_m(\phi)=\sqrt{\frac{1}{2\pi}}\,e^{im\phi}, \qquad m=0,\ \pm1,\ \pm2\ldots \tag{7.23}$$

Substituting (6.27) into (6.23) and canceling the functions $\Phi_m(\phi)$, we obtain an ordinary differential equation for $\Theta(\theta)$

$$\left\{\frac{1}{\sin\theta}\frac{d}{d\theta}\sin\theta\frac{d}{d\theta}-\frac{m^2}{\sin^2\theta}+\lambda\right\}\Theta(\theta)=0. \tag{7.24}$$

As shown in Sect. 5.6, the single-valued, finite solutions to (7.24) are known as *associated Legendre functions*. The parameters λ and m are restricted to the values

$$\lambda=l(l+1), \qquad l=0,\ 1,\ 2\ldots, \tag{7.25}$$

while

$$m=0,\ \pm1,\ \pm2\ldots\pm l, \qquad (2l+1\text{ values}). \tag{7.26}$$

Putting (7.25) into (7.20), the allowed energy levels for a particle on a sphere are found to be

$$E_\ell=\frac{\hbar^2}{2I}l(l+1). \tag{7.27}$$

Since the energy is independent of the second quantum number m, these levels are $(2l+1)$-fold degenerate.

The explicit forms for the spherical harmonics we will need is given in Eq. (5.112). The spherical harmonics constitute an orthonormal set satisfying the integral relations

$$\int_0^\pi \int_0^{2\pi} Y^*_{l'm'}(\theta,\phi) Y_{lm}(\theta,\phi) \sin\theta \, d\theta \, d\phi = \delta_{ll'}\delta_{mm'}. \tag{7.28}$$

7.4 Theory of angular momentum

Generalization of the energy-angular momentum relation (7.3) to three dimensions gives

$$E = \frac{L^2}{2I}. \tag{7.29}$$

Thus from (7.18) and (7.19) we can identify the operator for the square of total angular momentum

$$\hat{L}^2 = -\hbar^2 \left\{ \frac{1}{\sin\theta}\frac{\partial}{\partial\theta}\sin\theta\frac{\partial}{\partial\theta} + \frac{1}{\sin^2\theta}\frac{\partial^2}{\partial\phi^2} \right\}. \tag{7.30}$$

The functions $Y(\theta,\phi)$ are simultaneous eigenfunctions of $\hat{L}^2$ and $\hat{L}_z$ such that

$$\hat{L}^2 Y_{lm}(\theta,\phi) = l(l+1)\,\hbar^2\, Y_{lm}(\theta,\phi) \tag{7.31}$$

and

$$\hat{L}_z Y_{lm}(\theta,\phi) = m\,\hbar\, Y_{lm}(\theta,\phi). \tag{7.32}$$

But $Y_{lm}(\theta,\phi)$ is *not* an eigenfunction of either L_x and L_y (unless $\ell = 0$). Note that the magnitude of the total angular momentum $\sqrt{l(l+1)}\,\hbar$ is greater than its maximum observable component in any direction, namely $l\,\hbar$. The quantum-mechanical behavior of the angular momentum and its components can be represented by a vector model, illustrated in Fig. 7.2. The angular momentum vector **L**, with magnitude $\sqrt{l(l+1)}\,\hbar$, can be pictured as diffused around a cone about the z-axis, with only the z-component L_z having a definite value. The components L_x and L_y are indeterminate, corresponding to the fact that the system is not in an eigenstate of either. There are $2l+1$ different allowed values for L_z, with eigenvalues $m\,\hbar$ ($m = 0, \pm 1, \pm 2 \ldots \pm l$) equally spaced between $+l\,\hbar$ and $-l\,\hbar$.

The magnitude of the total angular momentum $|\vec{L}| = \sqrt{l(l+1)}\hbar$ is greater than its maximum observable component in any direction, namely $l\hbar$. The quantum-mechanical behavior of the angular momentum and its components can be represented by a vector model, as shown in Fig. 7.2. The angular momentum vector $\vec{L}$ can be pictured as precessing about the z-axis, with its

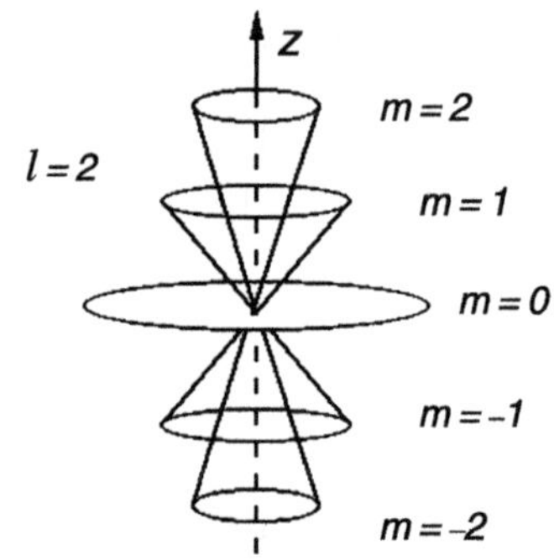

FIGURE 7.2 Space quantization of angular momentum, showing the case $l=2$.

z-component L_z constant. The components L_x and L_y fluctuate in the course of precession, mirroring the fact that the system is not in an eigenstate of either, as implied by the commutation relations. There are $2l+1$ possible values for L_z, with eigenvalues $m\hbar$ ($m=0,\pm1,\pm2,\ldots,\pm l$), equally spaced between $+l\hbar$ and $-l\hbar$. This discreteness in the allowed directions of the angular momentum vector is called *space quantization*. The existence of simultaneous eigenstates of $\hat{L}^2$ and any *one* component, conventionally $\hat{L}_z$, is consistent with the commutation relations derived in Chap. 4:

$$[\hat{L}^2, \hat{L}_z] = 0 \tag{7.33}$$

and

$$[\hat{L}_x, \hat{L}_y] = i\hbar\hat{L}_z, \qquad \textit{et cyc.} \tag{7.34}$$

7.5 Operator derivation of angular momentum eigenvalues

The eigenvalues of angular momentum can alternatively be determined by making use of ladder operators, analogous to the treatment of the harmonic oscillator in Sect. 6.4. As a consequence of (7.33), there must exist simultaneous eigenstates of L^2 and one component, generally chosen as L_z. Let us denote the corresponding eigenvectors by $|l,m\rangle$, and write the eigenvalue equations

$$L_z|l,m\rangle = m\hbar|l,m\rangle \quad \text{and} \quad L^2|l,m\rangle = \lambda\hbar^2|l,m\rangle, \tag{7.35}$$

with λ remaining to be determined.

At this point we introduce the operators

$$L^+ = L_x + iL_y, \qquad L^- = L_x - iL_y, \tag{7.36}$$

which, we will see in a moment, are *raising* and *lowering* operators for the eigenvalues of L_z. It is easy to see that L^2 commutes with L^+, since it commutes with both L_x and L_y. Therefore

$$L^2L^+|l,m\rangle = L^+L^2|l,m\rangle = \lambda\hbar^2 L^+|l,m\rangle, \tag{7.37}$$

using the eigenvalue equation for L^2, Eq. (7.35). We can then write: $L^2(L^+|l, m\rangle) = \lambda\hbar^2(L^+|l,m\rangle)$, which shows that $L^+|l,m\rangle$ is an eigenvector of L^2 with the *same eigenvalue* $\lambda\hbar^2$ as $|l,m\rangle$. Thus, the operator L^+ applied to $|l,m\rangle$ does *not* change the *magnitude* of the angular momentum.

Consider now the commutator $[L_z, L^+]$

$$[L_z, L^+] = [L_z, L_x] + i[L_z, L_y] = i\hbar L_y + i(-i\hbar)L_x = \hbar(iL_y + L_x) = \hbar L^+. \tag{7.38}$$

Operating on the eigenvector $|l,m\rangle$:

$$\begin{aligned} [L_z, L^+]|l,m\rangle &= L_z L^+|l,m\rangle - L^+L_z|l,m\rangle = \hbar L^+|l,m\rangle, \\ L_z(L^+|l,m\rangle) - m\hbar(L^+|l,m\rangle) &= \hbar(L^+|l,m\rangle), \\ L_z(L^+|l,m\rangle) &= (m+1)\hbar(L^+|l,m\rangle). \end{aligned} \tag{7.39}$$

Evidently, $L^+|l,m\rangle$ is an eigenvector of L_z with the eigenvalue $(m+1)\hbar$, thus the designation of L^+ as a *raising operator*, meaning that

$$L^+|l,m\rangle = \text{const}|l,m+1\rangle. \tag{7.40}$$

Analogously, L^- is a *lowering operator*, with

$$L^-|l,m\rangle = \text{const}|l,m-1\rangle. \tag{7.41}$$

Suppose that the quantum number l is now defined as the *maximum* allowed value of m for a given eigenvalue $\lambda\hbar^2$ of L^2. (Correspondingly, $-l$ would be the *minimum* value of m.) Since there is no higher possible value of m, the raising operator on $|l,l\rangle$ should annihilate the vector:

$$L^+|l,l\rangle = 0. \tag{7.42}$$

Now here is an identity you would never ordinarily think of:

$$L^-L^+ = L_x^2 + L_y^2 + i(L_xL_y - L_yL_x) = L^2 - L_z^2 + i(i\hbar)L_z. \tag{7.43}$$

Thus

$$L^2 = L^-L^+ + L_z^2 + \hbar L_z. \tag{7.44}$$

Applying this to $|l,l\rangle$

$$L^2|l,l\rangle = L^-L^+|l,l\rangle + L_z^2|l,l\rangle + \hbar L_z|l,l\rangle, \tag{7.45}$$

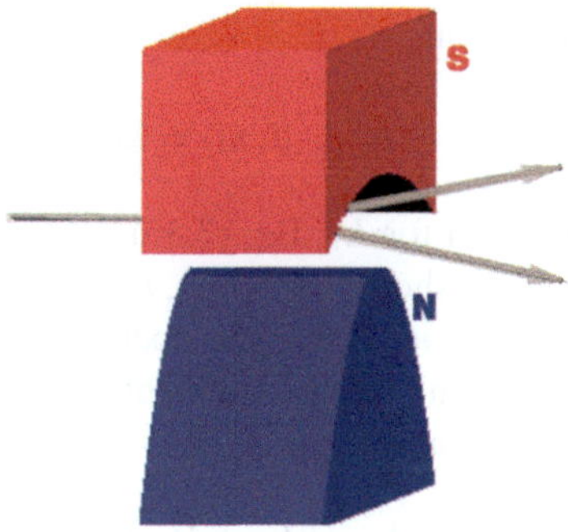

FIGURE 7.3 Schematic representation of Stern-Gerlach experiment.

giving

$$\lambda\hbar^2|l,l\rangle = 0 + l^2\hbar^2|l,l\rangle + l\hbar^2|l,l\rangle. \tag{7.46}$$

Finally, we can identify

$$\lambda = l^2 + l = l(l+1). \tag{7.47}$$

Thus we arrive at the eigenvalues and eigenvectors for orbital angular momentum:

$$\begin{aligned} L^2|l,m\rangle &= l(l+1)\hbar^2|l,m\rangle, \qquad l = 0,1,2,\ldots, \\ L_z|l,m\rangle &= m\hbar|l,m\rangle, \qquad m = 0,\pm1,\pm2,\ldots,\pm l. \end{aligned} \tag{7.48}$$

7.6 Electron spin

Many atomic spectral lines appear, under sufficiently high resolution, to be closely-spaced doublets, for example the 17.2 cm^{-1} splitting of the yellow sodium D lines. Uhlenbeck and Goudsmit proposed in 1925 that such doublets were due to an intrinsic angular momentum possessed by the electron (in addition to its orbital angular momentum) that could be oriented in just two possible ways. This property, known as *spin*, occurs as well in other elementary particles. Spin and orbital angular momenta are roughly analogous to the daily and annual motions, respectively, of the Earth around the Sun. In a classic experiment by Stern and Gerlach in 1922, shown in Fig. 7.3, a beam of silver atoms passed through an inhomogeneous magnetic field splits into two beams, corresponding to the two possible orientations of the magnetic moment of the single unpaired electron.

To distinguish the spin angular momentum from the orbital, we designate the quantum numbers as s and m_s, in place of l and m. For the electron, the quantum number s always has the value $\frac{1}{2}$, while m_s can have one of two values, $\pm\frac{1}{2}$, as shown in Fig. 7.4. The electron is said to be an elementary particle of spin $\frac{1}{2}$. The proton and neutron also have spin $\frac{1}{2}$ and belong to the classification of particles called *fermions*, which are governed by the Pauli exclusion principle. Other particles, including the photon, have integer values of spin and are classified as *bosons*. These do *not* obey the Pauli principle, so that an arbitrary number can occupy the same quantum state. A complete theory of spin requires relativistic quantum mechanics. For our purposes, it is sufficient to recognize the two possible internal states of the electron, which can be called *spin up* and *spin down*. These are designated, respectively, by α and β as factors in the electron wavefunction. Spins play an essential role in determining the possible electronic states of atoms and molecules.

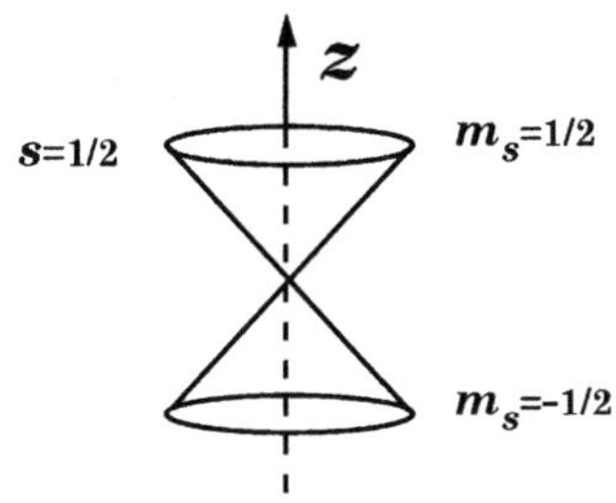

FIGURE 7.4 Electron spin.

Wolfgang Pauli

7.7 Pauli spin algebra

A more sophisticated way of representing a quantum system with an internal degree of freedom (such as electron spin) is to introduce a *spinor* wavefunction:

$$\Psi(\mathbf{r}) = \begin{pmatrix} \psi_1(\mathbf{r}) \\ \psi_2(\mathbf{r}) \end{pmatrix} \tag{7.49}$$

The spinorbital written $\psi(\mathbf{r})\alpha$ would then be given by

$$\Psi(\mathbf{r}) = \begin{pmatrix} \psi(\mathbf{r}) \\ 0 \end{pmatrix} = \psi(\mathbf{r}) \begin{pmatrix} 1 \\ 0 \end{pmatrix}, \tag{7.50}$$

while $\psi(\mathbf{r})\beta$ is

$$\Psi(\mathbf{r}) = \begin{pmatrix} 0 \\ \psi(\mathbf{r}) \end{pmatrix} = \psi(\mathbf{r}) \begin{pmatrix} 0 \\ 1 \end{pmatrix}. \tag{7.51}$$

The matrix operator

$$S_z = \frac{\hbar}{2} \begin{pmatrix} 1 & 0 \\ 0 & -1 \end{pmatrix} \tag{7.52}$$

represents the z-component of electron spin. There are two eigenstates,

$$|\alpha\rangle = \begin{pmatrix} 1 \\ 0 \end{pmatrix} \quad \text{and} \quad |\beta\rangle = \begin{pmatrix} 0 \\ 1 \end{pmatrix}, \tag{7.53}$$

such that

$$S_z \begin{pmatrix} 1 \\ 0 \end{pmatrix} = \frac{\hbar}{2} \begin{pmatrix} 1 \\ 0 \end{pmatrix} \tag{7.54}$$

and

$$S_z \begin{pmatrix} 0 \\ 1 \end{pmatrix} = -\frac{\hbar}{2} \begin{pmatrix} 0 \\ 1 \end{pmatrix}. \tag{7.55}$$

Operators for the x- and y-components of spin angular momentum can be represented by

$$S_x = \frac{\hbar}{2} \begin{pmatrix} 0 & 1 \\ 1 & 0 \end{pmatrix} \quad \text{and} \quad S_y = \frac{\hbar}{2} \begin{pmatrix} 0 & -i \\ i & 0 \end{pmatrix}, \tag{7.56}$$

such that the commutation relations

$$[S_x, S_y] = i\hbar S_z, \qquad et\ cyc, \tag{7.57}$$

analogous to (7.34) are satisfied. The magnitude of the spin angular momentum is given by

$$S^2 = S_x^2 + S_y^2 + S_z^2 = \frac{3}{4}\hbar^2 \begin{pmatrix} 1 & 0 \\ 0 & 1 \end{pmatrix}, \tag{7.58}$$

consistent with the value $s(s+1)\hbar^2$ for angular-momentum quantum number $s = 1/2$.

The three unit Hermitian matrices

$$\sigma_1 = \begin{pmatrix} 0 & 1 \\ 1 & 0 \end{pmatrix}, \qquad \sigma_2 = \begin{pmatrix} 0 & -i \\ i & 0 \end{pmatrix}, \qquad \sigma_3 = \begin{pmatrix} 1 & 0 \\ 0 & -1 \end{pmatrix} \tag{7.59}$$

called the *Pauli spin matrices* are important in relativistic quantum mechanics.

For future reference, we summarize the action of the spin operators on the two spin eigenstates:

$$\begin{aligned} S_z|\alpha\rangle &= \frac{\hbar}{2}|\alpha\rangle, & S_z|\beta\rangle &= -\frac{\hbar}{2}|\beta\rangle, & S_x|\alpha\rangle &= \frac{\hbar}{2}|\beta\rangle, \\ S_x|\beta\rangle &= \frac{\hbar}{2}|\alpha\rangle, & S_y|\alpha\rangle &= \frac{i\hbar}{2}|\beta\rangle, & S_y|\beta\rangle &= -\frac{i\hbar}{2}|\alpha\rangle, \\ & & S^2|\alpha\rangle &= \frac{3}{4}\hbar^2|\alpha\rangle, & S^2|\beta\rangle &= \frac{3}{4}\hbar^2|\beta\rangle. \end{aligned} \tag{7.60}$$

7.8 Addition of angular momenta

The total angular momentum of an atom or molecule is the vector sum of the angular momenta of its constituent parts, e.g., electrons. For example, the total orbital angular momentum of several electrons is given by

$$\mathbf{L} = \mathbf{l}_1 + \mathbf{l}_2 + \ldots, \tag{7.61}$$

while the total spin angular momentum is analogously

$$\mathbf{S} = \mathbf{s}_1 + \mathbf{s}_2 + \ldots. \tag{7.62}$$

These can combine to give a total electronic angular momentum

$$\mathbf{J} = \mathbf{L} + \mathbf{S}. \tag{7.63}$$

Later, we will also encounter angular momentum from molecular rotation and from nuclear spins. Consider the general case of vector addition of two angular momenta, which we will denote as $\mathbf{J}_1$ and $\mathbf{J}_2$:

$$\mathbf{J} = \mathbf{J}_1 + \mathbf{J}_2. \tag{7.64}$$

We can picture $\mathbf{J}_1$ and $\mathbf{J}_2$ as cones around their resultant $\mathbf{J}$, which is itself represented by a conical surface about some axis in space, as shown in Fig. 7.5. According to quantum theory, each component of angular momentum, as well as their resultant, has a magnitude given by $\sqrt{J(J+1)}\,\hbar$ with J having possible values 0, $\frac{1}{2}$, 1, $\frac{3}{2}$, 2 ..., now including the possibility of spins contributing multiples of $\frac{1}{2}$. The observable components of $\mathbf{J}$ are again given by $M\hbar$, with M running from $-J$ to $+J$ in integer steps. If $\mathbf{J}_1$ and $\mathbf{J}_2$ are described by quantum numbers J_1 and J_2, respectively, then the total angular momentum quantum number J has the possible values

$$J = |J_2 - J_1|,\ |J_2 - J_1| + 1,\ \ldots,\ J_2 + J_1, \tag{7.65}$$

again in integer steps. The value of J depends on the relative orientation of the components $\mathbf{J_1}$ and $\mathbf{J_2}$. For example, angular momenta 1 and $\frac{1}{2}$ can combine to give either $J = \frac{1}{2}$ or $J = \frac{3}{2}$.

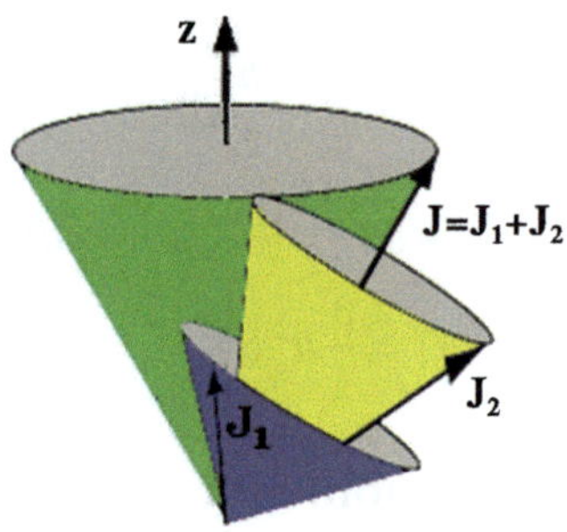

FIGURE 7.5 Vector addition of angular momenta.

Chapter 8

The hydrogen atom and atomic orbitals

The hydrogen atom has provided the most fundamental prototype at several levels in the advancement of quantum theory, beginning with the old quantum theory, through nonrelativistic, then relativistic, quantum mechanics and quantum field theory (the Lamb shift, etc.). It is the only real physical system that can be solved exactly, giving analytical solutions in closed-form (although this might also be said for the radiation field, as an assembly of harmonic oscillators).

8.1 Atomic spectra

When gaseous hydrogen in a glass tube is excited by a 5000-volt electrical discharge, four lines are observed in the visible part of the emission spectrum: red at 656.3 nm, blue-green at 486.1 nm, blue violet at 434.1 nm and violet at 410.2 nm:

FIGURE 8.1 Visible spectrum of atomic hydrogen.

Other series of lines have been observed in the ultraviolet and infrared regions. Rydberg (1890) found that all the lines of the atomic hydrogen spectrum could be fitted to a single formula

$$\frac{1}{\lambda} = \mathcal{R}\left(\frac{1}{n_1^2} - \frac{1}{n_2^2}\right), \qquad n_1, n_2 = 1, 2, 3\ldots, \qquad n_2 > n_1, \tag{8.1}$$

where $\mathcal{R}$, known as the Rydberg constant, has the value 109,677 cm^{-1} for hydrogen. The reciprocal of wavelength, in units of cm^{-1}, is in general use by spectroscopists. This unit is also designated *wavenumbers*, since it represents the number of wavelengths per cm. The Balmer series of spectral lines in the visible region, shown in Fig. 8.1, correspond to the values $n_1 = 2$, $n_2 = 3, 4, 5$ and 6. The lines with $n_1 = 1$ in the ultraviolet make up the Lyman series. The

Introduction to Quantum Mechanics. https://doi.org/10.1016/B978-0-12-822310-9.00016-1

line with $n_2 = 2$, designated Lyman alpha, has the longest wavelength (lowest wavenumber) in this series, with $1/\lambda = 82.258\ \text{cm}^{-1}$ or $\lambda = 121.57$ nm.

In the words of Arthur Schawlow, "The spectrum of the hydrogen atom has proved to be the Rosetta stone of modern physics: once this pattern of lines had been deciphered much else could also be understood."

Other atomic species have line spectra, which can be used as a "fingerprint" to identify the element. However, no atom other than hydrogen has a simple relation analogous to (8.1) for its spectral frequencies. Bohr in 1913 proposed that all atomic spectral lines arise from transitions between discrete energy levels, giving a photon such that

$$\Delta E = h\nu = \frac{hc}{\lambda}. \tag{8.2}$$

This is called the *Bohr frequency condition*. We now understand that the atomic transition energy ΔE is equal to the energy of a photon, as proposed earlier by Planck and Einstein.

8.2 The Bohr atom

The nuclear model proposed by Rutherford in 1911 pictures the atom as a heavy, positively-charged nucleus, around which much lighter, negatively-charged electrons circulate, much like planets in the Solar system. This model is however completely untenable from the standpoint of classical electromagnetic theory, for an accelerating electron (circular motion represents an acceleration) should radiate away its energy. In fact, a hydrogen atom should exist for no longer than 5×10^{-11} sec, time enough for the electron's death spiral into the nucleus. This is one of the worst quantitative predictions in the history of physics. It has been called the Hindenberg disaster on an atomic scale. (The Hindenberg, a hydrogen-filled dirigible, crashed and burned in a famous disaster in 1937.)

Bohr sought to avoid an atomic catastrophe by proposing that certain orbits of the electron around the nucleus could be exempted from classical electrodynamics and remain stable. The Bohr model was quantitatively successful for the hydrogen atom, as we shall now show.

Recall that the attraction between two opposite charges, such as the electron and proton, is given by Coulomb's law

$$F = \begin{cases} -\dfrac{e^2}{r^2} & \text{(gaussian units)} \\[2ex] -\dfrac{e^2}{4\pi\epsilon_0 r^2} & \text{(SI units).} \end{cases} \tag{8.3}$$

We prefer to use the gaussian system in applications to atomic phenomena. (In any event, it won't matter once we change to atomic units.) Since the Coulomb attraction is a central force (dependent only on r), the potential energy is related by

$$F = -\frac{dV(r)}{dr}. \tag{8.4}$$

We find therefore, for the mutual potential energy of a proton and electron,

$$V(r) = -\frac{e^2}{r}. \tag{8.5}$$

Bohr considered an electron in a circular orbit of radius r around the proton. To remain in this orbit, the electron must be experiencing a centripetal acceleration

$$a = -\frac{v^2}{r}, \tag{8.6}$$

where v is the speed of the electron.

By Newton's second law $F = ma$, we can write

$$\frac{e^2}{r^2} = \frac{mv^2}{r}, \tag{8.7}$$

where m is the mass of the electron. For simplicity, we assume that the proton mass is infinite (actually $m_p \approx 1836m_e$) so that the proton's position remains fixed. We will later correct for this approximation by introducing reduced mass. The energy of the hydrogen atom is the sum of the kinetic and potential energies:

$$E = T + V = \tfrac{1}{2}mv^2 - \frac{e^2}{r}. \tag{8.8}$$

Using Eq. (8.7), we see that

$$T = -\tfrac{1}{2}V \qquad \text{and} \qquad E = \tfrac{1}{2}V = -T. \tag{8.9}$$

This is the form of the virial theorem for a force law with r^{-2} dependence. Note that the energy of a bound atom is *negative*, since it is lower than the energy of the separated electron and proton, which is taken to be zero.

For further progress, we need some restriction on the possible values of r or v. This is where we can introduce the quantization of angular momentum $\mathbf{L} = \mathbf{r} \times \mathbf{p}$. Since $\mathbf{p}$ is perpendicular to $\mathbf{r}$, we can write simply

$$L = rp = mvr. \tag{8.10}$$

Using (8.7), we find that

$$r = \frac{L^2}{me^2}. \tag{8.11}$$

Assuming angular momentum quantization,

$$L = n\hbar, \qquad n = 1, 2 \ldots, \tag{8.12}$$

excluding $n = 0$, which would not give a circular orbit. The allowed orbital radii are then given by

$$r_n = n^2 a_0, \tag{8.13}$$

where

$$a_0 \equiv \frac{\hbar^2}{me^2} = 5.29 \times 10^{-11} \text{ m} = 0.529 \text{ Å}, \tag{8.14}$$

which is known as the *Bohr radius*. The corresponding energy is

$$E_n = -\frac{e^2}{2a_0 n^2} = -\frac{me^4}{2\hbar^2 n^2}, \qquad n = 1, 2 \ldots. \tag{8.15}$$

Rydberg's formula (8.1) can now be deduced from the Bohr model. We have

$$\frac{hc}{\lambda} = E_{n_2} - E_{n_1} = \frac{2\pi^2 me^4}{h^2}\left(\frac{1}{n_1^2} - \frac{1}{n_2^2}\right) \tag{8.16}$$

and the Rydbeg constant can be identified as

$$\mathcal{R} = \frac{2\pi^2 me^4}{h^3 c} \approx 109{,}737 \text{ cm}^{-1}. \tag{8.17}$$

The slight discrepancy with the experimental value for hydrogen (109,677) is due to the finite proton mass. This will be corrected by introducing reduced mass in Sect 8.8.

The Bohr model can be readily extended to hydrogenlike ions, systems in which a single electron orbits a nucleus of arbitrary atomic number Z. Thus $Z = 1$ for hydrogen, $Z = 2$ for He^+, $Z = 3$ for Li^{++}, and so on. The Coulomb potential (8.5) generalizes to

$$V(r) = -\frac{Ze^2}{r}, \tag{8.18}$$

while the radius of the orbit (8.13) becomes

$$r_n = \frac{n^2 a_0}{Z} \tag{8.19}$$

giving the energy (8.15)

$$E_n = -\frac{Z^2 e^2}{2a_0 n^2}. \tag{8.20}$$

De Broglie's proposal that electrons can have wavelike properties was actually inspired by the Bohr atomic model. Since

$$L = rp = n\hbar = \frac{nh}{2\pi}, \tag{8.21}$$

we find

$$2\pi r = \frac{nh}{p} = n\lambda. \tag{8.22}$$

Therefore, each allowed orbit traces out an integral number of de Broglie wavelengths.

Wilson (1915) and Sommerfeld (1916) generalized Bohr's formula for the allowed orbits to

$$\oint p\,dr = nh, \qquad n = 1, 2 \ldots . \tag{8.23}$$

The Sommerfeld-Wilson quantum conditions reduce to Bohr's results for circular orbits, but allow, in addition, elliptical orbits along which the momentum p is variable. According to Kepler's first law of planetary motion, the orbits of planets are ellipses with the Sun at one focus. Fig. 8.2 shows the generalization of the Bohr theory for hydrogen, including the elliptical orbits. The lowest energy state $n = 1$ is still a circular orbit. But $n = 2$ allows an elliptical orbit in addition to the circular one; $n = 3$ has three possible orbits, and so on. The energy still depends on n alone, so that the elliptical orbits represent degenerate states. Atomic spectroscopy shows in fact that energy levels with $n > 1$ consist of multiple states, as implied by the splitting of atomic lines by an electric field (Stark effect) or a magnetic field (Zeeman effect). The shapes of some of these orbits are drawn in Fig. 8.2.

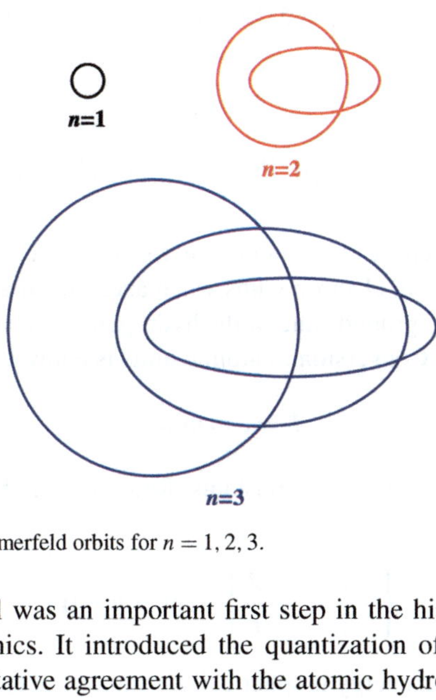

FIGURE 8.2 Bohr-Sommerfeld orbits for $n = 1, 2, 3$.

The Bohr model was an important first step in the historical development of quantum mechanics. It introduced the quantization of atomic energy levels and gave quantitative agreement with the atomic hydrogen spectrum. With the Sommerfeld-Wilson generalization, it accounted as well for the degeneracy of hydrogen energy levels. Although the Bohr model was able to sidestep the atomic "Hindenberg disaster," it cannot avoid what we might call the "Heisenberg disaster." By this we mean that the assumption of well-defined electronic orbits around a nucleus is completely contrary to the basic premises of quantum mechanics. Another flaw in the Bohr picture is that the angular momenta are all too large by one unit, for example, the ground state actually has zero orbital angular momentum (rather than $\hbar$).

8.3 Quantum mechanics of hydrogenlike atoms

In contrast to the particle in a box and the harmonic oscillator, the hydrogen atom is a *real* physical system that can be treated exactly by quantum mechanics. In addition to their inherent significance, these solutions suggest prototypes for atomic orbitals used in approximate computations on complex atoms and molecules.

For an electron in the field of a nucleus of charge $+Ze$, the Schrödinger equation can be written

$$\left\{-\frac{\hbar^2}{2m}\nabla^2 - \frac{Ze^2}{r}\right\}\psi(\mathbf{r}) = E\,\psi(\mathbf{r}). \tag{8.24}$$

It is convenient to introduce *atomic units* in which length is measured in bohrs:

$$a_0 = \frac{\hbar^2}{me^2} = 5.29 \times 10^{-11}\ \text{m} \equiv 1\ \text{bohr},$$

and energy in hartrees:

$$\frac{e^2}{a_0} = 4.358 \times 10^{-18}\ \text{J} = 27.211\ \text{eV} \equiv 1\,\text{hartree}.$$

Electron volts (eV) are a convenient unit for atomic energies, one eV being defined as the energy an electron gains when accelerated across a potential difference of 1 volt. The ground state of the hydrogen atom has an energy of $-1/2$ hartree or -13.6 eV. Conversion to atomic units is equivalent to setting

$$\hbar = e = m = 1$$

in all formulas containing these constants. Rewriting the Schrödinger equation in atomic units, we have

$$\left\{-\frac{1}{2}\nabla^2 - \frac{Z}{r}\right\}\psi(\mathbf{r}) = E\,\psi(\mathbf{r}). \tag{8.25}$$

Since the potential energy is spherically symmetrical (a function of r alone), it is obviously advantageous to treat this problem in spherical polar coordinates

r, θ, ϕ. Expressing the Laplacian operator in these coordinates (cf. Eq. (5.81)),

$$-\frac{1}{2}\left\{\frac{1}{r^2}\frac{\partial}{\partial r}r^2\frac{\partial}{\partial r}+\frac{1}{r^2\sin\theta}\frac{\partial}{\partial\theta}\sin\theta\frac{\partial}{\partial\theta}+\frac{1}{r^2\sin^2\theta}\frac{\partial^2}{\partial\phi^2}\right\}\psi(r,\theta,\phi)$$
$$-\frac{Z}{r}\psi(r,\theta,\phi)=E\,\psi(r,\theta,\phi). \tag{8.26}$$

Eq. (7.30) shows that the second and third terms in the Laplacian represent the angular momentum operator $\hat{L}^2$. Clearly, Eq. (8.26) will then have separable solutions of the form

$$\psi(r,\theta,\phi)=R(r)\,Y_{lm}(\theta,\phi). \tag{8.27}$$

Substituting this into (8.26) and using the angular momentum eigenvalue equation (7.31), we obtain an ordinary differential equation for the radial function $R(r)$:

$$\left\{-\frac{1}{2r^2}\frac{d}{dr}r^2\frac{d}{dr}+\frac{l(l+1)}{2r^2}-\frac{Z}{r}\right\}R(r)=E\,R(r). \tag{8.28}$$

Note that in the domain of the variable r, the angular momentum contribution $l(l+1)/2r^2$ acts as an effective addition to the potential energy. It can be identified with *centrifugal force*, which pulls the electron outward, in opposition to the Coulomb attraction. Carrying out the successive differentiations and simplifying, we obtain

$$\frac{1}{2}R''(r)+\frac{1}{r}R'(r)+\left[\frac{Z}{r}-\frac{l(l+1)}{2r^2}+E\right]R(r)=0, \tag{8.29}$$

another second-order linear differential equation with non-constant coefficients. It is again useful to explore the asymptotic solutions as $r\to\infty$. In the asymptotic approximation, the equation simplifies to

$$R''(r)-2|E|R(r)\approx 0, \tag{8.30}$$

having noted that the energy E is negative for bound states. Solutions to this equation are

$$R(r)\approx \text{const}\, e^{\pm\sqrt{2|E|}\,r}. \tag{8.31}$$

We reject the positive exponential as physically unacceptable, since $R(r)$ would approach infinity as $r\to\infty$, in violation of the requirement that the wavefunction must be finite everywhere. Choosing the negative exponential and setting $E=-Z^2/2$, the ground state energy in the Bohr theory (in atomic units), we obtain

$$R(r)\approx \text{const}\, e^{-Zr}. \tag{8.32}$$

It turns out that this asymptotic approximation is also an *exact* solution of the radial equation (8.29) with $l = 0$, just what happened for the harmonic-oscillator problem in Chap 6. The solutions are designated $R_{nl}(r)$, where the label n is known as the *principal quantum number*, as well as by the angular momentum l, which is a parameter in the radial equation. This solution corresponds to $R_{10}(r)$. It should be normalized according to the condition

$$\int_0^\infty [R_{10}(r)]^2 r^2 \, dr = 1. \tag{8.33}$$

Recalling the definite integral

$$\int_0^\infty r^n e^{-\alpha r} \, dr = \frac{n!}{\alpha^{n+1}}, \tag{8.34}$$

we find the normalized radial function

$$R_{10}(r) = 2Z^{3/2} e^{-Zr}. \tag{8.35}$$

Since this function is nodeless, we identify it with the ground state of the hydrogenlike atom. Multiplying by the spherical harmonic $Y_{00} = 1/\sqrt{4\pi}$, we obtain the complete wavefunction

$$\psi_{100}(r) = \sqrt{\frac{Z^3}{\pi}} e^{-Zr}. \tag{8.36}$$

This is conventionally designated as the $1s$ function $\psi_{1s}(r)$.

Making use of the simplified integration for a spherically-symmetrical function (cf. Eq. (5.80)), the normalization condition for the $1s$ wavefunction can be written

$$\int_0^\infty [\psi_{1s}(r)]^2 4\pi r^2 \, dr = 1. \tag{8.37}$$

8.4 Hydrogen-atom ground state

There are a number of different ways of representing hydrogen-atom wavefunctions graphically. We will illustrate some of these with the $1s$ ground state. In atomic units,

$$\psi_{1s}(r) = \frac{1}{\sqrt{\pi}} e^{-r} \tag{8.38}$$

is a decreasing exponential function of a single variable r, and is simply plotted in Fig. 8.3. Fig. 8.4 gives a somewhat more pictorial representation, a three-dimensional contour plot of $\psi_{1s}(r)$ as a function of x and y in the x, y-plane.

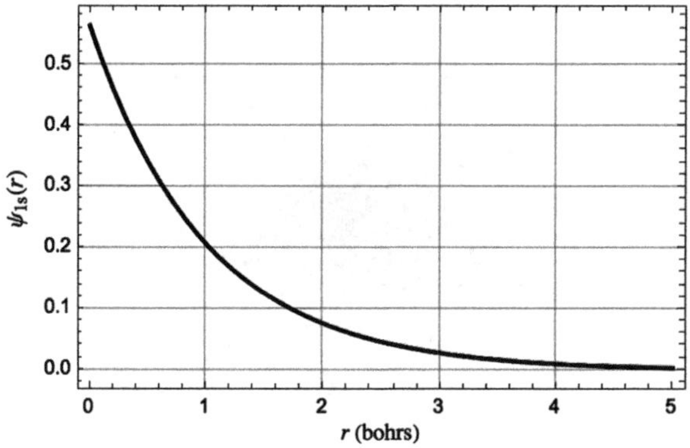

FIGURE 8.3 Plot of $1s$ orbital.

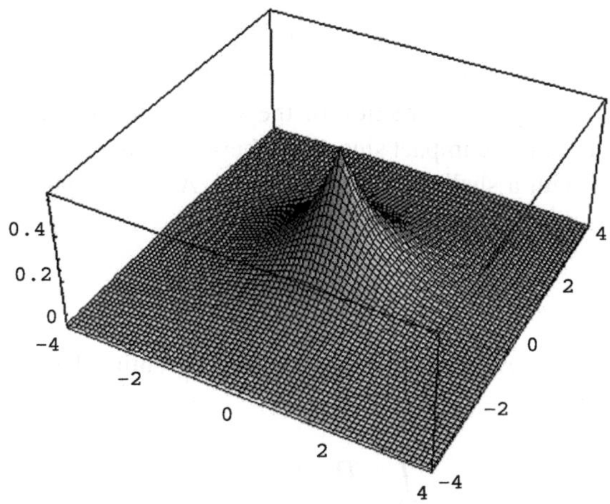

FIGURE 8.4 Contour map of $1s$ orbital in the xy-plane.

According to Born's interpretation of the wavefunction, the probability per unit volume of finding the electron at the point (r, θ, ϕ) is equal to the square of the normalized wavefunction

$$\rho_{1s}(r) = [\psi_{1s}(r)]^2 = \frac{1}{\pi} e^{-2r}. \tag{8.39}$$

This is represented in Fig. 8.5 by a scatter plot describing a possible sequence of observations of the electron position. Although results of individual measurements are not predictable, a statistical pattern does emerge after a sufficiently large number of measurements. The probability density is normalized such that

$$\int_0^\infty \rho_{1s}(r)\, 4\pi r^2\, dr = 1. \tag{8.40}$$

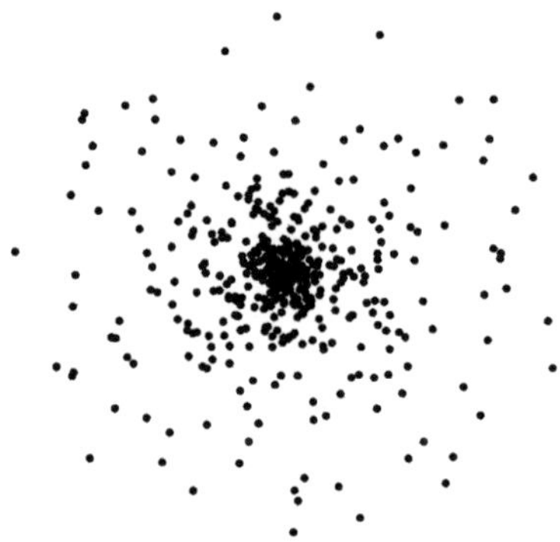

FIGURE 8.5 Scatter plot of 500 random electron position measurements on hydrogen 1s orbital.

In some ways $\rho(r)$ does *not* provide the best description of the electron distribution, since the region around $r = 0$, where the wavefunction has its largest values, is a relatively small fraction of the volume accessible to the electron. Larger radii r have more impact since, in spherical polar coordinates, a value of r is associated with a shell of volume $4\pi r^2 dr$. A more significant measure is therefore the *radial distribution function*

$$D_{1s}(r) = 4\pi r^2 [\psi_{1s}(r)]^2, \tag{8.41}$$

which represents the probability density within the entire shell of radius r, normalized such that

$$\int_0^\infty D_{1s}(r)\, dr = 1. \tag{8.42}$$

The functions $\rho_{1s}(r)$ and $D_{1s}(r)$ are both shown in Fig. 8.6. Remarkably, the 1s RDF has its maximum at $r = a_0 = 1$ bohr, which coincides with the radius of the first Bohr orbit.

8.5 Schrödinger equation for atomic orbitals

To construct a general solution for the radial function $R_{nl}(r)$ satisfying Eq. (8.29), we build in its limiting behavior both as $r \to \infty$ and as $r \to 0$. As we have seen (Eq. (8.29)), $R \approx e^{-\sqrt{2|E|}r}$ as $r \to \infty$. As $r \to 0$, the Coulomb potential $-Z/r$ becomes negligible compared to the centrifugal term $l(l+1/2r^2$. The limiting form as $r \to 0$ can then be determined from the solution of Laplace's equation in spherical coordinates (see Eq. (5.81)). Then term finite as $r \to 0$ is then given by $R \approx r^l$ (cf. Eq. (5.85)). The radial function can be written in the form

$$R(r) = r^l e^{-\sqrt{2|E|}r} F(r). \tag{8.43}$$

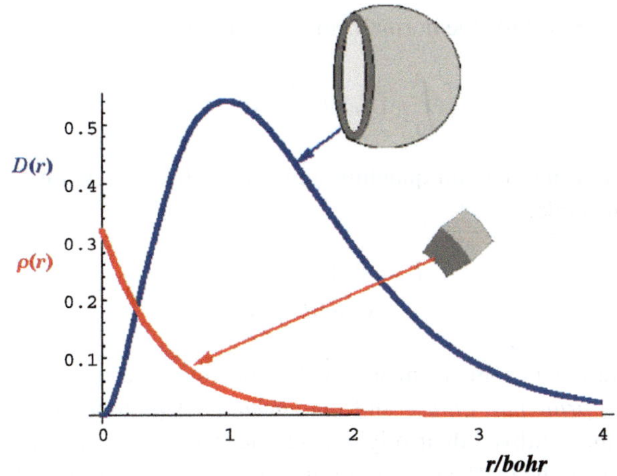

FIGURE 8.6 Density $\rho(r)$ and radial distribution function $D(r)$ for hydrogen $1s$ orbital.

At this point, we can make a very opportunistic move, since we already know, from the Bohr theory, that the energy eigenvalues in atomic units are given by

$$E_n = -\frac{Z^2}{2n^2}. \tag{8.44}$$

It is expedient to define a new variable

$$\rho = 2Zr/n, \tag{8.45}$$

so that the radial function can now be written

$$R(\rho) = \rho^l e^{-\rho/2} L(\rho). \tag{8.46}$$

The radial equation then reduces to the following equation for $L(\rho)$:

$$\rho L''(\rho) + (\beta + 1 - \rho)L'(\rho) + \alpha L(\rho) = 0, \tag{8.47}$$

where $\alpha = n + l$, $\beta = 2l + 1$. This is the known equation for associated Laguerre polynomials, which were introduced in Sect 5.7, with solutions $L_\alpha^\beta(\rho)$. Thus we arrive finally at the radial function for the hydrogenlike problem:

$$R_{nl}(r) = N_{nl}\rho^l e^{-\rho/2} L_{n+l}^{2l+1}(\rho), \tag{8.48}$$

where N_{nl} is a normalization constant

$$N_{nl} = -\left[\frac{(n-l-1)!}{2n[(n+l)!]^3}\right]^{1/2} \left(\frac{2Z}{n}\right)^{3/2}, \tag{8.49}$$

such that $R_{nl}(r)$ fulfills the normalization condition

$$\int_0^\infty [R_{nl}(r)]^2 r^2\, dr = 1. \tag{8.50}$$

The angular momentum quantum number l is by convention designated by the following code:

$$\begin{array}{cccccccc} l = & 0 & 1 & 2 & 3 & 4 & \ldots \\ & s & p & d & f & g & \ldots \end{array}$$

The first four letters come from an old classification scheme for atomic spectral lines: *sharp*, *principal*, *diffuse* and *fundamental*. Although these designations have long since outlived their original significance, they remain in general use. The solutions of the hydrogenic Schrödinger equation in spherical polar coordinates can now be written in full:

$$\psi_{nlm}(r,\theta,\phi) = R_{nl}(r)Y_{lm}(\theta,\phi),$$
$$n = 1,2,3\ldots, \quad l = 0,1\ldots n-1, \quad m = 0,\pm1,\pm2\cdots\pm l, \tag{8.51}$$

where Y_{lm} are the spherical harmonics tabulated in Eq. (5.112). In many applications, particularly in chemistry, *real* forms of the atomic orbitals are used. The $m = 0$ orbital functions are already real. For $m \neq 0$, the linear combinations of degenerate functions $\frac{1}{\sqrt{2}}(\psi_{nlm} \pm \psi_{nl-m})$ gives real atomic orbitals.

Table 8.1 on the next page enumerates all the hydrogenlike functions we will actually need. These are called *atomic orbitals*, in anticipation of their later applications to the structure of atoms and molecules. The energy levels for a hydrogenic system are given by

$$E_n = -\frac{Z^2}{2n^2} \quad \text{hartrees.} \tag{8.52}$$

The energy depends on the principal quantum number n alone. Considering all the allowed values of l and m, the level E_n has a degeneracy of n^2. Fig. 8.7 shows an energy level diagram for hydrogen ($Z = 1$). For $E \geq 0$, the energy is a continuum. The continuum represents states of an electron and proton in interaction, but not bound into a stable atom. Also shown are some of the transitions which make up the Lyman series in the ultraviolet and the Balmer series in the visible region.

The ns orbitals are all spherically symmetrical, being associated with a constant angular factor, the spherical harmonic $Y_{00} = 1/\sqrt{4\pi}$. They have $n - 1$ radial nodes—spherical shells on which the wavefunction equals zero. The $1s$ ground state is nodeless and the number of nodes increases with energy, in a pattern now familiar from our study of the particle-in-a-box and harmonic oscillator. For example, the $2s$ orbital has a radial node at $r = 2$ bohrs. The $3s$ orbital

TABLE 8.1 Real hydrogenic orbitals in atomic units.

$\psi_{1s} = \frac{Z^{3/2}}{\sqrt{\pi}} e^{-Zr}$	
$\psi_{2s} = \frac{Z^{3/2}}{2\sqrt{2\pi}} \left(1 - \frac{Zr}{2}\right) e^{-Zr/2}$	
$\psi_{2p_z} = \frac{Z^{5/2}}{4\sqrt{2\pi}} z e^{-Zr/2}$	ψ_{2p_x}, ψ_{2p_y} analogous
$\psi_{3s} = \frac{Z^{3/2}}{81\sqrt{3\pi}} (27 - 18Zr + 2Z^2r^2) e^{-Zr/3}$	
$\psi_{3p_z} = \frac{\sqrt{2}Z^{5/2}}{81\sqrt{\pi}} (6 - Zr) z e^{-Zr/3}$	ψ_{3p_x}, ψ_{3p_y} analogous
$\psi_{3d_{z^2}} = \frac{Z^{7/2}}{81\sqrt{6\pi}} (3z^2 - r^2) e^{-Zr/3}$	
$\psi_{3d_{zx}} = \frac{\sqrt{2}Z^{7/2}}{81\sqrt{\pi}} zx e^{-Zr/3}$	$\psi_{3d_{yz}}, \psi_{3d_{xy}}$ analogous
$\psi_{3d_{x^2-y^2}} = \frac{Z^{7/2}}{81\sqrt{\pi}} (x^2 - y^2) e^{-Zr/3}$	

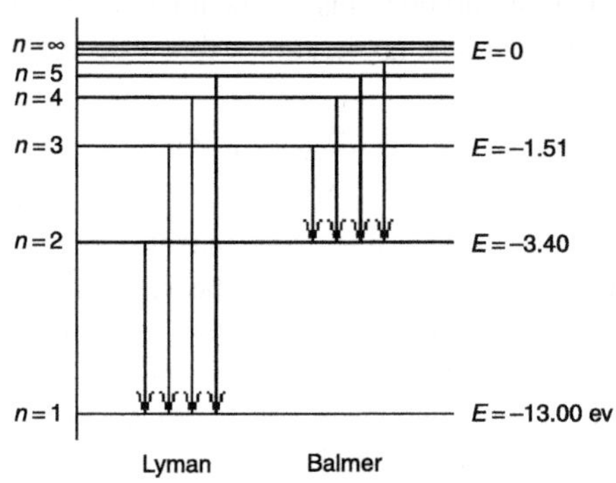

FIGURE 8.7 Energy levels and allowed transitions for atomic hydrogen.

has two radial nodes at $r = \frac{9}{2} - \frac{3\sqrt{3}}{2}$ and $\frac{9}{2} + \frac{3\sqrt{3}}{2}$. These orbitals are plotted in Fig. 8.8.

8.6 p- and d-orbitals

The lowest-energy solutions deviating from spherical symmetry are the $2p$-orbitals. With $R_{21}(r)$ and the three $l = 1$ spherical harmonics, we find three degenerate eigenfunctions:

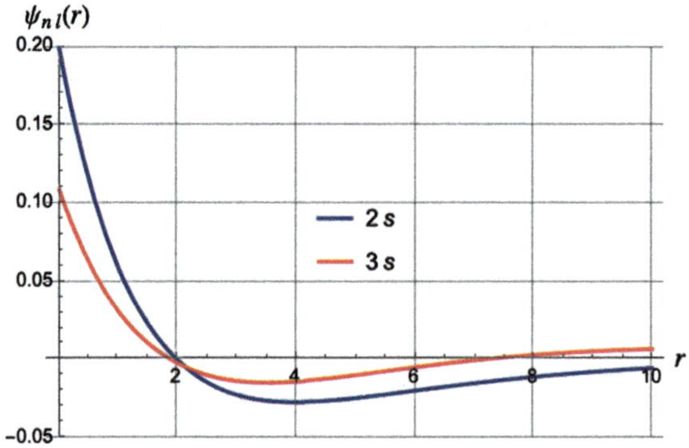

FIGURE 8.8 Plots of 2*s* and 3*s* orbitals. Scale units in bohrs.

$$\psi_{210}(r,\theta,\phi)=\frac{Z^{5/2}}{4\sqrt{2\pi}}\,r\,e^{-Zr/2}\cos\theta \tag{8.53}$$

and

$$\psi_{21\pm1}(r,\theta,\phi)=\mp\frac{Z^{5/2}}{4\sqrt{2\pi}}\,r\,e^{-Zr/2}\sin\theta\,e^{\pm i\phi}. \tag{8.54}$$

The function ψ_{210} is real and contains the factor $r\cos\theta$, which is equal to the cartesian variable z. In chemical applications, this is designated as a $2p_z$ orbital:

$$\psi_{2pz}=\frac{Z^{5/2}}{4\sqrt{2\pi}}\,z\,e^{-Zr/2}. \tag{8.55}$$

A contour plot is shown in Fig. 8.9. Note that this function is cylindrically-symmetrical about the z-axis with a node in the x, y-plane. The eigenfunctions $\psi_{21\pm1}$ are complex and not as easy to represent graphically, their angular dependence being that of the spherical harmonics $Y_{1\pm1}$. It has been shown that any linear combination of degenerate eigenfunctions is an equally-valid alternative eigenfunction. Making use of the Euler formulas for sine and cosine

$$\cos\phi=\frac{e^{i\phi}+e^{-i\phi}}{2}\qquad\text{and}\qquad\sin\phi=\frac{e^{i\phi}-e^{-i\phi}}{2i} \tag{8.56}$$

and noting that the combinations $\sin\theta\cos\phi$ and $\sin\theta\sin\phi$ correspond to the cartesian variables x and y, respectively, we can define the alternative $2p$ orbitals

$$\psi_{2px}=\frac{1}{\sqrt{2}}(\psi_{21-1}-\psi_{211})=\frac{Z^{5/2}}{4\sqrt{2\pi}}\,x\,e^{-Zr/2} \tag{8.57}$$

and

$$\psi_{2py}=-\frac{i}{\sqrt{2}}(\psi_{21-1}+\psi_{211})=\frac{Z^{5/2}}{4\sqrt{2\pi}}\,y\,e^{-Zr/2}. \tag{8.58}$$

Clearly, these have the same shape as the $2pz$-orbital, but are oriented along the x- and y-axes, respectively. The threefold degeneracy of the p-orbitals is very clearly shown by the geometric equivalence the functions $2px$, $2py$ and $2pz$, which is not obvious from the spherical harmonics. The functions listed in Table 8.1 are, in fact, the real forms for atomic orbitals. All higher p-orbitals have analogous functional forms $xf(r)$, $yf(r)$ and $zf(r)$ and are likewise 3-fold degenerate.

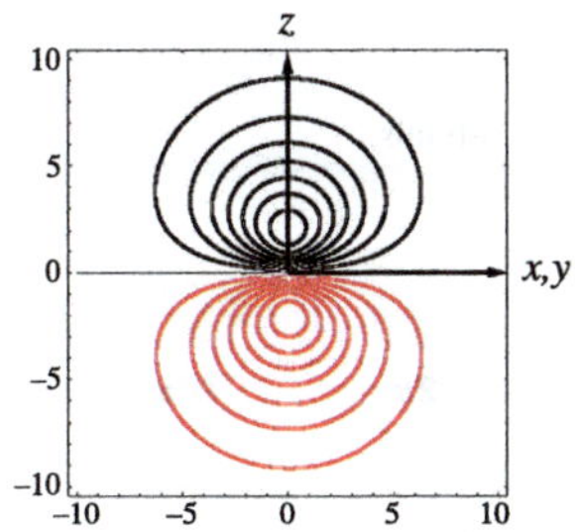

FIGURE 8.9 Contour plot of $2p_z$ orbital. Negative values are shown in red. Scale units in bohrs.

The orbital ψ_{320} is, like ψ_{210}, a real function. It is known in chemistry as the d_{z^2}-orbital and can be expressed as a cartesian factor times a function of r:

$$\psi_{3d_{z^2}} = \psi_{320} = (3z^2 - r^2)f(r). \tag{8.59}$$

A contour plot is shown in Fig. 8.10. This function is also cylindrically symmetric about the z-axis with *two* angular nodes—conical surfaces with $3z^2-r^2=0$. The remaining four $3d$ orbitals are complex functions containing the spherical harmonics $Y_{2\pm1}$ and $Y_{2\pm2}$ pictured in Fig. 6.4. We can again construct real functions from linear combinations, the result being four geometrically equivalent "four-leaf clover" functions with two perpendicular planar nodes. These orbitals are designated $d_{x^2-y^2}$, d_{xy}, d_{zx} and d_{yz}. Two of these are also shown in Fig. 8.10. The d_{z^2} orbital has a different shape. However, it can be expressed in terms of two non-standard d-orbitals, $d_{z^2-x^2}$ and $d_{y^2-z^2}$. The latter functions, along with $d_{x^2-y^2}$ add to zero and thus constitute a linearly *dependent* set. Only *two* combinations of these three functions can be chosen as independent eigenfunctions.

8.7 Summary on atomic orbitals

The atomic orbitals listed in Table 8.1 are illustrated in Fig. 8.11. Blue and yellow indicate, respectively, positive and negative regions of the wavefunctions (the radial nodes of the $2s$ and $3s$ orbitals are obscured). These pictures are intended as stylized representations of atomic orbitals and should *not* be interpreted as quantitatively accurate.

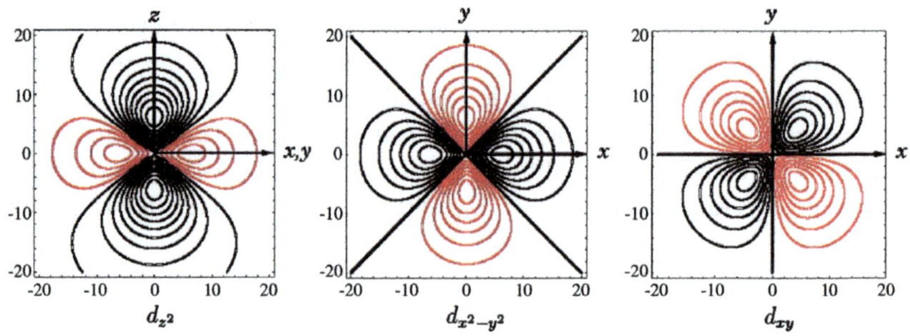

FIGURE 8.10 Contour plots of 3d orbitals.

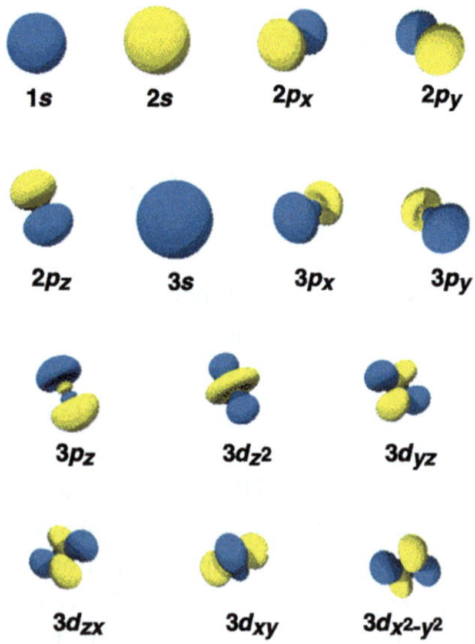

FIGURE 8.11 Hydrogenic atomic orbitals.

The electron charge distribution in an orbital $\psi_{nlm}(\mathbf{r})$ is given by

$$\rho(\mathbf{r}) = |\psi_{nlm}(\mathbf{r})|^2, \tag{8.60}$$

which for the s-orbitals is a function of r alone. The radial distribution function can be defined, even for orbitals containing angular dependence, by

$$D_{nl}(r) = 4\pi r^2[R_{nl}(r)]^2. \tag{8.61}$$

This represents the electron density in a shell of radius r, including all values of the angular variables θ, ϕ. Fig. 8.12 shows plots of the RDF for the first few hydrogen orbitals.

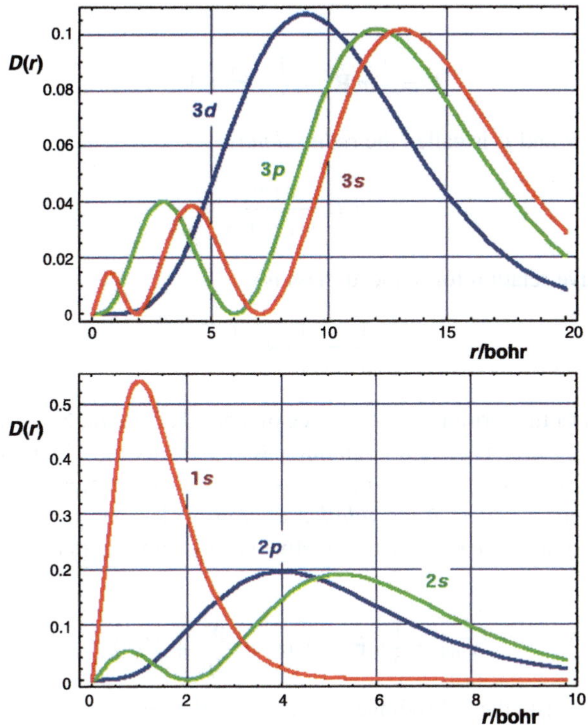

FIGURE 8.12 Some radial distribution functions.

8.8 Reduced mass

Consider a system of two particles of masses m_1 and m_2 interacting with a potential energy which depends only on the separation of the particles. The classical energy is given by

$$E = \frac{1}{2} m_1 \dot{\mathbf{r}}_1^2 + \frac{1}{2} m_2 \dot{\mathbf{r}}_2^2 + V(|\mathbf{r}_2 - \mathbf{r}_1|), \tag{8.62}$$

the dots signifying derivative wrt time. Introduce two new variables, the particle separation $\mathbf{r}$ and the position of the center of mass $\mathbf{R}$:

$$\mathbf{r} = \mathbf{r}_2 - \mathbf{r}_1, \qquad \mathbf{R} = \frac{m_1 \mathbf{r}_1 + m_2 \mathbf{r}_2}{m}, \tag{8.63}$$

where $m = m_1 + m_2$. In terms of the new coordinates

$$\mathbf{r}_1 = \mathbf{R} + \frac{m_2}{m}\mathbf{r}, \qquad \mathbf{r}_2 = \mathbf{R} - \frac{m_1}{m}\mathbf{r} \tag{8.64}$$

and

$$E = \frac{1}{2} m \dot{\mathbf{R}}^2 + \frac{1}{2} \mu \dot{\mathbf{r}}^2 + V(r), \tag{8.65}$$

where $r = |\mathbf{r}|$ and μ is called the *reduced mass*

$$\mu \equiv \frac{m_1 m_2}{m_1 + m_2}. \tag{8.66}$$

An alternative relation for reduced mass is

$$\frac{1}{\mu} = \frac{1}{m_1} + \frac{1}{m_2}, \tag{8.67}$$

reminiscent of the formula for resistance of a parallel circuit. Note that, if $m_2 \to \infty$, then $\mu \to m_1$. The term containing $\dot{\mathbf{R}}$ represents the kinetic energy of a single hypothetical particle of mass m located at the center of mass $\mathbf{R}$. The remaining terms represent the *relative* motion of the two particles. It has the appearance of a *single* particle of effective mass μ moving in the potential field $V(r)$.

$$E_{\text{rel}} = \frac{1}{2} \mu \dot{\mathbf{r}}^2 + V(r) = \frac{\mathbf{p}^2}{2\mu} + V(r). \tag{8.68}$$

We can thus write the Schrödinger equation for the relative motion

$$\left\{ -\frac{\hbar^2}{2\mu} \nabla^2 + V(r) \right\} \psi(\mathbf{r}) = E \psi(\mathbf{r}). \tag{8.69}$$

When we treated the hydrogen atom earlier, it was assumed that the nuclear mass was infinite. In that case we can set $\mu = m$, the mass of an electron. The Rydberg constant for infinite nuclear mass was calculated to be

$$\mathcal{R}_\infty = \frac{2\pi^2 m e^4}{h^3 c} = 109{,}737 \text{ cm}^{-1}. \tag{8.70}$$

If instead, we use the reduced mass of the electron-proton system

$$\mu = \frac{mM}{m + M} \approx \frac{1836}{1837} m \approx 0.999456 m. \tag{8.71}$$

This changes the Rydberg constant for hydrogen to

$$\mathcal{R}_{\text{H}} \approx 109{,}677 \text{ cm}^{-1} \tag{8.72}$$

in perfect agreement with experiment.

In 1931, H. C. Urey evaporated four liters of hydrogen down to one milliliter and measured the spectrum of the residue. The result was a set of lines displaced slightly from the hydrogen spectrum. This amounted to the discovery of deuterium, or heavy hydrogen, for which Urey was awarded in 1934 Nobel Prize in Chemistry. Estimating the mass of the deuteron, $^2H_1$, as twice that of the proton, gives

$$\mathcal{R}_D \approx 109{,}707 \text{ cm}^{-1}. \tag{8.73}$$

Another interesting example involves positronium, a short-lived combination of an electron and a positron—the electron's antiparticle. The electron and positron mutually annihilate with a half-life of approximately 10^{-7} sec and positronium decays into gamma rays. The reduced mass of positronium is

$$\mu = \frac{m \times m}{m + m} = \frac{m}{2} \tag{8.74}$$

half the mass of the electron. Thus the ionization energy equals 6.80 eV, half that of hydrogen atom. Positron emission tomography (PET) provides a sensitive scanning technique for functioning living tissue, notably the brain. A compound containing a positron-emitting radionuclide, for example, ^{11}C, ^{13}N, ^{15}O or ^{18}F, is injected into the body. The emitted positrons attach to electrons to form short-lived positronium (see Fig. 8.13), and the annihilation radiation is monitored.

FIGURE 8.13 Jens Zorn sculpture depicting positronium annihilation. Outside University of Michigan Physics Building.

Chapter 8. Exercises

8.1. Assume that each circular Bohr orbit for an electron in a hydrogen atom contains an integer number of de Broglie wavelengths, $n = 1, 2, 3$, etc. Show that the orbital angular momentum must then be quantized. Bohr's formula for the hydrogen energy levels follows from this.

8.2. Based on your knowledge of the first few hydrogenic eigenfunctions, deduce general formulas, in terms of n and l, for: (i) the number of radial nodes in an atomic orbital; (ii) the number of angular nodes; (iii) the total number of nodes.

8.3. Calculate the wavelength of the Lyman alpha transition ($1s \leftarrow 2p$) in atomic hydrogen and in He^+. Express the results in both nm and cm^{-1}.

8.4. Determine the maximum of the radial distribution function for the ground state of hydrogen atom. Compare this value with the corresponding radius in the Bohr theory.

8.5. The nodeless $1s$ ground state hydrogenlike eigenfunction is given by $\psi_{1s} = \frac{Z^{3/2}}{\sqrt{\pi}} e^{-Zr}$. Assuming only orthogonality and normalization, construct the $2s$ and $3s$ eigenfunctions, with 1 and 2 nodes, respectively, with the forms $\psi_{2s} = \text{const}(1 + \alpha r)e^{-Zr/2}$ and $\psi_{2s} = \text{const}(1 + \beta r + \gamma r^2)e^{-Zr/3}$, respectively.

8.6. The following reaction might occur in the interior of a star:

$$He^{++} + H \rightarrow He^+ + H^+.$$

Calculate the electronic energy change (in eV). Assume all species in their ground states.

8.7. Which of the following operators is *not* equal to the other four:
(i) $\partial^2/\partial r^2$ (ii) $r^{-2}\,\partial/\partial r\, r^2\, \partial/\partial r$ (iii) $r^{-1}\,\partial^2/\partial r^2\, r$ (iv) $(r^{-1}\,\partial/\partial r\, r)^2$
(v) $\partial^2/\partial r^2 + 2r^{-1}\partial/\partial r$.

8.8. Calculate the expectation values of r, r^2 and of r^{-1} in the ground state of the hydrogen atom. Give results in atomic units.

8.9. Calculate the expectation values of potential and kinetic energies for the $1s$ state of a hydrogenlike atom.

8.10. Verify that the $3d_{xy}$ orbital given in the table is a normalized eigenfunction of the hydrogenlike Schrödinger equation.

8.11. Show that the function

$$\psi(r, \theta, \phi) = \text{const}\left[1 - r\,\sin^2(\theta/2)\right] e^{-r/2}$$

is a solution of the Schrödinger equation for the hydrogen atom and find the corresponding eigenvalue (in atomic units).

8.12. For the ground state of a hydrogenlike atom, calculate the radius of the sphere enclosing 90% of the electron probability in the $1s$ state of hydrogen atom. (This involves a numerical computation with successive approximations.)

8.13. Consider as a variational approximation to the ground state of the hydrogen atom the wavefunction $\psi(r) = e^{-\alpha r}$. Calculate the corresponding energy $E(\alpha)$ then optimize with respect to the parameter α. Compare with the exact solution.

8.14. The electron-spin resonance hyperfine splitting for atomic hydrogen is given by

$$\Delta\nu = 532.65\left[\frac{8\pi}{3}|\psi(0)|^2 + \left\langle\frac{3\cos^2\theta - 1}{r^3}\right\rangle\right] \text{MHz}.$$

Calculate $\Delta\nu$ for the $1s$ and for the $2p_0$ states. The result is in MHz when the bracketed terms are expressed in atomic units. (Hint: In evaluating the expectation value, do the integral over angles first.)

Chapter 9

The helium atom

The second element in the periodic table provides our first example of a quantum-mechanical problem which *cannot* be solved exactly. Nevertheless, as we will show, approximation methods applied to helium can give accurate solutions in essentially perfect agreement with experimental results. In this sense, it can be concluded that quantum mechanics is correct for atoms more complicated than hydrogen. By contrast, the Bohr theory failed miserably in attempts to apply it beyond the hydrogen atom.

9.1 Experimental energies

The helium atom has two electrons bound to a nucleus with charge $Z = 2$. The successive removal of the two electrons can be diagrammed as

$$\text{He} \xrightarrow{I_1} \text{He}^+ + e^- \xrightarrow{I_2} \text{He}^{++} + 2e^-. \tag{9.1}$$

The *first ionization energy* I_1, the minimum energy required to remove the first electron from helium, is experimentally 24.5874 eV. The second ionization energy, I_2, is 54.4228 eV, which can be calculated exactly since He^+ is a hydrogenlike ion. We have

$$I_2 = -E_{1s}(\text{He}^+) = -\frac{Z^2}{2n^2} = -2 \text{ hartrees}, \tag{9.2}$$

which corresponds to 54.4228 eV. The energy of the three separated particles on the right side of Eq. (9.1) is, by definition, zero. Therefore the ground-state energy of helium atom is given by $E_0 = -(I_1 + I_2) = -79.0102 \text{ eV} = -2.9037$ hartrees. We will attempt to reproduce this value, as closely as possible, by theoretical analysis.

9.2 Schrödinger equation and variational calculations

The Schrödinger equation for He atom, again using atomic units and assuming infinite nuclear mass, can be written

$$\left\{-\frac{1}{2}\nabla_1^2 - \frac{1}{2}\nabla_2^2 - \frac{Z}{r_1} - \frac{Z}{r_2} + \frac{1}{r_{12}}\right\} \psi(\mathbf{r}_1, \mathbf{r}_2) = E\, \psi(\mathbf{r}_1, \mathbf{r}_2). \tag{9.3}$$

Introduction to Quantum Mechanics. https://doi.org/10.1016/B978-0-12-822310-9.00017-3

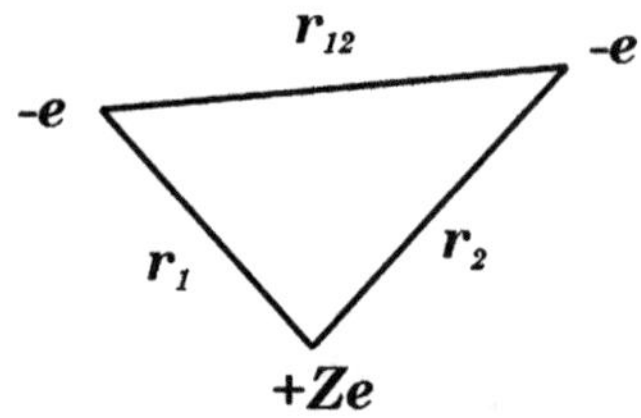

FIGURE 9.1 Coordinates for helium atom Schrödinger equation.

The five terms in the Hamiltonian represent, respectively, the kinetic energies of electrons 1 and 2, the nuclear attractions of electrons 1 and 2, and the repulsive interaction between the two electrons (see Fig. 9.1). It is this last contribution which prevents an exact solution of the Schrödinger equation and accounts for much of the complication in the theory. In seeking an approximation to the ground state, we might first work out the solution in the absence of the $1/r_{12}$-term. In the Schrödinger equation thus simplified, we can separate the variables $\mathbf{r}_1$ and $\mathbf{r}_2$ to reduce the equation to two independent hydrogenlike problems. The ground state wavefunction (not normalized) for this hypothetical helium atom would be

$$\psi(\mathbf{r}_1, \mathbf{r}_2) = \psi_{1s}(r_1)\psi_{1s}(r_2) = e^{-Z(r_1+r_2)} \tag{9.4}$$

and the energy would equal $2 \times (-Z^2/2) = -4$ hartrees, compared to the experimental value of -2.90 hartrees. Neglect of electron repulsion evidently introduces a very large error.

A significantly improved result can be obtained by keeping the functional form (9.4), but replacing Z by an adjustable parameter α. Using the function

$$\tilde{\psi}(r_1, r_2) = e^{-\alpha(r_1+r_2)} \tag{9.5}$$

in the variational principle [cf. Eq. (4.58)], we have

$$\tilde{E} = \frac{\int \tilde{\psi}(r_1, r_2)\, H\, \tilde{\psi}(r_1, r_2)\, d\tau_1\, d\tau_2}{\int \tilde{\psi}(r_1, r_2)\, \tilde{\psi}(r_1, r_2)\, d\tau_1\, d\tau_2}, \tag{9.6}$$

where H is the full Hamiltonian as in Eq. (9.3), including the $1/r_{12}$-term. The expectation values of the five parts of the Hamiltonian work out as follows:

$$\left\langle -\frac{1}{2}\nabla_1^2 \right\rangle = \left\langle -\frac{1}{2}\nabla_2^2 \right\rangle = \frac{\alpha^2}{2},$$
$$\left\langle -\frac{Z}{r_1} \right\rangle = \left\langle -\frac{Z}{r_2} \right\rangle = -Z\alpha, \qquad \left\langle \frac{1}{r_{12}} \right\rangle = \frac{5}{8}\alpha. \tag{9.7}$$

The sum of the integrals in (9.7) gives the variational energy

$$\tilde{E}(\alpha) = \alpha^2 - 2Z\alpha + \frac{5}{8}\alpha. \tag{9.8}$$

This will always be an upper bound for the true ground-state energy. We can optimize our result by finding the value of α which *minimizes* the energy (9.8). We find

$$\frac{d\tilde{E}}{d\alpha} = 2\alpha - 2Z + \frac{5}{8} = 0, \tag{9.9}$$

giving the optimal value

$$\alpha = Z - \frac{5}{16}. \tag{9.10}$$

This can be given a physical interpretation, noting that the parameter α in the approximate wavefunction (9.5) represents an *effective* nuclear charge. Each electron partially shields the other electron from the positively-charged nucleus by an amount equivalent to 5/16 of an electron charge. Substituting (9.10) into (9.8), we obtain the optimized approximation to the energy

$$\tilde{E} = -\left(Z - \frac{5}{16}\right)^2. \tag{9.11}$$

For helium ($Z = 2$), this gives -2.84765 hartrees, an error of about 2% (exact $E_0 = -2.90372$). Note that the inequality $\tilde{E} > E_0$ applies in an *algebraic* sense.

In the late 1920's, it was considered important to do the best possible helium computation, as a test of the validity of quantum mechanics for many electron atoms. The table below gives the results for a selection of variational computations on helium.

wavefunction	parameters	energy
$e^{-Z(r_1+r_2)}$	$Z = 2$	−2.75
$e^{-\alpha(r_1+r_2)}$	$\alpha = 1.6875$	−2.84765
$\psi(r_1)\psi(r_2)$	best $\psi(r)$	−2.86168
$e^{-\alpha(r_1+r_2)}(1 + cr_{12})$	best α, c	−2.89112
Hylleraas (1929)	10 parameters	−2.90363
Pekeris (1959)	1078 parameters	−2.90372

The third entry refers to the *self-consistent field* method, developed by Hartree. Even for the best possible choice of one-electron functions $\psi(r)$, there remains a considerable error. This is due to failure to include the variable r_{12} in the wavefunction. The effect is known as *electron correlation*. The fourth entry, containing a simple correction for correlation, gives a considerable improvement. Hylleraas (1929) extended this approach with a variational function of the form

$$\psi(r_1, r_2, r_{12}) = e^{-\alpha(r_1+r_2)} \times \text{polynomial in } r_1, r_2, r_{12} \tag{9.12}$$

and obtained a result in agreement with experiment using a function with 10 optimized parameters. More recently, using modern computers, results in essentially perfect agreement with experiment have been obtained.

9.3 Spinorbitals and the exclusion principle

The simpler wavefunctions for helium atom, for example (9.5), can be interpreted as representing two electrons in hydrogenlike $1s$ orbitals, designated as a $1s^2$ configuration. Pauli's exclusion principle, which states that no two electrons in an atom can have the same set of four quantum numbers, requires the two $1s$ electrons to have *different* spins: one spin-up or α, the other spin-down or β. A product of an orbital with a spin function is called a *spinorbital*. For example, electron 1 might occupy a spinorbital which we designate

$$\phi(1) = \psi_{1s}(1)\alpha(1) \quad \text{or} \quad \psi_{1s}(1)\beta(1). \tag{9.13}$$

Spinorbitals can be designated by a single subscript, for example, ϕ_a or ϕ_b, where the subscript stands for a *set* of four quantum numbers. In a two-electron system the occupied spinorbitals ϕ_a and ϕ_b must be different, meaning that at least one of their four quantum numbers must be unequal. A two-electron spinorbital function of the form

$$\Psi(1,2) = \frac{1}{\sqrt{2}}\Big(\phi_a(1)\phi_b(2) - \phi_b(1)\phi_a(2)\Big) \tag{9.14}$$

automatically fulfills the Pauli principle since it vanishes if $a = b$. Moreover, this function associates each electron equally with each orbital, which is consistent with the *indistinguishability* of identical particles in quantum mechanics. The factor $1/\sqrt{2}$ normalizes the two-particle wavefunction, assuming that ϕ_a and ϕ_b are normalized and mutually orthogonal. The function (9.13) is *antisymmetric* with respect to interchange of electron labels, meaning that

$$\Psi(2,1) = -\Psi(1,2). \tag{9.15}$$

This antisymmetry property is an elegant way of expressing the Pauli principle.

We note, for future reference, that the function (9.13) can be expressed as a 2×2 determinant:

$$\Psi(1,2) = \frac{1}{\sqrt{2}}\begin{vmatrix} \phi_a(1) & \phi_b(1) \\ \phi_a(2) & \phi_b(2) \end{vmatrix}. \tag{9.16}$$

For the $1s^2$ configuration of helium, the two orbital functions are the same and the total wavefunction including spin can be written

$$\Psi(1,2) = \psi_{1s}(1)\psi_{1s}(2) \times \frac{1}{\sqrt{2}}\Big(\alpha(1)\beta(2) - \beta(1)\alpha(2)\Big). \tag{9.17}$$

For two-electron systems (but *not* for three or more electrons), the wavefunction can be factored into an orbital function times a spin function. The two-electron spin function

$$\sigma_{0,0}(1,2) = \frac{1}{\sqrt{2}}\Big(\alpha(1)\beta(2) - \beta(1)\alpha(2)\Big) \tag{9.18}$$

represents the two electron spins in opposing directions (antiparallel) with a total spin angular momentum of zero. The two subscripts are the quantum numbers S and M_S for the total electron spin. Eq. (9.17) is called the *singlet* spin state since there is only a single orientation for a total spin quantum number of zero. It is also possible to have both spins in the *same* state, provided the orbitals are different. There are three possible states for two parallel spins:

$$\begin{aligned}\sigma_{1,1}(1,2) &= \alpha(1)\alpha(2),\\ \sigma_{1,0}(1,2) &= \frac{1}{\sqrt{2}}\Big(\alpha(1)\beta(2) + \beta(1)\alpha(2)\Big),\\ \sigma_{1,-1}(1,2) &= \beta(1)\beta(2).\end{aligned} \tag{9.19}$$

These make up the *triplet* spin states, which have the three possible orientations of an angular momentum of 1.

9.4 Excited states of helium

The lowest excitated states of helium have the electron configuration $1s\,2s$. The $1s\,2p$ configuration has higher energy, even though the $2s$ and $2p$ orbitals in hydrogen are degenerate. The repulsive interaction between $1s$ and $2s$ has a *lower* energy than that between $1s$ and $2p$ because $2s$ overlaps with the $1s$ *less* than $2p$ does, as is evident from the radial distribution functions plotted in Fig. 8.12. This is enhanced by the fact that $2s$ has a radial node in a region of space where the $1s$ density is significant. When electrons are in different orbitals, their spins can be either parallel or antiparallel. In order that the wavefunction satisfy the antisymmetry condition (9.14), the two-electron orbital and spin functions must have *opposite* behavior under exchange of electron labels. There are four possible states from the $1s\,2s$ configuration: a singlet state

$$\Psi^{+}(1,2) = \frac{1}{\sqrt{2}}\Big(\psi_{1s}(1)\psi_{2s}(2) + \psi_{2s}(1)\psi_{1s}(2)\Big)\,\sigma_{0,0}(1,2) \tag{9.20}$$

and three triplet states

$$\Psi^{-}(1,2) = \frac{1}{\sqrt{2}}\Big(\psi_{1s}(1)\psi_{2s}(2) - \psi_{2s}(1)\psi_{1s}(2)\Big)\begin{cases}\sigma_{1,1}(1,2)\\ \sigma_{1,0}(1,2)\\ \sigma_{1,-1}(1,2).\end{cases} \tag{9.21}$$

Using the Hamiltonian (9.3), we can compute the approximate energies

$$E^{\pm} = \int\int \Psi^{\pm}(1,2)\, H\, \Psi^{\pm}(1,2)\, d\tau_1\, d\tau_2. \tag{9.22}$$

After evaluating some fierce-looking integrals, this reduces to the form

$$E^{\pm} = I_{1s} + I_{2s} + J_{1s,2s} \pm K_{1s,2s}, \tag{9.23}$$

in terms of *one electron integrals*

$$I_a = \int \psi_a(\mathbf{r}) \left\{ -\frac{1}{2}\nabla^2 - \frac{Z}{r} \right\} \psi_a(\mathbf{r})\, d\tau, \tag{9.24}$$

Coulomb integrals

$$J_{a,b} = \int\int \psi_a(\mathbf{r}_1)^2 \frac{1}{r_{12}} \psi_b(\mathbf{r}_2)^2\, d\tau_1\, d\tau_2 \tag{9.25}$$

and *exchange integrals*

$$K_{a,b} = \int\int \psi_a(\mathbf{r}_1)\psi_b(\mathbf{r}_1) \frac{1}{r_{12}} \psi_a(\mathbf{r}_2)\psi_b(\mathbf{r}_2)\, d\tau_1\, d\tau_2. \tag{9.26}$$

The Coulomb integral represents the repulsive potential energy for two interacting charge distributions $[\psi_a(\mathbf{r}_1)]^2$ and $[\psi_b(\mathbf{r}_2)]^2$. The exchange integral, which has no classical analog, arises because of the exchange symmetry (or antisymmetry) requirement of the wavefunction. Both J and K can be shown to be positive quantities. Therefore the negative sign in (9.22) is associated with the state of lower energy. Thus the triplet state of the configuration $1s\,2s$ is lower in energy than the singlet state. This is an almost universal generalization and contributes to Hund's rule, to be discussed in the next Chapter.

The excitation energies of the $1s2s$ and $1s2p$ states of helium are shown in Fig. 9.2. On this scale the $1s^2$ ground state is the zero of energy. If the energies are approximated as in Eq. (9.22) by

$$\begin{aligned} E(^{1,3}S) &\approx I_{1s} + I_{2s} + J_{1s,2s} \pm K_{1s,2s}, \\ E(^{1,3}P) &\approx I_{1s} + I_{2p} + J_{1s,2p} \pm K_{1s,2p}, \end{aligned} \tag{9.27}$$

then the results are consistent with the following values of the Coulomb and exchange integrals: $J_{1s,2p} \approx 10.0$ eV, $K_{1s,2p} \approx 0.1$ eV, $J_{1s,2s} \approx 9.1$ eV and $K_{1s,2s} \approx 0.4$ eV.

Chapter 9. Exercises

9.1. For the optimized helium variational wavefunction

$$\psi(r_1, r_2) = e^{-\alpha(r_1+r_2)}$$

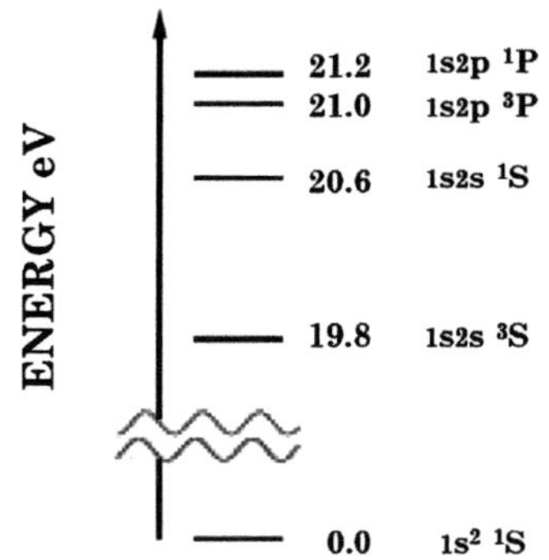

FIGURE 9.2 Lower excited states of helium atom.

calculate the expectation values of total kinetic and potential energies. Do these satisfy the virial theorem?

9.2. Using the same form of an optimized variational wavefunction

$$\psi(r_1, r_2) = e^{-\alpha(r_1+r_2)}$$

estimate the ground-state energy of Li^+.

9.3. Calculate the energy of the hypothetical $1s^3$ state of the Li atom using the optimized variational wavefunction

$$\psi(1, 2, 3) = e^{-\alpha(r_1+r_2+r_3)}.$$

Neglect electron spin, of course. Compare with the experimental ground-state energy, $E_0 = -7.478$ hartrees. Comment on the applicability of the variational theorem.

Chapter 10

Atomic structure and the periodic law

The discovery of the periodic structure of the elements by Mendeleev, shown in Fig. 10.1, must be ranked as one the greatest achievements in the history of science. And perhaps the most impressive conceptual accomplishment of quantum mechanics has been its rational account of the origin of the periodic table. Although accurate computations become increasingly more difficult as the number of electrons increases, the general patterns of atomic behavior can be predicted with remarkable accuracy. A modern version of the periodic table is shown in Fig. 10.2.

Dmitri Ivanovich Mendeleev

10.1 Slater determinants

According to the orbital approximation, which was introduced in the last Chapter, an N-electron atom contains N occupied spinorbitals, which can be designated $\phi_a, \phi_b \ldots \phi_n$. In accordance with the Pauli exclusion principle, no two of these spinorbitals can be identical. Also, every electron should be equally associated with every spinorbital. A very neat mathematical representation for these properties is a generalization of the two-electron wavefunction (9.15) called a *Slater determinant*

Introduction to Quantum Mechanics. https://doi.org/10.1016/B978-0-12-822310-9.00018-5

Reihen	Gruppe I. - R^2O	Gruppe II. - RO	Gruppe III. - R^2O^3	Gruppe IV. RH^4 RO^2	Gruppe V. RH^3 R^2O^5	Gruppe VI. RH^2 RO^3	Gruppe VII. RH R^2O^7	Gruppe VIII. - RO^4
1	H = 1							
2	Li = 7	Be = 9,4	B = 11	C = 12	N = 14	O = 16	F = 19	
3	Na = 23	Mg = 24	Al = 27,3	Si = 28	P = 31	S = 32	Cl = 35,5	
4	K = 39	Ca = 40	– = 44	Ti = 48	V = 51	Cr = 52	Mn = 55	Fe = 56, Co = 59 Ni = 59 Cu = 63
5	(Cu = 63)	Zn = 65	– = 68	– = 72	As = 75	Se = 78	Br = 80	
6	Rb = 85	Sr = 87	?Yt = 88	Zr = 90	Nb = 94	Mo = 96	– = 100	Ru = 104 Rh = 104 Pd = 106, Ag = 108
7	Ag = 108	Cd = 112	In = 113	Sn = 118	Sb = 122	Te = 125	J = 127	
8	Cs = 133	Ba = 137	?Di = 138	?Ce = 140	-	-	-	- - - -
9	(-)	-	-	-	-	-	-	
10	-	-	?Er = 178	?La = 180	Ta = 182	W = 184	-	Os = 195, Ir = 197, Pt = 198, Au = 199
11	(Au = 199)	Hg = 200	Tl = 204	Pb = 207	Bi = 208	-	-	
12	-	-	-	Th = 231	-	U = 240	-	- - - -

FIGURE 10.1 Mendeleev's original periodic table (1869). The elements in the gaps marked with red squares—now known to be Sc, Ga and Ge—were predicted by Mendeleev before they were actually discovered.

Group → / Period ↓	1	2	3	4	5	6	7	8	9	10	11	12	13	14	15	16	17	18
1	1 H																	2 He
2	3 Li	4 Be											5 B	6 C	7 N	8 O	9 F	10 Ne
3	11 Na	12 Mg											13 Al	14 Si	15 P	16 S	17 Cl	18 Ar
4	19 K	20 Ca	21 Sc	22 Ti	23 V	24 Cr	25 Mn	26 Fe	27 Co	28 Ni	29 Cu	30 Zn	31 Ga	32 Ge	33 As	34 Se	35 Br	36 Kr
5	37 Rb	38 Sr	39 Y	40 Zr	41 Nb	42 Mo	43 Tc	44 Ru	45 Rh	46 Pd	47 Ag	48 Cd	49 In	50 Sn	51 Sb	52 Te	53 I	54 Xe
6	55 Cs	56 Ba	57 La *	72 Hf	73 Ta	74 W	75 Re	76 Os	77 Ir	78 Pt	79 Au	80 Hg	81 Tl	82 Pb	83 Bi	84 Po	85 At	86 Rn
7	87 Fr	88 Ra	89 Ac **	104 Rf	105 Db	106 Sg	107 Bh	108 Hs	109 Mt	110 Ds	111 Rg	112 Cn	113 Nh	114 Fl	115 Mc	116 Lv	117 Ts	118 Og
			*	58 Ce	59 Pr	60 Nd	61 Pm	62 Sm	63 Eu	64 Gd	65 Tb	66 Dy	67 Ho	68 Er	69 Tm	70 Yb	71 Lu	
			**	90 Th	91 Pa	92 U	93 Np	94 Pu	95 Am	96 Cm	97 Bk	98 Cf	99 Es	100 Fm	101 Md	102 No	103 Lr	

FIGURE 10.2 Modern version of the periodic table extending through element 118.

$$\Psi(1,2\ldots N) = \frac{1}{\sqrt{N!}} \begin{vmatrix} \phi_a(1) & \phi_b(1) & \ldots & \phi_n(1) \\ \phi_a(2) & \phi_b(2) & \ldots & \phi_n(2). \\ & & \vdots & \\ \phi_a(N) & \phi_b(N) & \ldots & \phi_n(N) \end{vmatrix}. \tag{10.1}$$

Since interchanging any two rows (or columns) of a determinant multiplies it by -1, the antisymmetry property (9.14) is fulfilled, for *every* pair of electrons.

John C. Slater

The total number of electrons in an atom, the *cardinality*, can be known while the numbering of individual electrons, their *ordinality*, has no physical significance. This is something like making a deposit in your bank account. You know how many dollars you put in. But, when you make a withdrawal, you have no way of matching the *individual* dollars you take out with the dollars you put in.

The Hamiltonian for an atom with N electrons around a nucleus of charge Z can be written

$$H = \sum_{i=1}^{N} \left\{ -\frac{1}{2}\nabla_i^2 - \frac{Z}{r_i} \right\} + \sum_{i<j}^{N} \frac{1}{r_{ij}}. \tag{10.2}$$

The sum over electron repulsions is written so that each pair $\{i, j\}$ is counted just once. The energy of the state represented by a Slater determinant (10.1) is given by the multiple integral

$$E = \int\int \cdots \int \Psi(1, 2 \ldots N)\, H\, \Psi(1, 2 \ldots N)\, dx_1 dx_2 \ldots dx_N, \tag{10.3}$$

where x represents the four space and spin coordinates of each electron. The determinantal wavefunction Ψ contains $N!$ terms, while the Hamiltonian is the sum of N one-electron operators plus $N(N-1)/2$ two-electron contributions. Evaluation of the resulting sum of integrals, after a lengthy derivation, results in the expression:

$$\tilde{E} = \sum_a I_a + \frac{1}{2} \sum_{a,b} (J_{ab} - K_{ab}), \tag{10.4}$$

where the sums run over all occupied spinorbitals. The one-electron, Coulomb and exchange integrals have the same form as those defined for helium atom in Eqs. (9.23)-(9.25), except that they are now integrals over *both* space and spin coordinates. Thus the one-electron integrals are given by

$$I_a = \int \phi_a(\mathbf{r}) \left\{ -\frac{1}{2}\nabla^2 - \frac{Z}{r} \right\} \phi_a(\mathbf{r})\, dx = \int \psi_a(\mathbf{r}) \left\{ -\frac{1}{2}\nabla^2 - \frac{Z}{r} \right\} \psi_a(\mathbf{r})\, d\tau, \tag{10.5}$$

noting that integration the over spin gives 1. The Coulomb integrals are

$$J_{a,b} = \int\int \phi_a(\mathbf{r}_1)^2 \frac{1}{r_{12}} \phi_b(\mathbf{r}_2)^2\, dx_1\, dx_2 = \int\int \psi_a(\mathbf{r}_1)^2 \frac{1}{r_{12}} \psi_b(\mathbf{r}_2)^2\, d\tau_1\, d\tau_2, \tag{10.6}$$

again in agreement with the previous form. The exchange integrals, as now defined,

$$K_{a,b} = \int\int \phi_a(\mathbf{r}_1)\phi_b(\mathbf{r}_1) \frac{1}{r_{12}} \phi_a(\mathbf{r}_2)\phi_b(\mathbf{r}_2)\, dx_1\, dx_2 \tag{10.7}$$

will equal zero, by virtue of spin orthogonality, *unless* the spins of orbitals a and b are both α or both β. The factor 1/2 in Eq. (10.4) corrects for the double counting of pairs of spinorbitals in the second sum. The contributions with $a = b$ can be actually omitted since $J_{aa} = K_{aa}$. This effectively removes the Coulomb interaction of an orbital with itself, which is spurious.

The Hartree-Fock or self-consistent field (SCF) method is a procedure for optimizing the orbital functions in the Slater determinant (10.1), so as to minimize the energy (10.4). SCF computations have been carried out for all the atoms of the periodic table, with predictions of total energies and ionization energies generally accurate in the 1-2% range. Fig. 10.3 shows the electronic radial distribution function in the argon atom, obtained from a Hartree-Fock computation. The shell structure of the electron cloud is readily apparent.

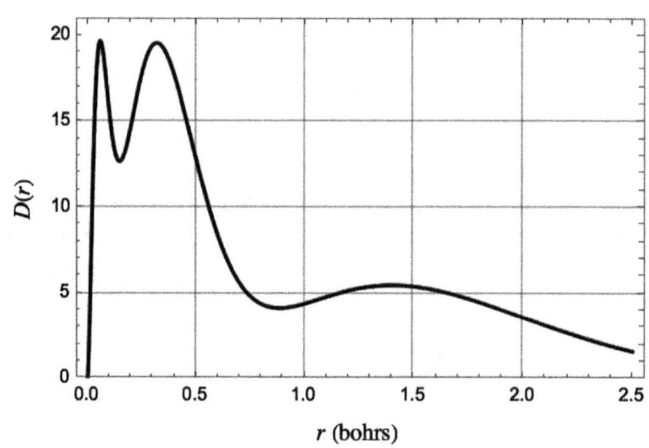

FIGURE 10.3 Radial distribution function for argon atom.

10.2 Self-consistent field theory

D. R. Hartree

V. A. Fock

Hartree in 1928 developed a method for treating many-electron atoms which has since evolved into several modern computational techniques for atomic and molecular structure. Consider first the simplest example, the two-electron helium atom. Electron #1 presumably moves in the combined field of the nucleus, taken as a point positive charge (Z=2), and electron #2, taken as a continuous negative charge distribution $\rho_b(\mathbf{r}_2) = -|\psi_b(\mathbf{r}_2)|^2$. The potential-energy for electron #1 is given by

$$V(\mathbf{r}) = V[\psi_b] = -\frac{Z}{r} + \int \frac{|\psi_b(\mathbf{r}')|^2}{|\mathbf{r} - \mathbf{r}'|} d\tau'. \tag{10.8}$$

The notation $V[\psi]$ indicates that V is a *functional* of ψ, meaning a quantity dependent on the functional form of ψ. The wavefunction for electron #1 is then determined by the one-particle Schrödinger equation

$$\left\{-\frac{1}{2}\nabla^2 + V[\psi_b]\right\}\psi_a(\mathbf{r}) = \epsilon_a \psi_a(\mathbf{r}). \tag{10.9}$$

By an analogous argument, the wavefunction for electron # 2 can be written

$$\left\{-\frac{1}{2}\nabla^2 + V[\psi_a]\right\}\psi_b(\mathbf{r}) = \epsilon_b \psi_b(\mathbf{r}). \tag{10.10}$$

These coupled integrodifferential equations can be compactly represented by

$$\begin{aligned} H_a^{\text{eff}}\psi_a(\mathbf{r}) &= \epsilon_a \psi_a(\mathbf{r}), \\ H_b^{\text{eff}}\psi_b(\mathbf{r}) &= \epsilon_b \psi_b(\mathbf{r}), \end{aligned} \tag{10.11}$$

known as the *Hartree equations*. They are coupled in the sense that the solution of the first equation enters into the second and vice versa. Practical solutions of these equations is accomplished by a successive approximation procedure. Initial "guesses" for $\psi_a(\mathbf{r})$ and $\psi_b(\mathbf{r})$ are used to calculate $V[\psi_a]$ and $V[\psi_b]$. The guesses are not that wild since we already know approximate forms such as $\psi(r) = e^{-\alpha r}$ with $\alpha = Z - 5/16$. The Hartree equations (10.11) are, at this stage, uncoupled and can be solved to obtain "first-improved" functions ψ_a and ψ_b. These, in turn, can be used to construct improved potentials $V[\psi_a]$ and $V[\psi_b]$ and the cyclic procedure is continued until input and output functions agree within a desired level of accuracy. The wavefunctions and potentials are then said to be *self-consistent*.

Extension of Hartree's method to an N-electron atom is straightforward, in principle. Each electron now moves in the field of the nucleus plus the overlapping charge clouds of the $N-1$ other electrons. N coupled Hartree equations are now to be solved:

$$\left\{-\frac{1}{2}\nabla^2 - \frac{Z}{r} + \sum_{b\neq a}\int \frac{|\psi_b(\mathbf{r}')|^2}{|\mathbf{r}-\mathbf{r}'|}\right\}\psi_a(\mathbf{r}) \equiv H_a^{\text{eff}}\psi_a(\mathbf{r}) = \epsilon_a\psi_a(\mathbf{r}), \qquad (10.12)$$

for the N orbitals $a, b \ldots n$. Again, an iterative (successive approximation) procedure can be applied until desired self-consistency is obtained. Each set of N one-electron functions—atomic orbitals—which satisfy the Hartree equations (10.12) can be identified with an electron configuration, for example $1s^2$, $1s2s$, etc. for helium atom. Each eigenvalue ϵ_a represents the energy of an electron occupying orbital a in the self-consistent field of all the other electrons. For bound electrons, the ϵ_a are negative numbers. Their magnitudes approximate the corresponding ionization energies, so $I_a \approx -\epsilon_a$. The N-electron Hartree wavefunction is a simple product of atomic orbitals

$$\Psi(\mathbf{r}_1, \mathbf{r}_2 \ldots \mathbf{r}_N) = \psi_a(\mathbf{r}_1)\psi_b(\mathbf{r}_2)\ldots\psi_n(\mathbf{r}_N) \qquad (10.13)$$

Conformity with the Pauli exclusion principle must be obtained "by hand", by allowing no more than two electrons to occupy any orbital.

As shown in the previous Section, the N-electron wavefunction is more correctly represented by a Slater determinant of spinorbitals (10.1), rather than a Hartree product of orbitals (10.13), thus accounting automatically for the exclusion principle and the indistinguishability of electrons. The *Hartree-Fock method* (1930) is a generalization of the self-consistent field based on Slater determinant wavefunctions. The *Hartree-Fock equations* (HF) for the spinorbitals ϕ_a have the form

$$\left\{-\frac{1}{2}\nabla^2 - \frac{Z}{r} + J[\phi_a, \phi_b \ldots \phi_n] - K[\phi_a, \phi_b \ldots \phi_n]\right\}\phi_a(x) = \epsilon_a\phi_a(x) \qquad (10.14)$$

where J and K are Coulomb and exchange functionals. The HF equations are quite similar to the Hartree equations (10.12), but now include exchange interactions. HF computations on atoms generally give agreement in the 1-2% range with experimental ionization and excitation energies. The residual error in HF computations is known as *electron correlation*. This originates from the approximation of instantaneous interelectronic interactions by those averaged over orbital distributions. Some electron correlation can be captured by *configuration interaction*, in which the N-electron wavefunction is extended to include a *sum* of Slater determinants. Another approach is density-functional theory, to be discussed in Chap. 14.

10.3 Aufbau principles

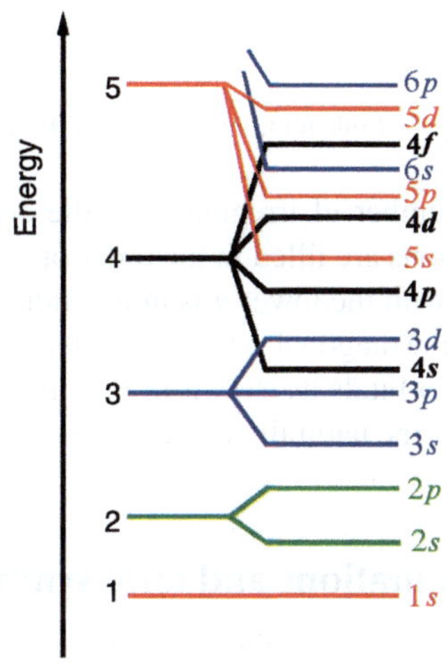

FIGURE 10.4 Approximate ordering of atomic orbital energy levels.

Aufbau means "building-up." Neils Bohr first proposed the pattern (*Aufbauprinzip*) in which atomic orbitals are filled as the atomic number is increased. For the hydrogen atom, the order of increasing orbital energy is given simply by $1s < 2s = 2p < 3s = 3p = 3d$, etc., determined by n alone. This degeneracy is removed however for orbitals in many-electron atoms. Thus $2s$ lies below $2p$, as already observed in helium. Similarly, $3s$, $3p$ and $3d$ increase energy in that order, and so on. The $4s$ is lowered sufficiently that it becomes comparable to $3d$. The general ordering of atomic orbitals is summarized by the following scheme:

$$\begin{aligned} &1s < 2s < 2p < 3s < 3p < 4s \sim 3d < 4p < 5s \sim 4d \\ &< 5p < 6s \sim 5d \sim 4f < 6p < 7s \sim 6d \sim 5f < 7p \end{aligned} \tag{10.15}$$

and illustrated in Fig. 10.4. This provides enough orbitals to fill the ground states of all the atoms in the periodic table. For orbitals designated as comparable in energy, e.g., $4s \sim 3d$, the actual order depends which other orbitals are occupied.

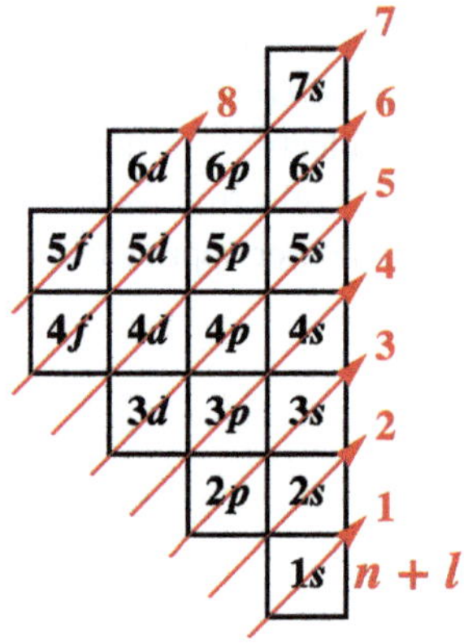

FIGURE 10.5 The Madelung $n+l$ rule for filling of atomic orbitals in atomic ground states.

With only a small number of irregularities, the orbitals in atomic ground-state electron configurations are filled in the order of increasing $n+l$. For equal $n+l$ values, the orbital with the lower n is most often filled first. This is known as the Madelung rule or the diagonal rule. The designation "diagonal rule" refers to the pattern of atomic orbitals as shown in Fig. 10.5. A sufficient number of orbitals is shown to cover the naturally occurring elements with atomic numbers Z from 1 to 92.

10.4 Atomic configurations and term symbols

Table 10.1 shows the ground-state electron configuration and term symbol for selected elements in the first part of the periodic table. From the term symbol, one can read off the total orbital angular momentum L and the total spin angular momentum S. The code for the total orbital angular momentum mirrors the one-electron notation, but using upper-case letters, as follows:

$$L = 0\ 1\ 2\ 3\ 4$$
$$\mathrm{S\ P\ D\ F\ G}$$

The vector sum of the orbital and spin angular momentum is designated

$$\mathbf{J} = \mathbf{L} + \mathbf{S}. \tag{10.16}$$

The possible values of the total angular momentum quantum number J runs in integer steps between $|L-S|$ and $L+S$. The J value is appended as a subscript on the term symbol, e.g., ${}^1\mathrm{S}_0$, ${}^2\mathrm{P}_{1/2}$, ${}^2\mathrm{P}_{3/2}$. The energy differences between J

states is a result of *spin-orbit interaction*, an interaction between the magnetic moments associated with orbital and spin angular momenta. For electron shells less than half filled, lower values of J have lower energy. In carbon, for example, $L = 1$ and $S = 1$, so that the possible values of J are 0, 1 and 2. The lowest energy is $J = 0$, thus the ground-state term is $^3P_0$. For electron shells *more* than half filled, the order of J levels is reversed. Thus oxygen, also with $L = 1$ and $S = 1$, has its lowest energy for $J = 2$. For atoms of low atomic number, the spin-orbit coupling is a relatively small correction to the energy, but it can become increasingly significant for heavier atoms. The total spin S is designated, somewhat indirectly, by the spin multiplicity $2S + 1$ written as a superscript *before* the S, P, D... symbol. For example ^{1}S (singlet S), ^{1}P (singlet P)... mean $S = 0$; ^{2}S (doublet S), ^{2}P (doublet P)... mean $S = 1/2$; ^{3}S (triplet S), ^{3}P (triplet P)... mean $S = 1$, and so on. Please do not confuse the spin quantum number S with the orbital designation S.

We will next consider in some detail the Aufbau of ground electronic states starting at the beginning of the periodic table. Hydrogen has one electron in an s-orbital so its total orbital angular momentum is also designated S. The single electron has $s = 1/2$, thus $S = 1/2$. The spin multiplicity $2S + 1$ equals 2, thus the term symbol is written ^{2}S. In helium, a second electron can occupy the $1s$ shell, provided it has the opposite spin. The total spin angular momentum is therefore zero, as is the total orbital angular momentum. The term symbol is ^{1}S, as it will be for all other atoms with complete electron shells. In determining the total spin and orbital angular moments, we need consider only electrons outside of closed shells. Therefore lithium and beryllium are a reprise of hydrogen and helium. The angular momentum of boron comes from the single $2p$ electron, with $l = 1$ and $s = 1/2$, giving a ^{2}P state.

To build the carbon atom, we add a second $2p$ electron. Since there are three degenerate $2p$ orbitals, the second electron can go into either the already-occupied $2p$ orbital or one of the unoccupied $2p$ orbitals. Clearly, two electrons in different $2p$ orbitals will have less repulsive energy than two electrons crowded into the same $2p$ orbital. In terms of the Coulomb integrals, we would expect, for example

$$J(2px, 2py) < J(2px, 2px) \tag{10.17}$$

This is fairly obvious from a simple schematic drawing showing the overlapping charge densities of two interacting p-orbitals:

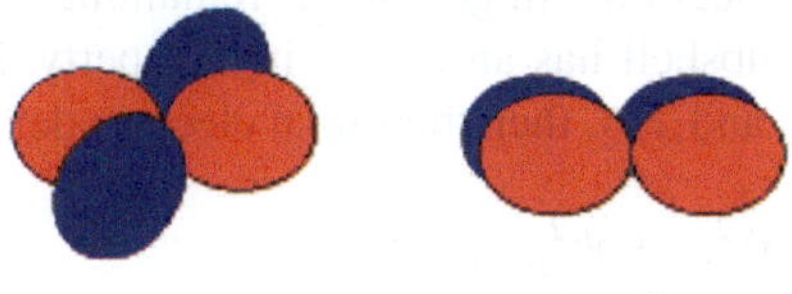

$2p_x\,2p_y$ $\qquad$ $2p_x^2$

TABLE 10.1 Atomic ground-state configurations.

Atom	Z	Electron Configuration	Term Symbol
H	1	$1s$	$^2S_{1/2}$
He	2	$1s^2$	$^1S_0$
Li	3	$[He]2s$	$^2S_{1/2}$
Be	4	$[He]2s^2$	$^1S_0$
B	5	$[He]2s^22p$	$^2P_{1/2}$
C	6	$[He]2s^22p^2$	$^3P_0$
N	7	$[He]2s^22p^3$	$^4S_{3/2}$
O	8	$[He]2s^22p^4$	$^3P_2$
F	9	$[He]2s^22p^5$	$^2P_{3/2}$
Ne	10	$[He]2s^22p^6$	$^1S_0$
Na	11	$[Ne]3s$	$^2S_{1/2}$
Cl	17	$[Ne]3s^23p^5$	$^2P_{3/2}$
Ar	18	$[Ne]3s^23p^6$	$^1S_0$
K	19	$[Ar]4s$	$^2S_{1/2}$
Ca	20	$[Ar]4s^2$	$^1S_0$
Sc	21	$[Ar]4s^23d$	$^2D_{3/2}$
Ti	22	$[Ar]4s^23d^2$	$^3F_2$
V	23	$[Ar]4s^23d^3$	$^4F_{3/2}$
Cr	24	$[Ar]4s3d^5$	$^7S_3$
Mn	25	$[Ar]4s^23d^5$	$^6S_{5/2}$
Fe	26	$[Ar]4s^23d^6$	$^5D_4$
Co	27	$[Ar]4s^23d^7$	$^4F_{9/2}$
Ni	28	$[Ar]4s^23d^8$	$^3F_4$
Cu	29	$[Ar]4s3d^{10}$	$^2S_{1/2}$
Zn	30	$[Ar]4s^23d^{10}$	$^1S_0$
Ga	31	$[Ar]4s^23d^{10}4p$	$^2P_{1/2}$
Br	35	$[Ar]4s^23d^{10}4p^5$	$^2P_{3/2}$
Kr	36	$[Ar]3d^{10}4s^24p^6$	$^1S_0$

For nitrogen atom, with three $2p$ electrons, we expect, by the same line of reasoning, that the third electron will go into the remaining unoccupied $2p$ orbital. The half-filled $2p^3$ subshell has an interesting property. If the three occupied orbitals are $2p_x, 2p_y$ and $2p_z$, then their total electron density is given by

$$\begin{aligned}\rho_{2p} &= \psi_{2p_x}^2 + \psi_{2p_y}^2 + \psi_{2p_z}^2 \\ &= (x^2 + y^2 + z^2) \times \text{function of r} = \text{function of r},\end{aligned} \tag{10.18}$$

noting that $x^2 + y^2 + z^2 = r^2$. But spherical symmetry implies zero angular momentum, like an s-orbital. In fact, any half filled subshell, such as p^3, d^5, f^7, will contribute zero angular momentum. The same is, of course true as well for *completely filled* subshells, such as p^6, d^{10}, f^{14}. These are all S terms. Another way to understand this vector cancelation of angular momentum is to consider the alternative representation of the degenerate $2p$-orbitals: $2p_{-1}$, $2p_0$ and $2p_1$. Obviously, the z-components of angular momentum now add to zero, and since only this component is observable, the total angular momentum must also be zero.

Returning to our interrupted consideration of carbon, the $2p^2$ subshell can be regarded, in concept, as a half-filled $2p^3$ subshell plus an electron "hole." The advantage of this picture is that the total orbital angular momentum can be identified as that of the hole, with magnitude $l = 1$, as shown below:

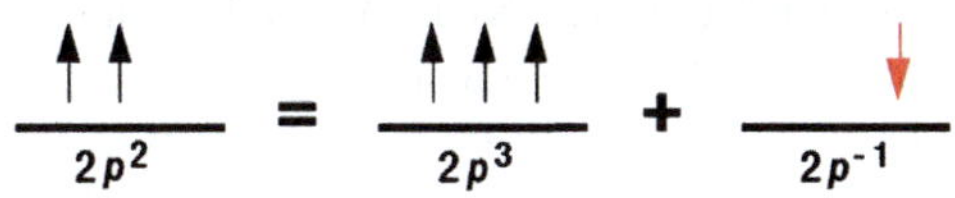

Thus the term symbol for the carbon ground state is P. It remains to determine the total spins of these subshells. Recall that exchange integrals K_{ab} are non-zero only if the orbitals a and b have the same spin. Since exchange integrals enter the energy formula (10.4) with negative signs, the more non-vanishing K integrals, the lower the energy. This is achieved by having the maximum possible number of electrons with *unpaired* spins. We conclude that $S = 1$ for carbon and $S = 3/2$ for nitrogen, so that the complete term symbols are ^{3}P and ^{4}S, respectively.

The allocation electrons among degenerate orbitals can be formalized by *Hund's rule*: For an atom in its ground state, the term with the highest multiplicity has the lowest energy.

Resuming Aufbau of the periodic table, oxygen with four $2p$ electrons must have one of the $2p$-orbitals doubly occupied. But the remaining two electrons will choose unoccupied orbitals with parallel spins. Thus oxygen has, like carbon, a ^{3}P ground state. Fluorine can be regarded as a complete shell with an electron hole, thus a ^{2}P ground state. Neon completes the $2s2p$ shells, thus term symbol ^{1}S. The chemical stability and high ionization energy of all the noble-gas atoms can be attributed to their electronic structure of complete shells. The third row of the periodic table is filled in complete analogy with the second row. The similarity of the outermost electron shells accounts for the periodicity of chemical properties. Thus, the alkali metals Na and K belong in the same family as Li, the halogens Cl and Br are chemically similar to F, and so forth.

The transition elements, atomic numbers 21 to 30, present further challenges to our understanding of electronic structure. A complicating factor is that the energies of the $4s$ and $3d$ orbitals are very close, so that interactions among occupied orbitals often determines the electronic state. Ground-state electron configurations can be deduced from spectroscopic and chemical evidence, and confirmed by accurate self-consistent field computations. The $4s$ orbital is the first to be filled in K and Ca. Then come $3d$ electrons in Sc, Ti and V. A discontinuity occurs at Cr. The ground-state configuration is found to be $4s3d^5$, instead of the extrapolated $4s^23d^4$. This can be attributed to the enhanced stability of a half-filled $3d^5$-shell. All six electrons in the valence shells have parallel spins, maximizing the number of stabilizing exchange integrals and giving the observed ^{6}S term. An analogous discontinuity occurs for copper, in which the $4s$ subshell is again raided to complete the $3d^{10}$ subshell.

The order in which orbitals are filled is not necessarily consistent with the order in which they are removed. Thus, in all the positive ions of the first transition series, the two $4s$-electrons are removed first. The inadequacy of any simple generalizations about orbital energies is demonstrated by comparing these three ground-state electron configurations: Ni $4s^23d^8$, Pd $5s^04d^{10}$ and Pt $6s5d^9$. The incomplete shells of d-orbitals in the transition elements enables them to absorb radiation in the visible region, producing many brightly-colored compounds of the transition metals. Because of their easily-removable d-electrons from their ions, most of the transition elements have variable valence and several possible oxidation states. Manganese, for example, has compounds with every oxidation state between $+1$ and $+7$.

10.5 Periodicity of atomic properties

The periodic structure of the elements is evident for many physical and chemical properties, including chemical valence, atomic radius, electronegativity, melting point, density, and hardness. Two classic prototypes for periodic behavior are the variations of the first ionization energy and the atomic radius with atomic number. These are plotted in Figs. 10.6 and 10.7.

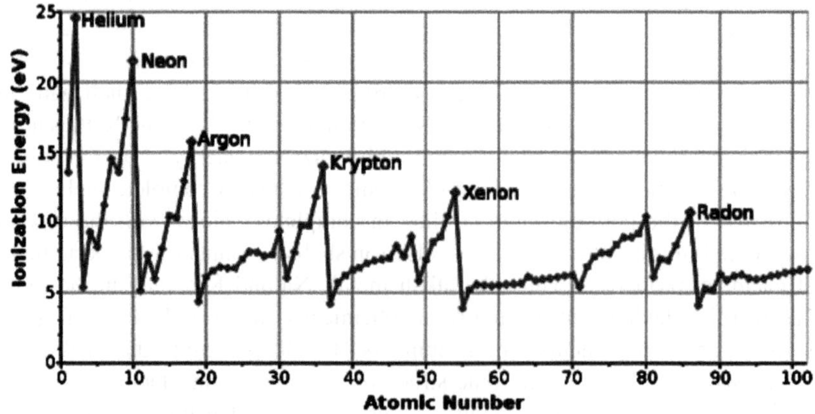

FIGURE 10.6 Periodic trends in ionization energy.

The general tendency is for ionization potential to increase and for atomic radius to decrease going across the periodic table from each alkali metal to the corresponding noble gas. Both trends can be attributed to the enhanced attraction to the increasing nuclear charge outweighing the mutual repulsions among the added electrons in the valence shell. The slopes flatten noticeably across the transition-metal blocks, Z=21–30, 39–48 and 71–80. This can be attributed to the larger and more diffuse d-orbitals, which are weakly attracted to the nucleus and also less mutually repulsive. The atomic radii in each d-block are very nearly constant. The atoms in second transition series are slightly larger than those in the first since additional shells of electrons are being occupied. Remarkably, the radii in the third transition series *do not* increase as expected. They are practically equal to the corresponding radii in the second transition series. This is a result of the *lanthanide contraction.*

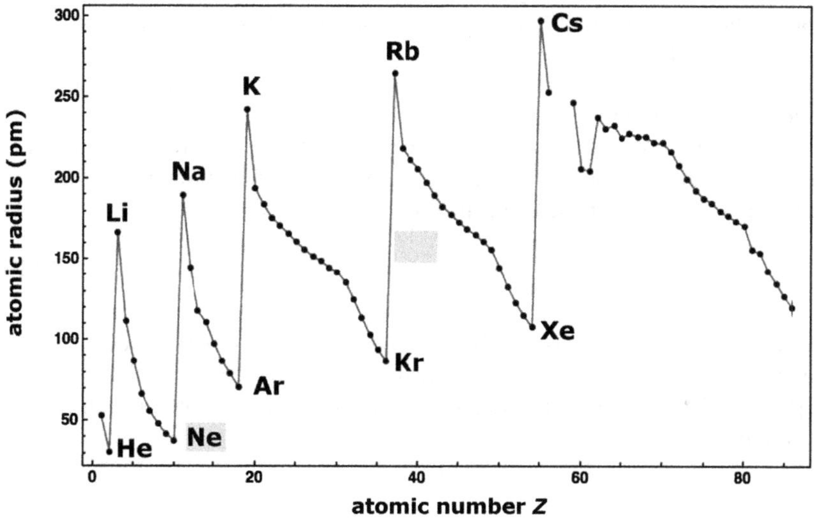

FIGURE 10.7 Periodic trends in atomic radii.

Between La ($Z = 57$) and Lu ($Z = 71$) the $4f$ orbitals are sequentially filled. The very diffuse f-orbitals are relatively ineffective at shielding nuclear charge, even more so than d-electrons. The increased effective nuclear charge in the third row transition metals causes them to be smaller than expected and virtually identical in size to those in the second row. Thus the $6s$, $6p$, and $5d$ orbitals in the third row are shrunken to approximately the same size as the $5s$, $5p$, and $4d$ in the second row. This accounts for the close chemical similarity of corresponding elements in the two groups. In fact, the chemistry of hafnium (Hf, $Z = 40$) and zirconium (Zr, $Z = 72$) is more similar than that of any other pair of elements in the periodic table.

In addition to making the third-row transition metals smaller, the lanthanide contraction also makes them less reactive because the valence electrons are relatively close to the nucleus and less susceptible to chemical reactions. This accounts for the relative inertness—or *nobility*—of these metals, particularly gold and platinum. Moreover, the third-row transition metals are the densest known elements, having the about same atomic size as the second-row transition metals but twice the atomic weight. The densest element is iridium (Ir, $Z = 77$), at 22.65 g/cm^3.

10.6 Relativistic effects

It has been realized in recent years that the lanthanide contraction is only part of the explanation for the behavior of the heavier elements. An equally-important factor is relativity. On a fundamental level, relativity actually plays an integral role in quantum theory, beginning with the symmetries between space and time, momentum and energy which suggested the form of the time-dependent Schrödinger equation (cf. Sect 2.3). Electron spin and the Pauli exclusion principle are, in fact, implications of relativistic quantum mechanics. One might validly claim therefore that the periodic structure of the elements itself is a consequence of relativity!

According to Einstein's special theory of relativity, the effective mass of an electron increases whenever its speed becomes a significant fraction of the speed of light, $c \approx 3 \times 10^8$ m/sec. Specifically,

$$m = \frac{m_0}{\sqrt{1 - v^2/c^2}}, \tag{10.19}$$

where m_0 is the *rest mass* of the electron (which we otherwise denote as m, equal to 9.1×10^{-31} kg). As an order of magnitude estimate of this effect, the speed an electron in the first Bohr orbit around a nucleus of atomic number Z is given by

$$\frac{v}{c} = Z\alpha, \tag{10.20}$$

where $\alpha = e^2/\hbar c \approx 1/137$, the fine-structure constant. For $Z = 1$ (hydrogen), $v/c \approx 1/137 = 0.0073$, so that $m = 1.00003m_0$ and relativistic corrections are negligible. But for $Z = 80$ (mercury), v/c increases to 0.58, so that $m = 1.23m_0$, which lowers energy levels quite significantly. Since electron velocities are larger closer to the nucleus (from the virial theorem $mv^2 = Ze^2/r$), this effect is most pronounced for 1s- and 2s-orbitals. Higher s-orbitals are also drawn inward to maintain their orthogonality to the inner shells. The contraction is somewhat smaller, but still significant, for p-orbitals. The most notable effect of relativity therefore is to lower the energies of s- and p-orbitals and to decrease their average radii. A secondary effect, caused by the more effective shielding of the nucleus by inner s- and p-shells, is to make outer d- and f-orbitals more diffuse and higher in energy.

Spin-orbit coupling is another effect which has its origin in relativity—as is, in fact, true for all magnetic phenomena. In the first part of the periodic table, the orbital and spin angular momenta in an atom are governed by *Russell-Saunders* or *LS*-coupling. According to this scheme, the net orbital angular momentum **L** adds to the net spin angular momentum **S** to give the total atomic angular momentum **J**. Spin-orbit coupling then causes splitting of the energy levels of a given term, for example ^{3}P, to the states $^3P_0$, $^3P_1$ and $^3P_2$, as determined by the value of J. With increasing atomic number, there is a gradual transition to what is known as jj-coupling. Here it becomes necessary to first add the orbital and spin angular momenta of each individual electron to give

$$\mathbf{j} = \mathbf{l} + \mathbf{s} \tag{10.21}$$

and then add the **j**'s to get the total atomic angular momentum **J**. Except for s-orbitals, which have zero orbital angular momentum, the spin-orbit interaction divides a shell of given l into two subshells with total angular momentum $j = l - 1/2$ and $j = l + 1/2$. Relativistic calculations yield different energies and radii for each subshell. Accordingly, atomic orbitals can now be reclassified as $s_{1/2}$, $p_{1/2}$, $p_{3/2}$, $d_{3/2}$, $d_{5/2}$, etc. For heavier atoms, the difference between the two j states of a given l can become as significant as those between different shells.

The most significant chemical effects of relativity are therefore contraction of s- and p-orbitals, expansion of d- and f-orbitals and spin orbit splitting. Heavy transition metals exhibit the strongest relativistic effects. Perhaps the most dramatic example is the comparative chemical behavior of silver and gold. The non-relativistic predictions of the energy levels of silver and gold atoms show no striking differences, and one would expect the two elements to behave very similarly chemically and spectroscopically. It is observed, however, that gold absorbs strongly in the visible region, thereby giving the metal its famous yellow lustre. Moreover, gold is much less reactive chemically—more of a noble metal. Referring to Fig. 10.8, these phenomena can be explained by the energy-level modifications ascribed to relativity. The "relativistic" results represent real life. The nonrelativistic analogs are hypothetical, obtained in principle by imagining the speed of light $c \to \infty$. Silver (Ag, $Z = 47$) has the ground state electron configuration [Kr]$4d^{10}5s$ while gold (Au, $Z = 79$) has [Xe]$4f^{14}5d^{10}6s$. The relativistic stabilization of the $6s$-orbital and the destabilization of the $5d$ narrows their energy gap sufficiently that the $6s \leftarrow 5d$ transition in the solid state is shifted into the visible part of the spectrum. By contrast, the $5s \leftarrow 4d$ transition in silver lies in the ultraviolet. Absorption of blue and violet wavelengths accounts for the color of gold. Moreover the lower energy of the $6s$ in Au (higher ionization potential), compared to the $5s$ in Ag, is related to the greater stability of elemental gold. Silver tarnishes in air, not gold. (As Chaucer even noted, "if Golde ruste, what shal Iren doo?")

The relativistic contraction of the filled $6s^2$ shell of mercury (Hg, $Z = 80$) makes it more inert, "almost a rare gas." This is also known as the *inert pair effect* because the possibility for hybridization with $6p$ is suppressed. The re-

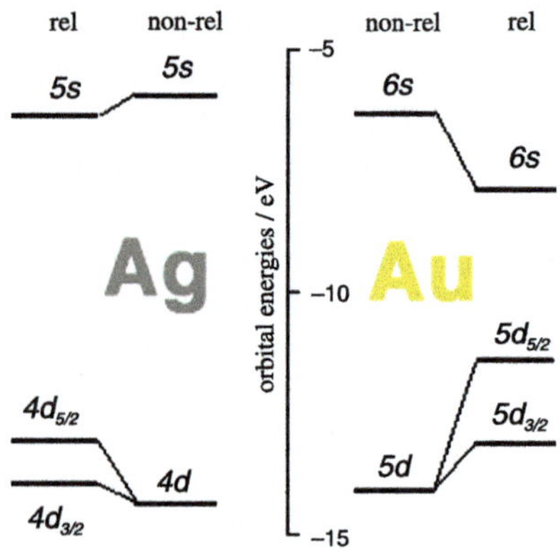

FIGURE 10.8 Valence-shell orbital energy levels for Ag and Au, showing actual (relativistic) values and hypothetical nonrelativistic counterparts.

duction in interatomic attraction is evidently sufficient to make Hg a liquid at room temperature.

Recent computations (2011) have found that about 80% of the voltage of the lead-acid battery, which starts your car, can be attributed to relativity. A cell of the lead-acid battery consists of lead and lead dioxide plates immersed in sulfuric acid, which produces a voltage difference of 2.1 volts between the two plates. The analogous battery using tin, with $Z = 50$, rather than lead with $Z = 82$, would produce a voltage only about 20% of this value. It might thereby be surmised that relativistic effects do not necessarily require near-light speeds, but can be apparent even at highway speeds.

10.7 Spiral form of the periodic table

A novel graphical representation of the sequence of atomic orbitals is due to Theodor Benfey (1960). In this spiral form of the periodic table, reproduced in Fig. 10.9, the transition elements and lanthanides appear as spokes from a spiral rather than as detached horizontal rows. This has the advantage that elements with successive atomic numbers are actually adjacent to one another.

Chapter 10. Exercises

10.1. Recall that a filled or half-filled p, d or f shell has spherical symmetry. Accordingly, go through the periodic table up to $Z = 54$ and predict which atomic ground states should have spherically-symmetrical electronic distributions (multiplet-S term symbols).

10.2. Give the electronic configurations and term symbols of the first excited electronic states of the atoms up to $Z = 10$.

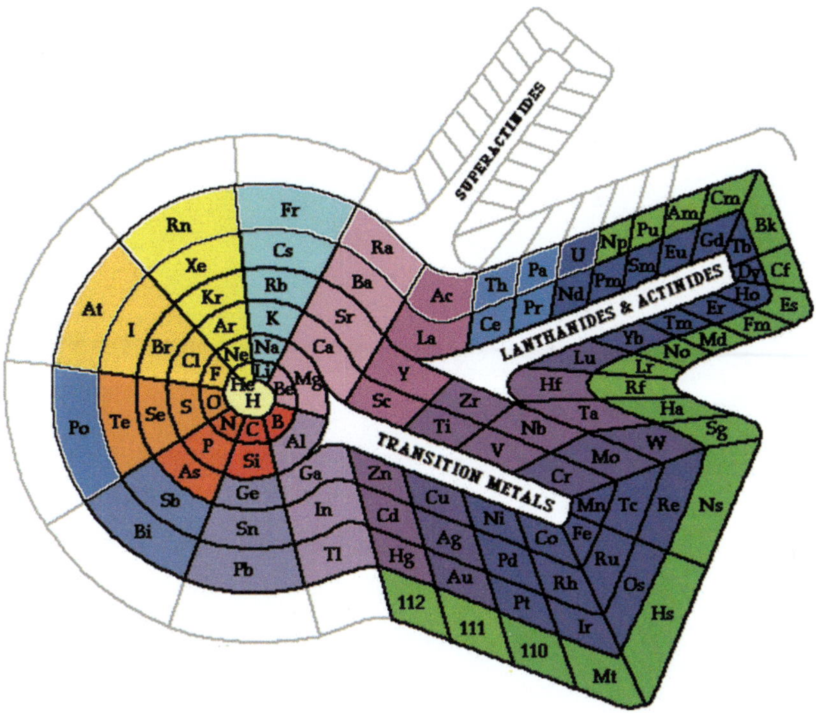

FIGURE 10.9 Benfey's spiral form of the periodic table.

10.3. Find an excited state of carbon atom which is spherically symmetrical. It will turn out that a similar state is responsible for the tetravalence of carbon.

10.4. Find at least one exception to the ordering predicted by Madelung's $n+l$ rule. Show that this can be remedied by an alternative rule based on the ordering $n+\frac{1}{2}(l+\delta_{l,3})$. Predict the ground-state electronic configuration of Rn ($Z=86$) on the basis of each rule.

10.5. Show that the speed of an electron in the lowest Bohr orbit of an atom with atomic number Z is given by

$$\frac{v}{c}=Z\alpha$$

where $\alpha=e^2/\hbar c\approx 1/137$. Find the energy of this orbit. Calculate the numerical value for Hg ($Z=80$) and compare to the nonrelativistic ($c=\infty$) result.

Chapter 11

The chemical bond

The nature of the chemical bond and the principles of molecular structure were formulated a long time ago to systematize an immense body of chemical knowledge. With the advent of quantum mechanics, it became possible to actually *derive* the concepts of chemical bonding from more fundamental laws governing matter on the atomic scale. Remarkably, many of the intuitive concepts developed by chemists have remained valid when reexpressed in terms of quantum-mechanical principles.

11.1 The hydrogen molecule

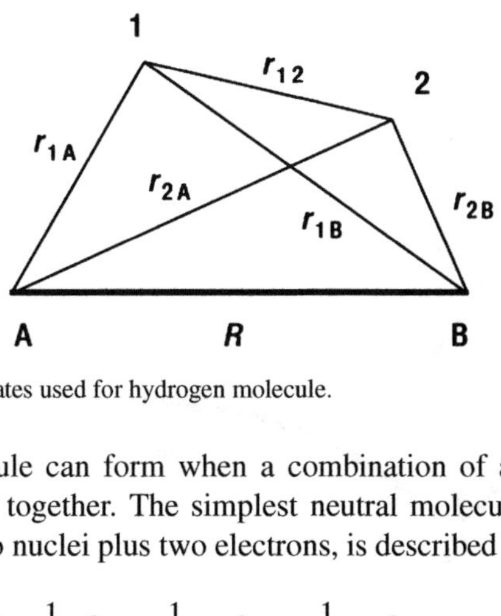

FIGURE 11.1 Coordinates used for hydrogen molecule.

A stable molecule can form when a combination of atoms can lower its energy by bonding together. The simplest neutral molecule is H_2. This four-particle system, two nuclei plus two electrons, is described by the Hamiltonian

$$H = -\frac{1}{2}\nabla_1^2 - \frac{1}{2}\nabla_2^2 - \frac{1}{2M_A}\nabla_A^2 - \frac{1}{2M_B}\nabla_B^2 - \frac{1}{r_{1A}} - \frac{1}{r_{2B}} - \frac{1}{r_{2A}} - \frac{1}{r_{1B}} + \frac{1}{r_{12}} + \frac{1}{R}, \tag{11.1}$$

using atomic units and the coordinates shown in Fig. 11.1: We note first that the masses of the nuclei are much greater than those of the electrons, $M_{\text{proton}} = 1836$ atomic units, compared to $m_{\text{electron}} = 1$ atomic unit. Therefore nuclear kinetic energies will be negligibly small compared to those of the electrons. Typically, the amplitude of nuclear vibration is of the order of 1% the spread of

Introduction to Quantum Mechanics. https://doi.org/10.1016/B978-0-12-822310-9.00019-7

an electron's probability distribution. In accordance with the *Born-Oppenheimer approximation*, we can first consider the electronic Schrödinger equation

$$H_{\text{elec}}\psi(\mathbf{r}_1, \mathbf{r}_2, R) = E_{\text{elec}}(R)\,\psi(\mathbf{r}_1, \mathbf{r}_2, R), \tag{11.2}$$

where

$$H_{\text{elec}} = -\frac{1}{2}\nabla_1^2 - \frac{1}{2}\nabla_2^2 - \frac{1}{r_{1A}} - \frac{1}{r_{2B}} - \frac{1}{r_{2A}} - \frac{1}{r_{1B}} + \frac{1}{r_{12}} + \frac{1}{R}. \tag{11.3}$$

The internuclear separation R occurs as a parameter in this equation so that the Schrödinger equation must, in concept, be solved for each value of the internuclear distance R. A typical result for the energy of a diatomic molecule as a function of R is shown in Fig. 11.2. For a bound state, the energy minimum occurs at for $R = R_e$, known as the *equilibrium internuclear distance*. The depth of the potential well at R_e is called the *binding energy* or *dissociation energy* D_e. For the H_2 molecule, $D_e = 4.746$ eV and $R_e = 1.400$ bohr $= 0.7406$ Å. Note that as $R \to 0$, $E(R) \to \infty$, since the $1/R$ nuclear repulsion will become dominant.

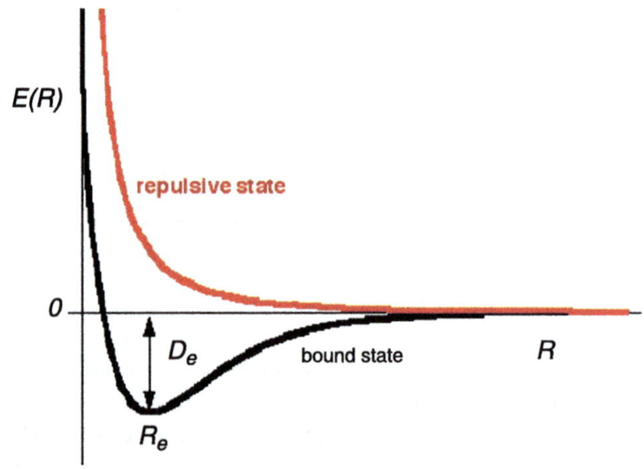

FIGURE 11.2 Energy curves for a diatomic molecule.

The more massive nuclei move much more slowly than the electrons. From the viewpoint of the nuclei, the electrons adjust almost instantaneously to any changes in the internuclear distance. The electronic energy $E_{\text{elec}}(R)$ therefore plays the role of a *potential energy* in the Schrödinger equation for nuclear motion

$$\left\{-\frac{1}{2M_A}\nabla_A^2 - \frac{1}{2M_B}\nabla_B^2 + V(R)\right\}\chi(\mathbf{r}_A, \mathbf{r}_B) = E\,\chi(\mathbf{r}_A, \mathbf{r}_B), \tag{11.4}$$

where

$$V(R) = E_{\text{elec}}(R), \tag{11.5}$$

from solution of Eq. (11.2). Solutions of Eq. (11.4) determine the vibrational and rotational energies of the molecule. These will be considered further in Chap. 16. For the present, we are interested in the obtaining electronic energy from Eqs. (11.2) and (11.3). We will thus drop the subscript "elec" on H and $E(R)$ for the remainder this Chapter.

The first quantum-mechanical account of chemical bonding is due to Heitler and London in 1927, only one year after the Schrödinger equation was proposed. They reasoned that, since the hydrogen molecule H_2 was formed from a combination of hydrogen atoms A and B, a first approximation to its electronic wavefunction might be

$$\psi(\mathbf{r}_1, \mathbf{r}_2) = \psi_{1s}(r_{1A})\psi_{1s}(r_{2B}). \tag{11.6}$$

Using this function into the variational integral

$$\tilde{E}(R) = \frac{\int \psi\, H\, \psi\, d\tau}{\int \psi^2\, d\tau}, \tag{11.7}$$

the value $R_e \approx 1.7$ bohr was obtained, indicating that the hydrogen atoms can indeed form a molecule. However, the calculated binding energy $D_e \approx 0.25$ eV, is much too small to account for the strongly-bound H_2 molecule. Heitler and London proposed that it was necessary to take into account the *exchange* of electrons, in which the electron labels in (11.6) are reversed. The properly symmetrized function

$$\psi(\mathbf{r}_1, \mathbf{r}_2) = \psi_{1s}(r_{1A})\psi_{1s}(r_{2B}) + \psi_{1s}(r_{1B})\psi_{1s}(r_{2A}) \tag{11.8}$$

gave a much more realistic binding energy value of 3.20 eV, with $R_e = 1.51$ bohr. We have already used exchange symmetry (and antisymmetry) in our treatment of the excited states of helium in Chap. 9. The variational function (11.8) was improved (Wang, 1928) by replacing the hydrogen $1s$ functions e^{-r} by $e^{-\zeta r}$. The optimized value $\zeta = 1.166$ gave a binding energy of 3.782 eV. The quantitative breakthrough was the computation of James and Coolidge (1933). Using a 13-parameter function of the form

$$\begin{gathered} \psi(\mathbf{r}_1, \mathbf{r}_2) = e^{-\alpha(\xi_1+\xi_2)} \times \text{polynomial in } \{\xi_1, \xi_2, \eta_1, \eta_2, \rho\}, \\ \xi_i \equiv \frac{r_{iA} + r_{iB}}{R}, \qquad \eta_i \equiv \frac{r_{iA} - r_{iB}}{R}, \qquad \rho \equiv \frac{r_{12}}{R}, \end{gathered} \tag{11.9}$$

they obtained $R_e = 1.40$ bohr, $D_e = 4.720$ eV. This result "proved" the validity of quantum mechanics for molecules, in the same sense that Hylleraas' computation on helium was a proof for the multi-electron atom.

11.2 Valence bond theory

G. N. Lewis Linus Pauling

The basic idea of the Heitler-London model for the hydrogen molecule can be extended to chemical bonds between any two atoms. The orbital function (10.8) must be associated with the singlet spin function $\sigma_{0,0}(1, 2)$ in order that the overall wavefunction be antisymmetric. This is a quantum-mechanical realization of the concept of an electron-pair bond, first proposed by G. N. Lewis in 1916. It is also now explained why the electron spins must be paired, i.e., antiparallel. It is also permissible to combine an antisymmetric orbital function with a triplet spin function but this will, in most cases, give a repulsive state, such as the one shown in red in Fig. 11.2.

According to valence-bond theory, unpaired orbitals in the valence shells of two adjoining atoms can combine to form a chemical bond if they overlap significantly and are symmetry compatible. As emphasized by Linus Pauling, a measure of the bonding potential of two orbitals is the *overlap integral*

$$S_{ab} = \int \psi_a(\mathbf{r})\psi_b(\mathbf{r})\, d\tau \tag{11.10}$$

For an effective σ-bond, S is typically in the range of 0.2 or 0.3. A σ-bond is cylindrically symmetrical about the axis joining the atoms. Two s AOs, two p_z AOs or an s and a p_z can contribute to a σ-bond, as shown in Fig. 10.3. The z-axis is chosen along the internuclear axis. Two p_x or two p_y AO's can form a π-bond, which has a nodal plane containing the internuclear axis. Examples of symmetry-incompatible AO's would be an s with a p_x or a p_x with a p_y. In such cases the overlap integral would vanish because of cancelation of positive and negative contributions. Some possible combinations of AO's forming σ and π bonds are shown in Fig. 11.3.

Bonding in the HCl molecule can be attributed to a combination of a hydrogen $1s$ with an unpaired $3p_z$ on chlorine. In Cl_2, a σ-bond is formed between the

$3p_z$ AO's on each chlorine. As a first approximation, the other doubly-occupied AO's on chlorine—the inner shells and the valence-shell lone pairs—are left undisturbed.

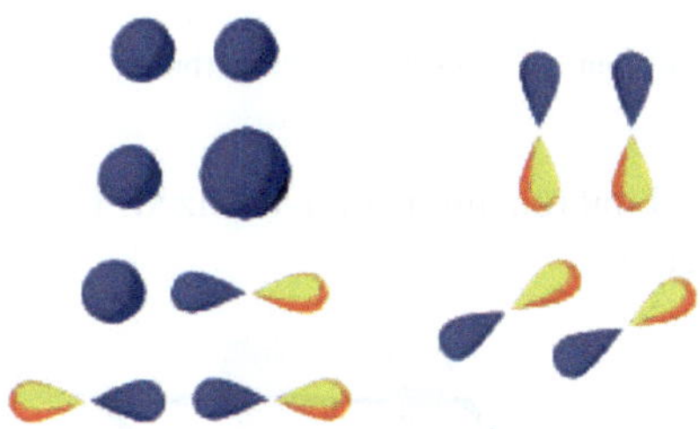

FIGURE 11.3 Combinations of atomic orbitals to form σ bonds (left) and π bonds (right).

The oxygen atom has two unpaired $2p$-electrons, say $2p_x$ and $2p_y$. Each of these can form a σ-bond with a hydrogen $1s$ to make a water molecule. It would appear from the geometry of the p-orbitals that the HOH bond angle would be 90°. It is actually around 104.5°. We will resolve this discrepancy shortly. The nitrogen atom, with three unpaired $2p$ electrons can form three bonds. In NH_3, each $2p$-orbital forms a σ-bond with a hydrogen $1s$. Again 90° HNH bond angles are predicted, compared with the experimental 107°. The diatomic nitrogen molecule has a triple bond between the two atoms, one σ bond from combining $2p_z$ AO's and two π bonds from the combinations of $2p_x$'s and $2p_y$'s, respectively.

11.3 Hybrid orbitals and molecular geometry

To understand the bonding of carbon atoms, we must introduce additional elaborations of valence-bond theory. We can write the valence shell configuration of carbon atom as $2s^2 2p_x 2p_y$, signifying that two of the $2p$ orbitals are unpaired. It might appear that carbon would be divalent, and indeed the species CH_2 (carbene or methylene radical) does have a transient existence. But the chemistry of carbon is dominated by tetravalence. Evidently it is a good investment for the atom to promote one of the $2s$ electrons to the unoccupied $2p_z$ orbital to give the hypothetical *valence state* configuration $2s2p_x 2p_y 2p_z$. The gain in stability attained by formation of four bonds more than compensates for the small excitation energy. It can thus be understood why the methane molecule CH_4 exists. The molecule has the shape of a regular tetrahedron, which is the result of *hybridization*, mixing of the s and three p orbitals to form four sp^3 hybrid atomic orbitals. Hybrid orbitals can overlap more strongly with neighboring atoms, thus producing stronger bonds. The result is four C–H σ-bonds, identical except for orientation in space, with 109.5° H-C-H bond angles, as shown in Fig. 11.4.

Other carbon compounds make use of two alternative hybridization schemes. The s AO can form hybrids with *two* of the p AO's to give three sp^2 hybrid

FIGURE 11.4 Formation of four sp^3 tetrahedral hybrid orbitals.

orbitals, with one p-orbital remaining unhybridized. This accounts for the structure of ethylene (ethene):

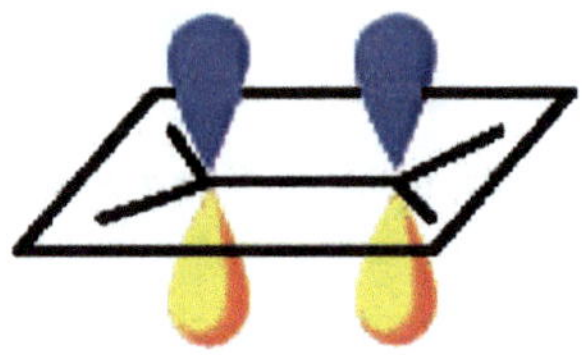

The C–H and C–C σ-bonds are all trigonal sp^2 hybrids, with 120° bond angles. The two unhybridized p-orbitals form a π-bond, which gives the molecule its rigid planar structure. The two carbon atoms are connected by a double bond, consisting of one σ and one π. The third canonical form of sp-hybridization occurs in C–C triple bonds, for example, acetylene (ethyne). Here, two of the p AO's on each carbon remain unhybridized and can form two π-bonds, in addition to two σ-bonds. Acetylene H–C≡C–H is a linear molecule, as shown below, since the sp-hybrids are oriented 180° apart.

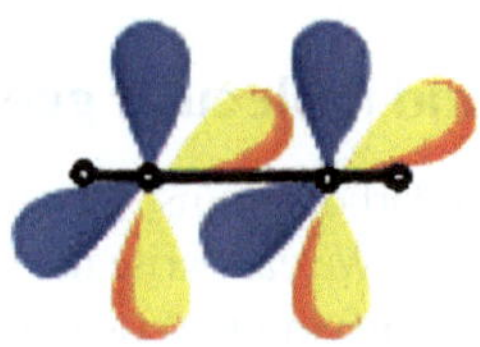

The canonical sp^n hybrid orbitals can be expressed in terms of their component atomic orbitals as follows:

Linear sp hybrids:

$$\psi_1^{\text{lin}} = \frac{1}{\sqrt{2}}\left(\psi_s + \psi_{p_z}\right), \qquad \psi_2^{\text{lin}} = \frac{1}{\sqrt{2}}\left(\psi_s - \psi_{p_z}\right). \tag{11.11}$$

Trigonal sp^2 hybrids:

$$\psi_1^{\text{trig}} = \frac{1}{\sqrt{3}}\psi_s + \sqrt{\frac{2}{3}}\psi_{p_x},$$
$$\psi_2^{\text{trig}} = \frac{1}{\sqrt{3}}\psi_s - \frac{1}{\sqrt{6}}\psi_{p_x} + \frac{1}{\sqrt{2}}\psi_{p_y},$$

$$\psi_3^{\text{trig}} = \frac{1}{\sqrt{3}}\psi_s - \frac{1}{\sqrt{6}}\psi_{p_x} - \frac{1}{\sqrt{2}}\psi_{p_y}. \tag{11.12}$$

Tetrahedral sp^3 hybrids:

$$\begin{aligned} \psi_1^{\text{tet}} &= \frac{1}{2}\left(\psi_s + \psi_{p_x} + \psi_{p_y} + \psi_{p_z}\right), \quad \psi_2^{\text{tet}} = \frac{1}{2}\left(\psi_s + \psi_{p_x} - \psi_{p_y} - \psi_{p_z}\right), \\ \psi_3^{\text{tet}} &= \frac{1}{2}\left(\psi_s - \psi_{p_x} + \psi_{p_y} - \psi_{p_z}\right), \quad \psi_4^{\text{tet}} = \frac{1}{2}\left(\psi_s - \psi_{p_x} - \psi_{p_y} + \psi_{p_z}\right). \end{aligned} \tag{11.13}$$

Each set can be shown to be orthonormalized, provided that the s- and p-functions are normalized and orthogonal. Fig. 11.5 shows a contour plot of a tetrahedral sp^3 hybrid. The strong directional character of this and other hybrid orbitals enhances the overlap with neighboring orbitals, thus contributing to stronger bonds.

FIGURE 11.5 Contour plot of tetrahedral sp^3 hybrid orbital.

The deviations of the bond angles in H_2O and NH_3 from 90° can be attributed to fractional hybridization. The angle H-O-H in water is 104.5° while H-N-H in ammonia is 106.6°. It is rationalized that the p-orbitals of the central atom acquire some s-character and increase their angles towards the tetrahedral value of 109.5°. Correspondingly, the lone pair orbitals must also become hybrids. Apparently, for both water and ammonia, a model based on tetrahedral orbitals on the central atoms would be closer to the actual behavior than the original selection of s- and p-orbitals. The hybridization is evidently driven by repulsions between the electron densities of neighboring bonds.

11.4 Hypervalent compounds

Hypervalent molecules have been defined as those formed by elements in groups V to VIII of the periodic table which exhibit valences greater than those which conform to the octet rule of conventional Lewis diagrams. Two representative examples are the molecules PF_5 and SF_6. A once-popular explanation (which we cited in the First Edition of this book) was that the additional bonds could be attributed to hybrids involving the unoccupied $3d$-orbitals of phosphorus and sulfur. For example, $3s\,3p^3\,3d$ hybrids might account for the trigonal bipyramidal configuration of PF_5, while $3s\,3p^3\,3d^2$ hybrids might explain octahedral SF_6. Accurate electronic-structure computations have since indicated, however, that $3d$ orbitals make only negligible contribution to the occupied molecular orbitals for second-row atoms, so that the d-orbital model is not a good explanation.

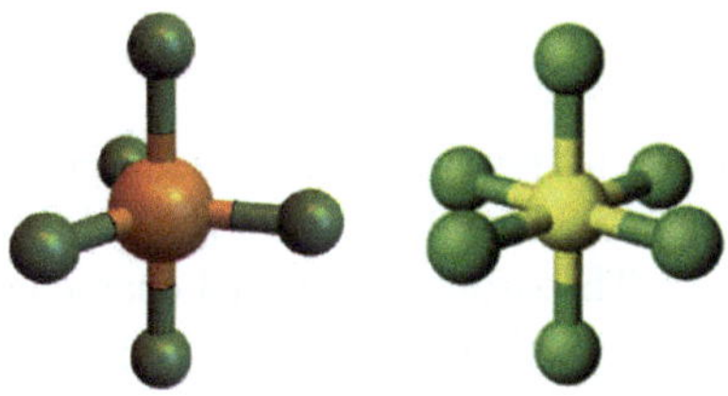

PF_5 SF_6

Rundle and Pimentel (ca 1951) advanced the idea of a three-center four-electron bond, which can explain the bonding in some hypervalent compounds. As shown in Fig. 11.6, three atomic orbitals: a p AO on the central atom plus two AOs from atoms on the opposite sides of the central atom combine to form three molecular orbitals. The highest energy combination is antibonding (repulsive), while the central one is nonbonding. Only the lowest-energy combination promotes bonding to the central atom. Two of the four electrons in the 3c-4e bond occupy this bonding orbital, while the remaining two electrons are in the nonbonding orbital, and remain associated with the outer atoms.

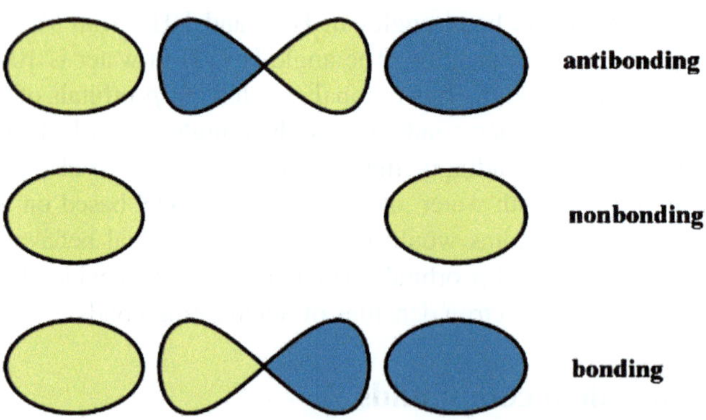

FIGURE 11.6 Model for the three-center four-electron bond.

For the case of hexavalent sulfur, one might picture the $3s^2 3p^4$ ground-state electron configuration transformed into a *valence state* $3s^0 3p^6$, in which the

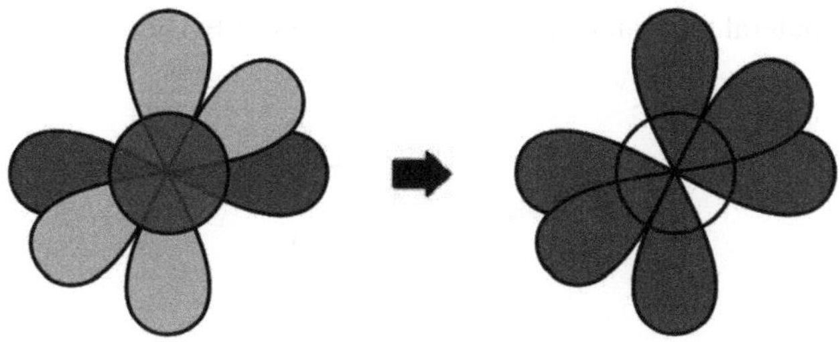

FIGURE 11.7 Schematic representation of the formation of the hexavalent valence state of sulfur in SF_6. Unoccupied orbitals are shown as transparent, singly-occupied orbitals as light gray and doubly occupied orbitals as darker gray.

$3p_x$, $3p_y$ and $3p_z$ become doubly occupied. This is shown schematically in Fig. 11.7. Each of these p-orbitals would then form a 4-electron bond with unoccupied AOs of two fluorine atoms, resulting in three perpendicular 3c-4e bonds in a octahedral SF_6 molecule.

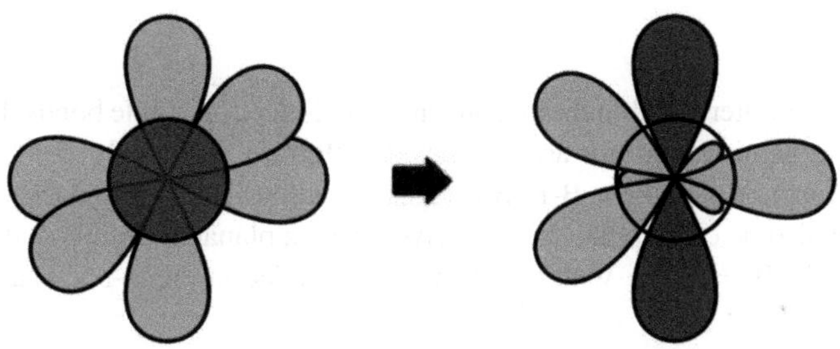

FIGURE 11.8 Schematic representation of the formation of the pentavalent valence state of phosphorus in PF_5. Unoccupied orbitals are shown as transparent, singly-occupied orbitals as light gray and doubly occupied orbitals as darker gray.

Analogously, as shown in Fig. 11.8, for pentavalent phosphorus, the $3s^2 3p^3$ ground-state configuration can be pictured as transforming into three singly-occupied sp^2 hybrids, directed to the vertices of an equilateral triangle, plus one doubly-occupied p-orbital, which takes part in 3c-4e bonding with two fluorine atoms to form the two axial bonds of the trigonal bipyramidal PF_5 molecule.

11.5 Boron hydrides

Boron hydrides or polyboranes are a group of *electron-deficient* molecules which exhibit some unusual non-classical bonding behavior. The stereochemistry of these compounds was first revealed in the 1950s by X-ray diffraction techniques. The simplest of these compounds is borane BH_3. This has the expected equilateral triangular structure, with three sp^2 B-H bonds

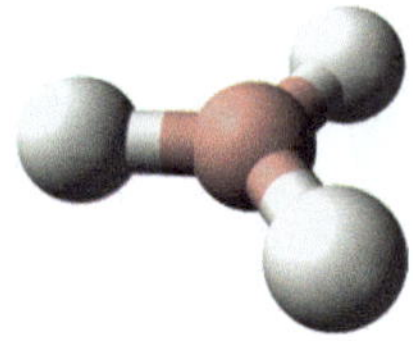

Because boron is electron deficient, there are only six electrons associated with the central atom, a technical violation of the octet rule. Next, we consider diborane B_2H_6. Although it has a formula analogous to ethane, is found to have the following structure:

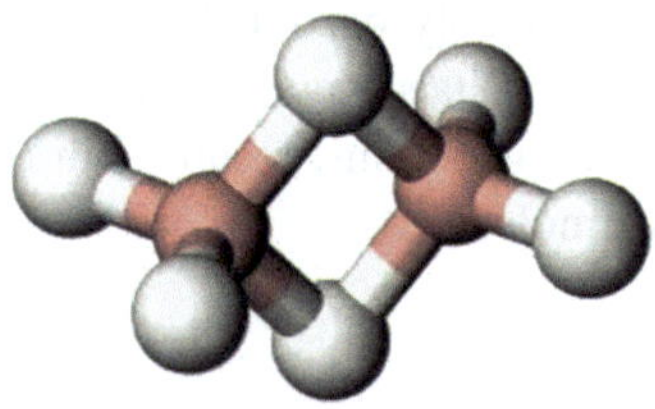

The four outer B-H bonds are conventional 2-electron single bonds. However the two remaining H atoms are connected to the two B atoms by *bridge bonds*. It is conjectured that these B-H-B bridges constitute 3-center 2-electron bonds. This can also be conceptualized as derived from a planar structure analogous to ethylene: $H_2B^-{=}B^-H_2$ with two protons embedded in the π-bonds above and below the plane:

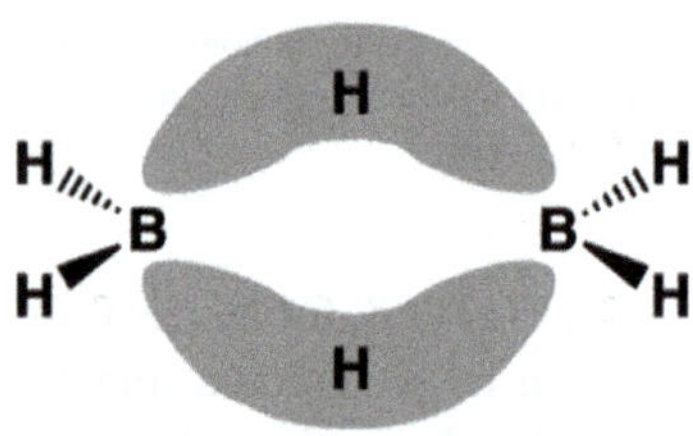

The protonated π-bonds are sometimes referred to as "banana bonds."

Higher boranes, we show B_4H_{10}, B_5H_9 and $B_{10}H_{14}$:

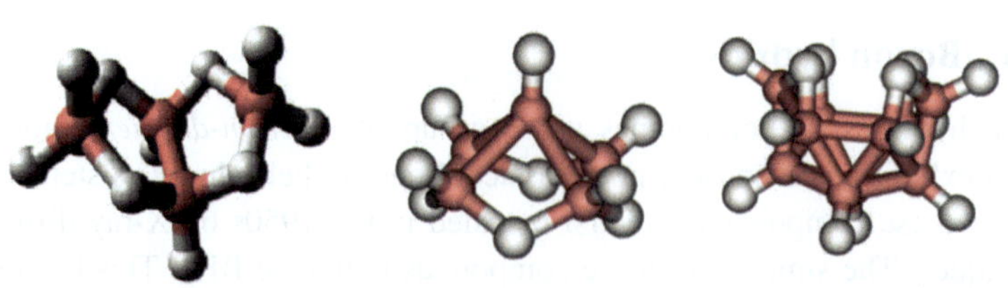

have complex cagelike structures containing multiple 3-center bonds, including those with B in the center.

11.6 Valence-shell model

An elementary, but quite successful, model for determining the shapes of molecules is the *valence-shell electron-pair repulsion* (VSEPR) theory, first proposed by Sidgewick and Powell and popularized by Gillespie. The local arrangement of atoms around each multivalent center in the molecule can be represented by $AX_{n-k}E_k$, where X is ligand atom and E is a nonbonding or lone pair of electrons. The geometry around the central atom is then determined by the arrangement of the n electron pairs (bonding plus nonbonding) which minimizes their mutual repulsion. The following geometric configurations are the most favorable for the given numbers of ligands:

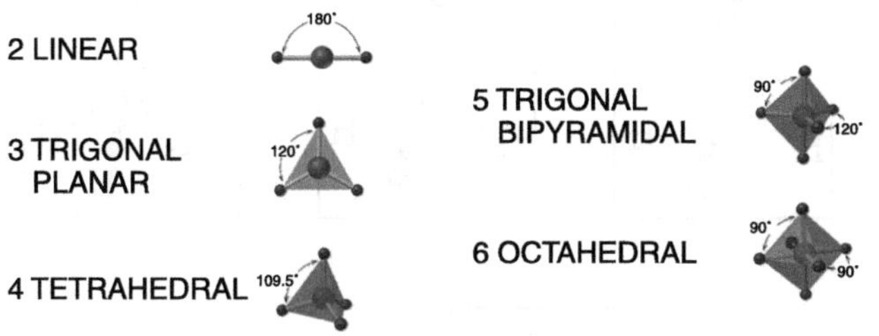

The basic geometry will be distorted if the n surrounding pairs are not identical. Also, since a lone pair is not attracted to a ligand, its distribution will tend to be more spread out and it will repel other electron pairs more strongly. The relative strength of repulsion between pairs follows the order E–E > E–X > X–X. The 4-coordinate molecule CH_4 will be a perfect tetrahedron. Ammonia, which is NH_3E, will be tetrahedral to a first approximation. But the lone pair E will repel the N–H bonds more than they repel one another. Thus the E–N–H angle will increase from the tetrahedral value of 109.5°, causing the H–N–H angles to decrease slightly. The observed value of 106.6° is quite well accounted for. For water, OH_2E_2, the opening of the E–O–E angle will cause an additional closing of H–O–H, to 104.5°.

The electron-deficient molecule BF_3 has just 3 electron pairs in the valence shell. It will therefore have a trigonal planar configuration. Likewise, the carbene radical CH_2, a rare case of divalent carbon, will be a bent molecule since the lone pair occupies one of the vertices.

The 6-coordinated compound SF_6 has a regular octahedral configuration. A lone pair, as in BrF_5, occupies one octahedral vertex, leaving a square pyramidal molecular framework. In xenon tetrafluoride, the 8 valence electrons from Xe form four Xe–F bonds, leaving two lone pairs. Since the lone pairs are maximilly repulsive, they will be as far apart as possible—on opposite sides of the octahedron. This leaves the XeF_4 molecule in a square planar configuration.

The 5-coordination compounds show some more exotic possibilities. PF_5 has a trigonal bipyramidal shape with inequivalent axial and equatorial positions. The lone pair in SF_4 chooses an equatorial position since it can do less "damage" there—it make a 90° angle with two F atoms, whereas an axial position would make 90° angles with three F atoms. This leaves SF_4 with a shape resembling a distorted see-saw. The two lone pairs in ClF_3 are most stable when located in two equatorial positions, separated by 120°. This leaves ClF_3 in a distorted tee shape. The complex that forms between I^- and I_2 in aqueous solution is a linear ion. Consider I^- as the central atom, with 8 valence electrons. Two electrons form bonds to other I atoms, leaving three lone pairs, which prefer to occupy all three equatorial positions, leaving the bonding pairs in the axial positions.

Following are illustrations of the molecules we have described:

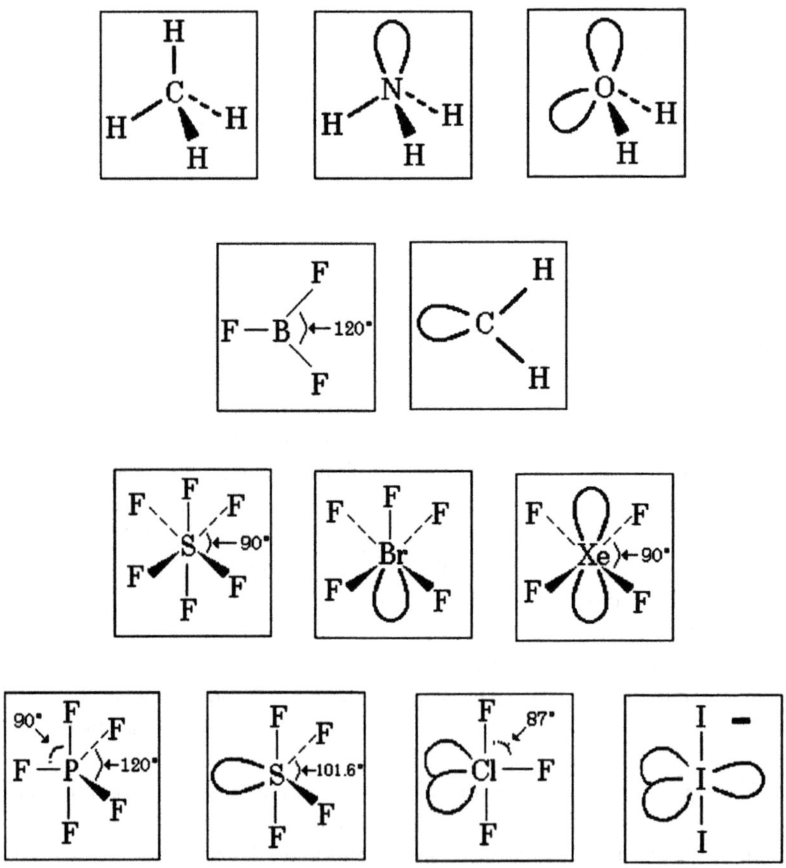

Multiple bonds behave very similarly to single-bond pairs, apart from being slightly more repulsive. CO_2 is a linear molecule with the two double bonds accounting for all the valence electrons. Double-bonded carbon $\left(=C\langle\right)$ forms a trigonal structure, as expected for sp^2 hybrids. Triple-bonded carbon ($\equiv$C–) forms a linear structure, as expected for sp hybrids. Sulfate anion SO_4^{2-} has two single bonds and two double bonds (individual bonds resonating between these). Consistent with the valence-shell model, the four bonds are in a tetrahedral configuration.

Unpaired electrons, as in free radicals, behave much like lone pairs, but are slightly *less* repulsive. The bonding in NO_2 can be described by contributing resonance structures including $O{=}\overset{\circ}{N}{=}O$ and $O{=}\overset{\circ}{N}{}^{+}{-}O^{-}$. The valence shell model predicts a bent molecule with the structure:

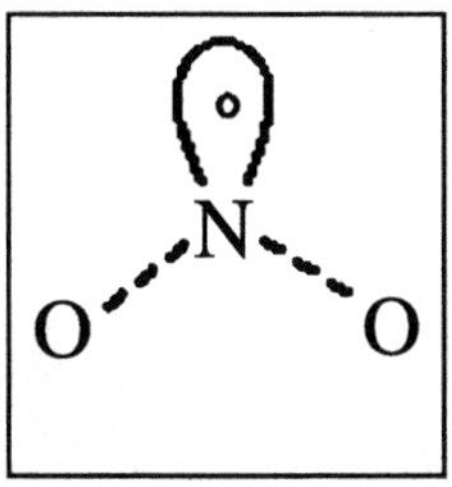

For more than six electron pairs, the geometries of valence-shell theory are no longer as simple to determine. For $n = 7$, two reasonable alternatives are a pentagonal bipyramid and a capped or distorted octahedron. Iodine heptafluoride, IF_7, has a pentagonal bipyramidal geometry. Xenon hexafluoride, XeF_6, also has seven electron pairs—six bonding pairs plus one lone pair. The structure of is believed to be a distorted octahedron, complicated by time-dependence of the fluorine positions—what is known as a *fluxional* molecule. For $n = 8$, a possible geometry is a square antiprism, obtained by twisting one face of a cube by 45° with respect to the opposite face. This is most likely the shape of the ion XeF_8^{-2}.

11.7 Transition metal complexes

Transition metal ions, with their unoccupied d-orbitals, are excellent electron acceptors or *Lewis acids*. Molecules or ions known as *ligands* can donate electron pairs to the metal ion, thus acting as *Lewis bases*. These are said to *coordinate* to the Lewis acid via *coordinate covalent bond* to form a *coordination compound*. Most transition metal ions have valence-shell d-electrons in excess of the number needed for bonding to the ligands. The geometry of the complex ion is determined by the most stable arrangement of its ligands—an octahedron for 6 ligands, a trigonal bipyramid for 5 and a tetrahedron for 4 (or, in certain cases, a square planar configuration). *Crystal field theory* was proposed by Hans Bethe in 1929 (with later elaborations by J. H. Van Vleck) to describe the nonbonding d-electrons on the central atom. This is a simple electrostatic model in which the ligands are idealized as negative point charges. In the absence of the crystal field, the five d-orbitals are degenerate. When ligands coordinate to the metal, however, the individual d-orbitals are shifted in energy. In the case of an octahedral crystal field, the metal atom is surrounded by six identical ligands in a regular octahedron, oriented as shown in Fig. 11.9. The d_{z^2} and the $d_{x^2-y^2}$ orbitals point directly at ligands along cartesian axes, and are thus *raised* in energy by the repulsive interactions. By contrast the d_{xy}, $d_{y^2-z^2}$ and $d_{z^2-x^2}$ orbitals are more successful at avoiding the ligands by pointing at 45° angles to the cartesian axes. As a result, the five-fold degeneracy of the d-orbitals is partially resolved by the octahedral field into two levels, as shown in Fig. 11.10. The doubly-degenerate upper level and the triply-degenerate lower level are designated e_g and t_{2g}, respectively, in group-theoretical notation (which will be explained in Chap. 15). The *crystal field splitting* is denoted by Δ (or, in older notation $10Dq$).

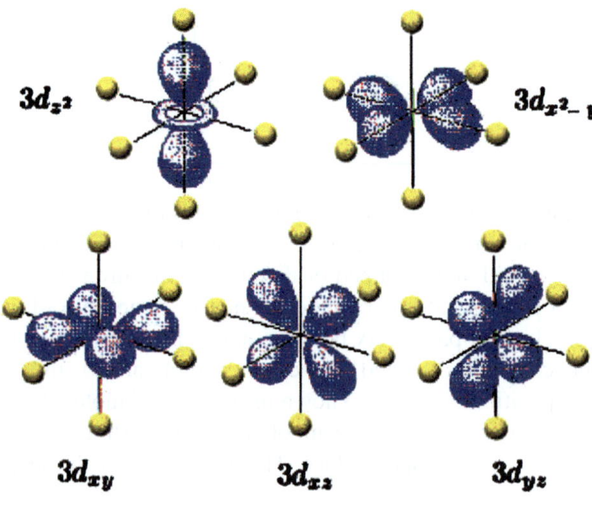

FIGURE 11.9 The five $3d$ orbitals in an octahedral crystal field.

There can be anywhere from 12 to 22 valence electrons in an octahedral complex. The first twelve of these are used for bonding to the six ligands. The remaining ones occupy atomic d-orbitals which interact with the crystal field. A simple example is the hexaquatitanium(III) octahedral complex $[Ti(H_2O)_6]^{3+}$, which has a single $3d$ electron after bonding to the ligands has been accounted for. In the ground state, this electron occupies one of the degenerate t_{2g} orbitals. The violet color of this complex is the result of absorption of light in excitation to the e_g level. The transition $e_g \leftarrow t_{2g}$ occurs at a maximum at 20,300 cm^{-1}, which is evidently the approximate value of Δ. Crystal-field splittings are typically in the range of 7000–30,000 cm^{-1}. (The visible region is approximately 14,000–24,000 cm^{-1}.) Extensive chemical experience has led to a *spectrochemical series*, an ordering of ligands by the strength of their crystal-field splittings. Following is a partial listing of some common ligands:

$$CO > CN^- > NO_2^- > en > NH_3 > H_20 > OH^- > F^- > Cl^-$$

Here "en" stands for ethylenediamine $NH_2CH_2CH_2NH_2$, a bidentate ligand. The dividing line between strong- and weak-field ligands is often considered to be between NH_3 and H_20. The variability of color depending on the crystal-field environment for some octahedral cobalt (III) complexes is illustrated by the following list:

Complex Ion	λ_{max} (nm)	Color
$[Co(CN)_6]^{3-}$	310	pale yellow
$[Co(en)_3]^{3+}$	340, 470	yellow-orange
$[Co(NH_3)_6]^{3+}$	437	yellow-orange
$[Co(H_2O)_6]^{3+}$	400, 600	deep blue
$[CoF_6]^{3-}$	700	green

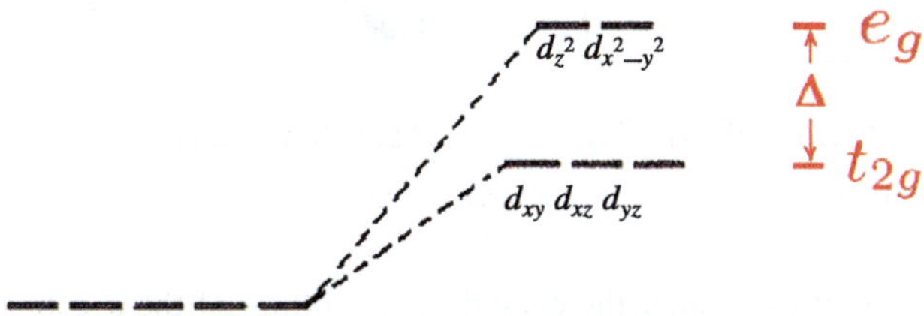

FIGURE 11.10 Splitting of d-orbitals in octahedral crystal field.

For d-shells containing between 4 and 7 electrons, the magnitude of Δ influences the magnetic properties of the complex. Consider for example two different octahedral complexes of ferric ion Fe(III) with five d-electrons. For $[Fe(CN)_6]^{3-}$, $\Delta \approx 30,000\ \text{cm}^{-1}$ while for $[FeF_6]^{3-}$, $\Delta \approx 10,000\ \text{cm}^{-1}$. The first of these is found to be a *high-spin complex* with spin $S = 5/2$, while the second is a *low spin complex* with $S = 1/2$. This behavior can be rationalized on the basis of Aufbau principles and Hund's rules, as shown in the schematic energy-level diagrams below. For $[FeF_6]^{3-}$, the e_g and t_{2g} energies are close enough together that the increased stability due to exchange integrals among 5 parallel electron spins more than compensates for the small excitation energy. This is analogous to what was encountered for atomic chromium (see Sect. 10.4), where the $4s3d^5$ ground-state electron configuration was preferred over the $4s^23d^4$.

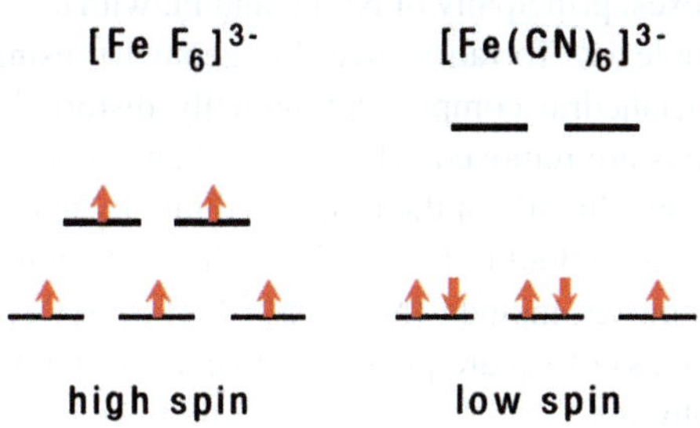

High-spin and low-spin states are readily distinguished by the magnetic properties of a complex. Since the environment of the central atom is *not* spherically symmetrical, the orbital angular momentum is not conserved and makes no contribution to magnetic behavior. The orbital angular momentum is said to be *quenched* by the crystal field.

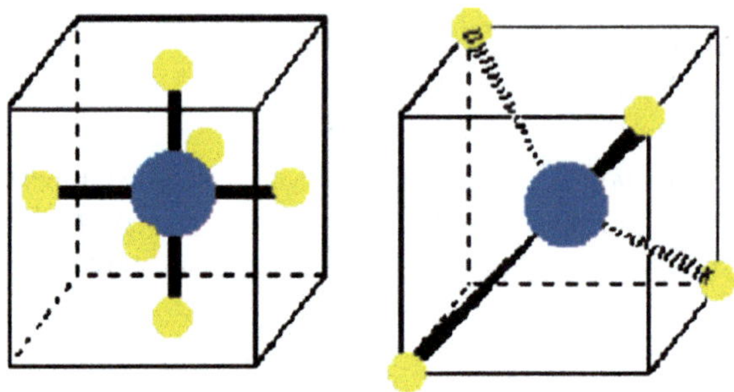

FIGURE 11.11 Octahedral and tetrahedral arrangements of ligands inscribed in a cube.

We will briefly consider the crystal field splitting of the d-orbitals in four-coordinate, tetrahedral complexes. The cube, octahedron and tetrahedron are related geometrically. Octahedral coordination results when ligands are placed in the centers of cube faces while tetrahedral coordination results when ligands are placed on alternate corners of a cube, as shown in Fig. 11.11. Four-coordinate complexes are most likely to occur when the total number of valence electrons—from metal plus ligands—is less than 18. In a tetrahedral complex, none of the five d-orbitals point directly at ligands, but the d_{xy}, d_{xz} and d_{yz} orbitals do come close. Thus these three orbitals will be raised in energy. The d_{z^2} and the $d_{x^2-y^2}$ orbitals point toward the sides of the cube, farther away from the attached ligands, so these two are lower in energy. This pattern is *inverted* compared to octahedral complexes. The two-fold degenerate lower level is designated e, while the three-fold degenerate upper level is t_2. The subscript g (*gerade* or even) is not relevant because the tetrahedron does not have a center of symmetry.

Several complexes, principally of Ni, Pd and Pt, with d^8 configurations form square planar complexes. To rationalize this geometry using crystal field theory, imagine an octahedral complex tetragonally distorted such that the two ligands on the z-axis are removed. The $d_{x^2-y^2}$ then becomes a uniquely "bad" orbital in that it points directly at the four remaining ligands. A good "strategy" for the d^8-complex is to doubly fill the four other d-orbitals. Some well-know square planar complexes include $[Ni(CN)_4]^{2-}$, $[Pt(NH_3)_4]^{2+}$, $[Cu(NH_3)_4]^{2+}$ and $[PdCl_4]^{2-}$. A class of square planar complexes of major biological significance are the porphyrins.

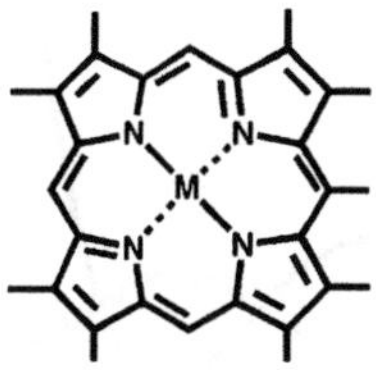

Porphyrins have a planar structure consisting of four pyrrole units linked by four methine bridges. The outer periphery is an aromatic system containing 18 delocalized π-electrons, fulfilling Hückel's $4N + 2$ rule. The four nitrogen atoms are in a perfect configuration to act as ligands. Heme proteins are iron-porphyrin complexes. They are the prosthetic groups in hemoglobins and myoglobins, responsible for oxygen transport and storage. The heme group is also found in cytochromes, which are vital electron-transporting molecules. Chlorophylls, central to photosynthesis in green plants, are magnesium-porphyrin complexes. Vitamin B-12, essential to the metabolism of proteins, carbohydrates and fats, is a cobalt-porphyrin structure.

The geometry of coordination compounds can not always be determined systematically. Ni(II), for example, can form octahedral complexes such as $[Ni(H_2O)_6]^{2+}$, tetrahedral complexes such as $[Ni(CO)_4]^{2+}$, square planar complexes such as $[Ni(CN)_4]^{2-}$, and even two different isomeric forms of $[Ni(CN)_5]^{3-}$, one a trigonal bipyramid, the other a square pyramid.

The chemistry and spectroscopy of transition-metal ions is a vast subject, and we have only scratched the surface. For a more extended treatment, refer, for example, to Ballhausen and Gray, *Molecular Orbital Theory*, (Benjamin, New York, 1965).

11.8 The hydrogen bond

Hydrogen usually forms a single electron-pair covalent bond. If, however, H is bonded to one highly electronegative atom (F, O or N) and is in close proximity to another one, it can form a three-atom bridge known as a hydrogen bond. This can be represented as a resonance hybrid between the two structures comprising one covalent bond (length ≈ 0.97 Å) and one electrostatic attraction (separation ≈ 1.79 Å):

$$\mathrm{X—H\cdots Y} \quad \text{and} \quad \mathrm{X\cdots H—Y}$$

with the three atoms in a linear configuration. As quantum-mechanical model for the hydrogen bond, one can consider a proton moving in a double-well potential, as shown in Fig. 11.12.

Hydrogen bonds are typically about 5% as strong as covalent bonds, but their collective effect can be quite significant. The fact that water is a liquid a room temperature (while the heavier compound H_2S remains a gas) is attributed to hydrogen bonding involving the two unshared electron pairs on each oxygen

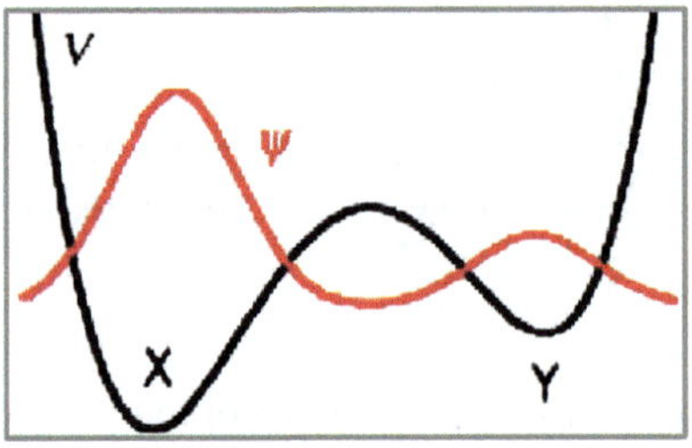

FIGURE 11.12 Double-well potential V representing hydrogen bond. Atom X is shown as more electronegative than atom Y. The proton wavefunction ψ is show in red.

atom. Both liquid and solid water contain networks of covalent and hydrogen bonds, as shown in Fig. 11.13:

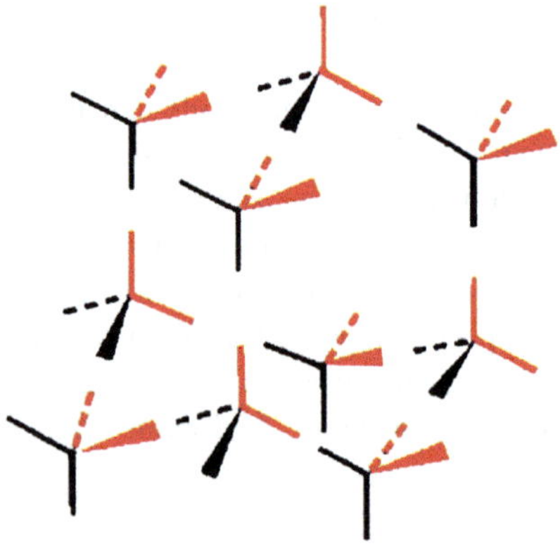

FIGURE 11.13 Tetrahedral network of covalent (black) and hydrogen (red) bonds in water. Each tetrahedron has O at center, H at each vertex.

Each oxygen atom is associated *on average* with two hydrogen atoms. An oxygen instantaneously connected with one more or one less hydrogen produces an H_3O^+ or an OH^- ion, respectively. In ice, the optimal arrangement is a hexagonal structure which *less dense* than the liquid.

Without exaggeration, life as we know it could not exist without hydrogen bonding. Recall that proteins are built up of amino-acid units connected by peptide bonds, as shown here:

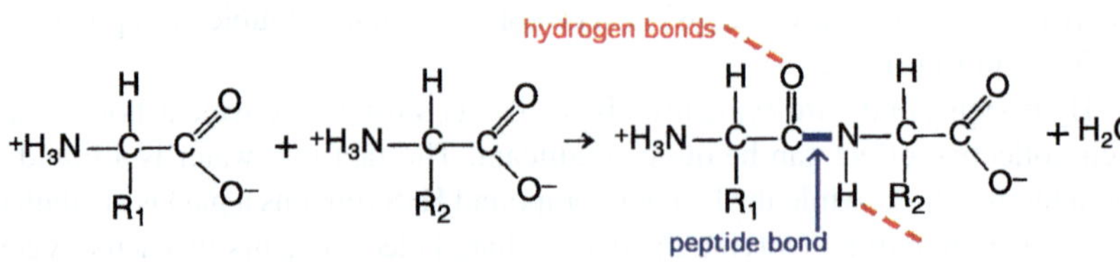

While the sequence of amino acids determines the primary structure of a protein, hydrogen bonding determines the secondary structure. Linus Pauling deduced that the α-helix structure results when each peptide carbonyl is hydrogen bonded to the amino group of the fourth peptide along the chain, as shown in Fig. 11.14. An alternative protein structure, the β-sheet, is based on hydrogen bonding between neighboring peptide chains.

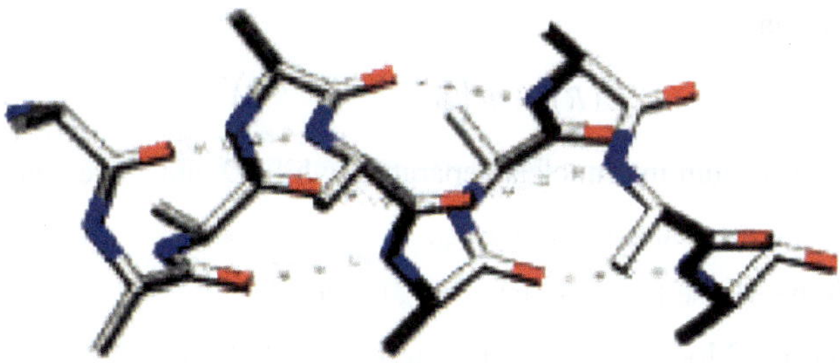

FIGURE 11.14 Schematic representation of alpha helix. Hydrogen bonds (dotted) connect carbonyl oxygens (red) to amino nitrogens (blue) four amino-acid units down the chain.

The genetic code carried by the DNA double-helix depends on the specificity of the base pairings thymine (T) to adenine (A) and cytosine (C) to guanine (G), which is determined by optimal configurations of hydrogen bonds, as shown in Fig. 11.15.

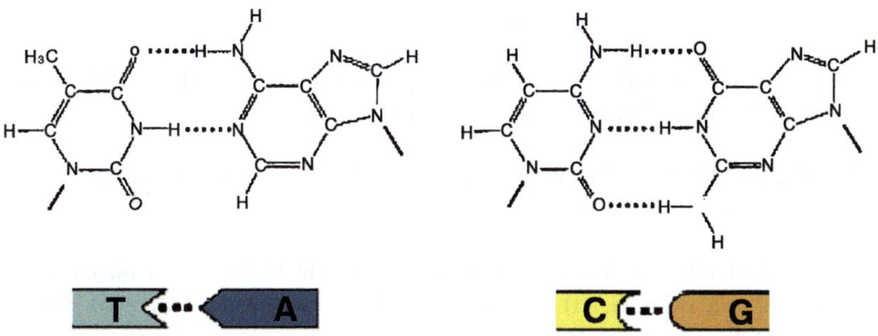

FIGURE 11.15 Base pairing in DNA via hydrogen bonds.

11.9 Critique of valence-bond theory

Valence-bond theory is over 90% successful in explaining much of the descriptive chemistry of ground states. VB theory is therefore particularly popular among chemists since it makes use of familiar concepts such as chemical bonds between atoms, resonance hybrids and the like. It can perhaps be characterized as a theory which "explains but does not predict." VB theory fails to account for the triplet ground state of O_2 or for the bonding in electron-deficient molecules such as diborane, B_2H_6. It is not very useful in consideration of excited states, hence for spectroscopy. Many of these deficiencies are remedied by molecular orbital theory, which we take up in the next two Chapters.

Chapter 11. Exercises

11.1. The electronic energy of a diatomic molecule can be approximated by the Morse function:

$$E(R) = D\left(1 - e^{-\beta(R-R_e)}\right)^2.$$

R_e is the equilibrium internuclear separation while D and β are constants.

(i) Identify the dissociation energy D_e.

(ii) Sketch the Morse function, labeling D_e and R_e.

(iii) Expand the Morse function up to terms quadratic in $(R - R_e)$. Show that this approximates a harmonic oscillator potential and identify the force constant k.

11.2. The allene molecule $CH_2{=}C{=}CH_2$ is known to have a linear geometry for the three carbon atoms. Rationalize this on the basis of hybridization of carbon AO's.

11.3. Show that the total electron distribution in the acetylene molecule H–C≡C–H is cylindrically symmetrical.

11.4. Applying the valence-shell model, predict the shapes of each of the following molecules: H_2S, SF_4, XeF_4, SF_6, BrF_5, IF_7.

11.5. Predict the ground-state geometry of each of the following species: CH_2, SO_2, SO_3, NO_3^-, XeO_3.

11.6. Consider the structure of hydrogen peroxide H_2O_2 based on (i) unhybridized atomic orbitals, (ii) the valence shell model. The experimental H–O–O angle is 94.8°.

11.7. As an alternative to the VSEPR approach, the following model for hypervalent molecules has been proposed [Pimentel, Rundle, Coulson, ca 1951]:

> *(i) Consider the ground-state electron configuration of the central atom, e.g., S($s^2p_x^2p_yp_z$), Br($s^2p_x^2p_y^2p_z$), Xe($s^2p_x^2p_y^2p_z^2$). (No promotion, hybridization or d-orbitals!)*
>
> *(ii) Singly-occupied p-orbitals form electron-pair bonds with ligand atoms in the usual way.*
>
> *(iii). Doubly-occupied p-orbitals form electron-pair bonds with two ligand atoms, located on opposite sides of the central atom.*

On the basis of the above model, predict the geometry of SF_4, BrF_3, BrF_5, XeF_2, XeF_4 and XeF_6. Compare with the corresponding valence-shell predictions.

11.8. Determine the d-electron configurations of the following octahedral complexes: $[Fe(H_2O)_6]^{3+}$, $[Cr(CN)_6]^{4-}$, $[Co(NH_3)_6]^{3+}$, $[Cu(H_2O)_6]^{2+}$. Note whether each is a high-spin or low-spin configuration.

Chapter 12

Molecular orbital theory of diatomic molecules

Molecular orbital theory is a conceptual extension of the orbital model, which was so successfully applied to atomic structure. As has been playfully remarked, "a molecule is nothing more than an atom with more nuclei." This may be overly simplistic, but we do attempt, as far as possible, to exploit analogies with atomic structure. Our understanding of atomic orbitals began with the exact solutions of a prototype problem—the hydrogen atom. We will begin our study of homonuclear diatomic molecules beginning with another exactly solvable prototype, the hydrogen molecule-ion H_2^+.

12.1 The hydrogen molecule-ion

The simplest conceivable molecule would be made of two protons and one electron, namely H_2^+. This species actually has a transient existence in electrical discharges through hydrogen gas and has been detected by mass spectrometry. It also has been detected in outer space. The Schrödinger equation for H_2^+ can be solved exactly within the Born-Oppenheimer approximation. For fixed internuclear distance R, this reduces to a problem of one electron in the field of two protons, designated A and B. We can write

$$\left\{-\frac{1}{2}\nabla^2 - \frac{1}{r_A} - \frac{1}{r_B} + \frac{1}{R}\right\}\psi(\mathbf{r}) = E\,\psi(\mathbf{r}), \tag{12.1}$$

where r_A and r_B are the distances from the electron to protons A and B, respectively. This equation was first solved by Burrau (1927), after separating the variables in prolate spheroidal coordinates. We will write down these coordinates but give only a pictorial account of the solutions. The three prolate spheroidal coordinates are designated ξ, η, ϕ. The first two are defined by

$$\xi = \frac{r_A + r_B}{R}, \qquad \eta = \frac{r_A - r_B}{R}, \tag{12.2}$$

while ϕ is the angle of rotation about the internuclear axis. The surfaces of constant ξ and η are, respectively, confocal ellipsoids and hyperboloids of revolution with foci at A and B. The two-dimensional analog should be familiar from analytic geometry, an ellipse being the locus of points such that the sum of the distances to two foci is a constant. Analogously, a hyperbola is the locus whose

Introduction to Quantum Mechanics. https://doi.org/10.1016/B978-0-12-822310-9.00020-3

difference is a constant. Fig. 12.1 shows several surfaces of constant ξ, η and ϕ. The ranges of the three coordinates are: $1 \leq \xi \leq \infty$, $-1 \leq \eta \leq 1$, $0 \leq \phi \leq 2\pi$. The prolate-spheroidal coordinate system conforms to the natural symmetry of the H_2^+ problem in the same way that spherical polar coordinates were the appropriate choice for the hydrogen atom.

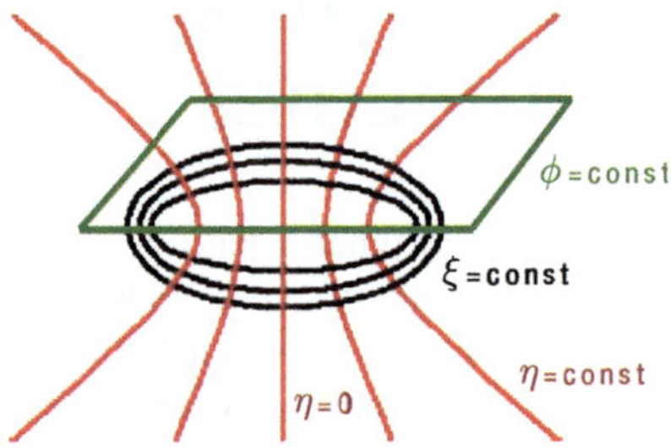

FIGURE 12.1 Prolate spheroidal coordinates.

The first few solutions of the H_2^+ Schrödinger equation are sketched in Fig. 12.3, roughly in order of increasing energy. The ϕ-dependence of the wavefunction is contained in a factor

$$\Phi(\phi) = e^{i\lambda\phi}, \qquad \lambda = 0, \pm 1, \pm 2 \ldots, \tag{12.3}$$

which is identical to the ϕ-dependence in atomic orbitals. In fact, the quantum number λ represents the component of orbital angular momentum along the internuclear axis, the only component which has a definite value in systems with axial (cylindrical) symmetry. The quantum number λ determines the basic shape of a diatomic molecular orbital, in the same way that ℓ did for an atomic orbital. An analogous code is used: σ for $\lambda = 0$, π for $\lambda = \pm 1$, δ for $\lambda = \pm 2$, and so on. We are already familiar with σ- and π-orbitals from valence-bond theory. A second classification of the H_2^+ eigenfunctions pertains to their symmetry with respect to inversion through the center of the molecule, also known as *parity* (see Fig. 12.2). If $\psi(-\mathbf{r}) = +\psi(\mathbf{r})$, the function is classified *gerade* or even parity, and the orbital designation is given a subscript g, as in σ_g or π_g. If $\psi(-\mathbf{r}) = -\psi(\mathbf{r})$, the function is classified as *ungerade* or odd parity, and we write instead σ_u or π_u. Atomic orbitals can also be classified by inversion symmetry. However, all s and d AO's are g, while all p and f orbitals are u, so no further designation is necessary.

The MO's of a given symmetry are numbered in order of increasing energy, for example, $1\sigma_g$, $2\sigma_g$, $3\sigma_g$. The lowest-energy orbital, as we have come to expect, is nodeless. It must obviously have cylindrical symmetry ($\lambda = 0$) and inversion symmetry (g). It is designated $1\sigma_g$ since it is the first orbital of this classification. The next higher orbital has a nodal plane, with $\eta = 0$, perpendicular to the axis. This function still has cylindrical symmetry (σ) but now changes sign upon inversion (u). It is designated $1\sigma_u$, as the first orbital of this type. The

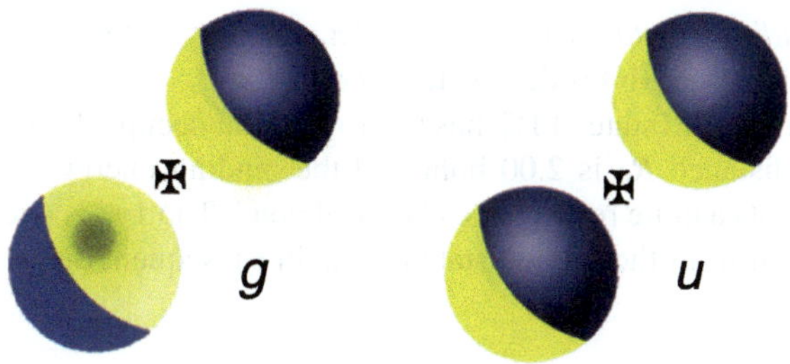

FIGURE 12.2 Inversion symmetry classification.

next higher orbital has an inner ellipsoidal node. It has the same symmetry as the lowest orbital and is designated $2\sigma_g$. Next comes the $2\sigma_u$ orbital, with both planar and ellipsoidal nodes. Two degenerate π-orbitals come next, each with a nodal plane containing the internuclear axis, with $\phi = \text{const}$. Their classification is $1\pi_u$. The second $1\pi_u$-orbital, not shown in Fig. 12.3, has the same shape rotated by 90°. The $3\sigma_g$ orbital has two hyperboloidal nodal surfaces, where $\eta = \pm\text{const}$. The $1\pi_g$, again doubly-degenerate, has two nodal planes, $\eta = 0$ and $\phi = \text{const}$. Finally, the $3\sigma_u$, the last orbital we consider, has three nodal surfaces where $\eta = \text{const}$.

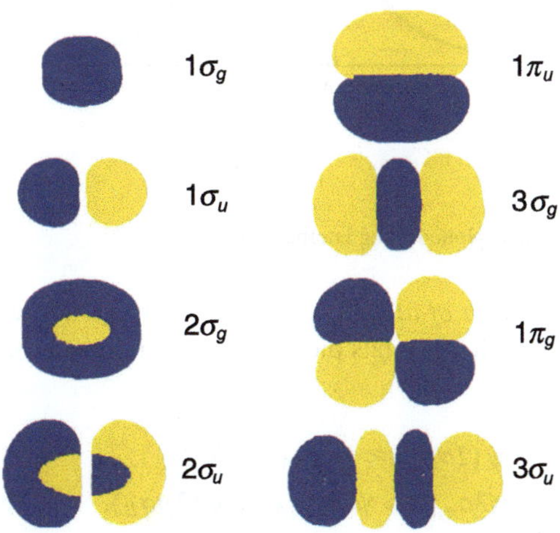

FIGURE 12.3 H_2^+ molecular orbitals. Wavefunctions are positive in blue regions, negative in yellow regions.

An MO is classified as a *bonding orbital* if it promotes the bonding of the two atoms. Generally a bonding MO, as a result of constructive interference between AO's, has a significant accumulation of electron charge in the region between the nuclei and thus reduces their mutual repulsion. The $1\sigma_g$, $2\sigma_g$, $1\pi_u$ and $3\sigma_g$ are evidently bonding orbitals. An MO which, because of destructive interference, does *not* significantly contribute to nuclear shielding is classified as an *antibonding orbital*. The $1\sigma_u$, $2\sigma_u$, $1\pi_g$ and $3\sigma_u$ belong in this category. Often an antibonding MO is designated by σ^* or π^*.

The actual ground state of H_2^+ has the $1\sigma_g$ orbital occupied. The equilibrium internuclear distance R_e is 2.00 bohr and the binding energy D_e is 2.79 eV, which represents a quite respectable chemical bond. The $1\sigma_u$ is a repulsive state and a transition from the ground state results in subsequent dissociation of the molecule.

12.2 The LCAO approximation

In Fig. 12.4, the $1\sigma_g$ and $1\sigma_u$ orbitals are plotted as functions of z, along the internuclear axis. Both functions have cusps, discontinuities in slope, at the positions of the two nuclei A and B. The $1s$ orbitals of hydrogen atoms have these same cusps. The shape of the $1\sigma_g$ and $1\sigma_u$ suggests that they can be approximated by a sum and difference, respectively, of hydrogen $1s$ orbitals, such that

$$\psi(1\sigma_{g,u}) \approx \psi(1s_A) \pm \psi(1s_B). \tag{12.4}$$

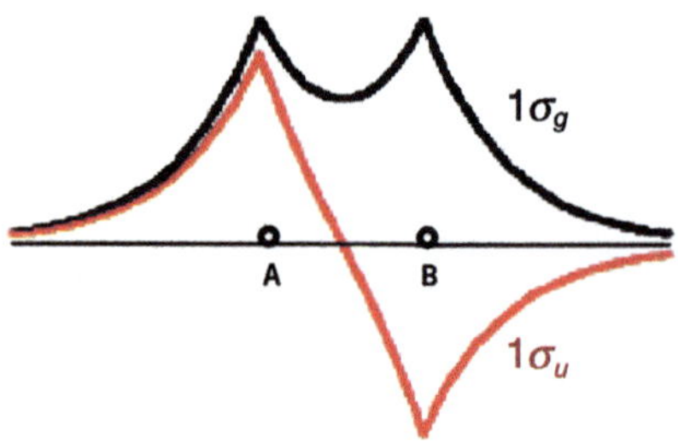

FIGURE 12.4 H_2^+ orbitals plotted along internuclear axis.

This *linear combination of atomic orbitals* is the basis of the so-called LCAO approximation. The other orbitals pictured in Fig. 12.3 can likewise be approximated as follows:

$$\begin{aligned}\psi(2\sigma_{g,u}) &\approx \psi(2s_A) \pm \psi(2s_B),\\ \psi(3\sigma_{g,u}) &\approx \psi(2p\sigma_A) \pm \psi(2p\sigma_B),\\ \psi(1\pi_{u,g}) &\approx \psi(2p\pi_A) \pm \psi(2p\pi_B).\end{aligned} \tag{12.5}$$

The $2p\sigma$ atomic orbital refers to $2p_z$, which has the axial symmetry of a σ-bond. Likewise $2p\pi$ refers to $2p_x$ or $2p_y$, which are positioned to form π-bonds. An alternative notation for diatomic molecular orbitals which specifies their atomic origin and bonding/antibonding character is the following:

$1\sigma_g$	$1\sigma_u$	$2\sigma_g$	$2\sigma_u$	$3\sigma_g$	$3\sigma_u$	$1\pi_u$	$1\pi_g$
$\sigma 1s$	$\sigma^* 1s$	$\sigma 2s$	$\sigma^* 2s$	$\sigma 2p$	$\sigma^* 2p$	$\pi 2p$	$\pi^* 2p$

Almost all applications of molecular-orbital theory are based on the LCAO approach, since the exact H_2^+ functions are too complicated to work with. The relationship between MO's and their constituent AO's can be represented in a correlation diagram, show in Fig. 12.5.

LCAO involves *addition* of orbital functions. By contrast, in Chap. 9 we *multiplied* orbitals to construct two-electron wavefunctions. Confusion can be avoided by keeping in mind this simple rule:

Orbital functions are added when they belong to the same electron, multiplied when they belong to different electrons.

Composite many-electron wavefunctions, such as Slater determinants, contain *sums of products*, as required for antisymmetry.

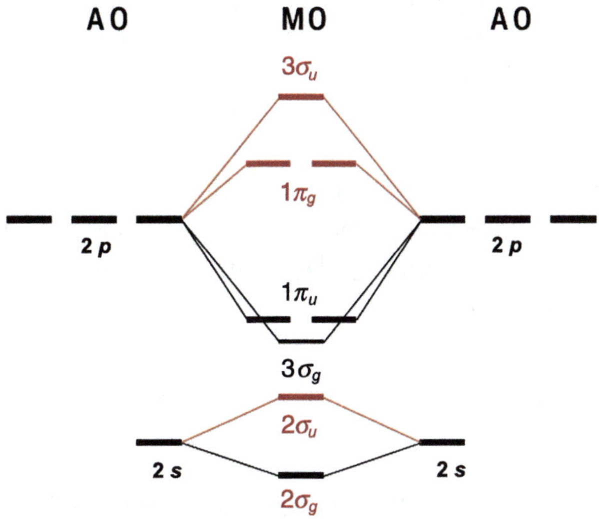

FIGURE 12.5 Molecular-orbital correlation diagram. The $1s \to 1\sigma_g$, $1\sigma_u$ is similar to the $2s$ correlations.

12.3 MO theory of homonuclear diatomic molecules

A sufficient number of orbitals is available for Aufbau of the ground states of all homonuclear diatomic species from H_2 to Ne_2. Table 12.1 summarizes the results. The most likely order in which the MO's are filled is given by

$$1\sigma_g < 1\sigma_u < 2\sigma_g < 2\sigma_u < 3\sigma_g \sim 1\pi_u < 1\pi_g < 3\sigma_u$$

The relative order of $3\sigma_g$ and $1\pi_u$ depends on which other MO's are occupied, much like the situation involving $4s$ and $3d$ atomic orbitals. The results of photoelectron spectroscopy indicate that $1\pi_u$ is lower up to and including N_2, but $3\sigma_g$ is lower thereafter.

TABLE 12.1 Homonuclear diatomic molecules.

Molecule	Electron configuration	Bond order	D_e/eV	R_e/Å
H_2^+	$1\sigma_g \ {}^2\Sigma_g^+$	0.5	2.79	1.06
H_2	$1\sigma_g^2 \ {}^1\Sigma_g^+$	1	4.75	0.741
He_2	$1\sigma_g^2 1\sigma_u^2 \ {}^1\Sigma_g^+$	0	0.0009[a]	3.0
	$1\sigma_g^2 1\sigma_u 2\sigma_g \ {}^3\Sigma_u^{+b}$	1	2.6	1.05
He_2^+	$1\sigma_g^2 1\sigma_u \ {}^2\Sigma_u^+$	0.5	2.5	1.08
Li_2	$1\sigma_g^2 1\sigma_u^2 2\sigma_g^2 \ {}^1\Sigma_g^+$	1	1.07	2.67
Be_2	$1\sigma_g^2 1\sigma_u^2 2\sigma_g^2 2\sigma_u^2 \ {}^1\Sigma_g^+$	0	0.1	2.5
B_2	$\ldots 1\pi_u^2 \ {}^3\Sigma_g^{-c}$	1	3.0	1.59
C_2	$\ldots 1\pi_u^4 \ {}^1\Sigma_g^+$	2	6.3	1.24
N_2	$\ldots 1\pi_u^4 3\sigma_g^2 \ {}^1\Sigma_g^+$	3	9.91	1.10
N_2^+	$\ldots 1\pi_u^4 3\sigma_g \ {}^2\Sigma_g^+$	2.5	8.85[d]	1.12
O_2	$\ldots 3\sigma_g^2 1\pi_u^4 1\pi_g^2 \ {}^3\Sigma_g^{-c,e}$	2	5.21	1.21
O_2^+	$\ldots 3\sigma_g^2 1\pi_u^4 1\pi_g \ {}^2\Pi_g$	2.5	6.78[d]	1.12
F_2	$\ldots 1\pi_u^4 3\sigma_g^2 1\pi_g^4 \ {}^1\Sigma_g^+$	1	1.66	1.41
Ne_2	$\ldots 1\pi_u^4 3\sigma_g^2 1\pi_g^4 3\sigma_u^2 \ {}^1\Sigma_g^+$	0	0.0036[a]	3.1

[a] *Van der Waals bonding.*
[b] *Lifetime* $\approx 10^{-4}$ *sec.*
[c] *Note application of Hund's rules.*
[d] *Compare effect of ionization on binding energy.*
[e] *Paramagnetism of* O_2 *predicted by MO theory.*

The term symbol Σ, Π, $\Delta \ldots$, analogous to the atomic S, P, D… symbolizes the axial component of the total orbital angular momentum. When a π-shell is filled (4 electrons) or half-filled (2 electrons), the orbital angular momentum cancels to zero and we find a Σ term. The spin multiplicity is completely analogous to the atomic case. The total parity is again designated by a subscript g or u. Since the many electron wavefunctions are made up of products of individual MO's, the total parity is odd only if the molecule contains an *odd* number of u orbitals. Thus a σ_u^2 or a π_u^2 subshell transforms like g.

For Σ terms, the superscript $\pm$ denotes the sign change of the wavefunction under a reflection in a plane containing the internuclear axis. This is equivalent to a sign change in the variable $\phi \to -\phi$. This symmetry is needed when we deal with spectroscopic selection rules. In a spin-paired π_u^2 subshell the triplet spin function is symmetric so that the orbital factor must be antisymmetric, of the form

$$\frac{1}{\sqrt{2}}\Big[\pi_x(1)\pi_y(2) - \pi_y(1)\pi_x(2)\Big]. \tag{12.6}$$

This will change sign under the reflection, since $x \to x$ but $y \to -y$. We need only remember that a π_u^2 subshell will give the term symbol $^3\Sigma_g^-$.

The net bonding effect of the occupied MO's is determined by the *bond order*, half the excess of the number bonding minus the number antibonding. This definition brings the MO results into correspondence with the Lewis (valence-bond) concept of single, double and triple bonds. It is also possible in MO theory to have a bond order of 1/2, for example, in H_2^+ which is held together by a single bonding orbital. A bond order of zero generally indicates no stable chemical bond, although helium and neon atoms can still form clusters held together by much weaker van der Waals forces. Molecular-orbital theory successfully accounts for the transient stability of a $^3\Sigma_u^+$ excited state of He_2, in which one of the antibonding electrons is promoted to an excited bonding orbital. This species has a lifetime of about 10^{-4} sec, until it emits a photon and falls back into the unstable ground state. Another successful prediction of MO theory concerns the relative binding energy of the positive ions N_2^+ and O_2^+, compared to the neutral molecules. Ionization weakens the N–N bond since a bonding electron is lost, but it strengthens the O–O bond since an antibonding electron is lost.

One of the earliest triumphs of molecular orbital theory was the prediction that the oxygen molecule is paramagnetic. Fig. 12.6 shows that liquid O_2 is a magnetic substance, attracted to the region between the poles of a permanent magnet. The paramagnetism arises from the half-filled $1\pi_g^2$ subshell. According to Hund's rules the two electrons singly occupy the two degenerate $1\pi_g$ orbitals with their spins aligned *parallel*. The term symbol is $^3\Sigma_g^-$ and the molecule thus has nonzero spin angular momentum and a net magnetic moment, which interacts with an external magnetic field. Linus Pauling invented the paramagnetic oxygen analyzer, which was for a time extensively used in medical technology.

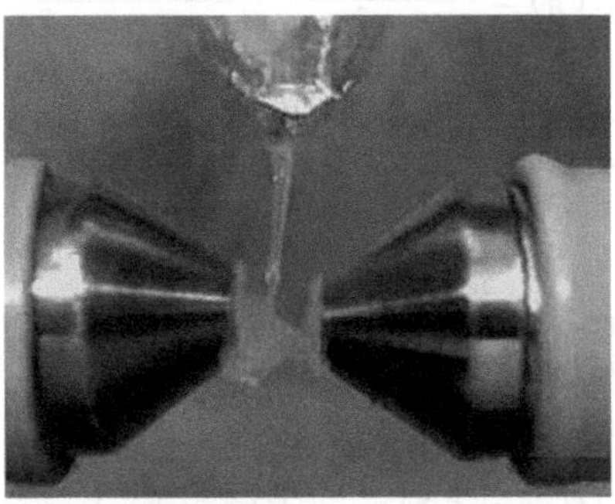

FIGURE 12.6 Demonstration showing blue liquid O_2 attracted to the poles of a permanent magnet.

12.4 Variational computation of molecular orbitals

Thus far we have approached MO theory from a mainly descriptive point of view. To begin a more quantitative treatment, recall the LCAO approximation to the H_2^+ ground state, Eq. (12.4), which can be written

$$\psi = c_A \psi_A + c_B \psi_B. \tag{12.7}$$

Using this as a trial function in the variational principle, we have

$$E(c_A, c_B) = \frac{\int \psi H \psi \, d\tau}{\int \psi^2 \, d\tau}, \tag{12.8}$$

where H is the Hamiltonian from Eq. (12.1). In fact, these equations can be applied more generally to construct *any* molecular orbital, not just solutions for H_2^+. In the general case, H will represent an *effective* one-electron Hamiltonian determined by the molecular environment of a given orbital. The energy expression involves some complicated integrals, but can be simplified somewhat by expressing it in a standard form. Hamiltonian matrix elements are defined by:

$$H_{AA} = \int \psi_A H \psi_A \, d\tau, \qquad H_{BB} = \int \psi_B H \psi_B \, d\tau,$$
$$H_{AB} = H_{BA} = \int \psi_A H \psi_B \, d\tau, \tag{12.9}$$

while the overlap integral is given by

$$S_{AB} = \int \psi_A \psi_B \, d\tau. \tag{12.10}$$

Presuming the functions ψ_A and ψ_B to be normalized, the variational energy (12.8) reduces to

$$E(c_A, c_B) = \frac{c_A^2 H_{AA} + 2c_A c_B H_{AB} + c_B^2 H_{BB}}{c_A^2 + 2c_A c_B S_{AB} + c_B^2}. \tag{12.11}$$

To optimize the MO, we find the minimum of E wrt variation in c_A and c_B, as determined by the two conditions:

$$\frac{\partial E}{\partial c_A} = 0, \qquad \frac{\partial E}{\partial c_B} = 0. \tag{12.12}$$

The result is a *secular equation* determining two values of the energy:

$$\begin{vmatrix} H_{AA} - E & H_{AB} - ES_{AB} \\ H_{AB} - ES_{AB} & H_{BB} - E \end{vmatrix} = 0. \tag{12.13}$$

For the case of a homonuclear diatomic molecule, for example H_2^+, the two Hamiltonian matrix elements H_{AA} and H_{BB} are equal, say to α. Setting $H_{AB} = \beta$ and $S_{AB} = S$, the secular equation reduces to

$$\begin{vmatrix} \alpha - E & \beta - ES \\ \beta - ES & \alpha - E \end{vmatrix} = (\alpha - E)^2 - (\beta - ES)^2 = 0, \tag{12.14}$$

with the two roots

$$E^{\pm} = \frac{\alpha \pm \beta}{1 \pm S}. \tag{12.15}$$

The calculated integrals α and β are usually negative, thus for the bonding orbital

$$E^{+} = \frac{\alpha + \beta}{1 + S} \qquad \text{(bonding)}, \tag{12.16}$$

while for the antibonding orbital

$$E^{-} = \frac{\alpha - \beta}{1 - S} \qquad \text{(antibonding)}. \tag{12.17}$$

Note that $(E^{-} - \alpha) > (\alpha - E^{+})$, thus the energy increase associated with antibonding is slightly greater than the energy decrease for bonding. For historical reasons, α is called a *Coulomb integral* and β, a *resonance integral.*

12.5 Heteronuclear molecules

The variational computation leading to Eq. (12.13) can be applied as well to the heteronuclear case in which the orbitals ψ_A and ψ_B are *not* equivalent. The matrix elements H_{AA} and H_{BB} are approximately equal to the energies of the atomic orbitals ψ_A and ψ_B, respectively, say E_A and E_B with $E_A > E_B$. It is generally true that $|E_A|$, $|E_B| \gg |H_{AB}|$. With these simplifications, secular equation can be written

$$\begin{vmatrix} E_A - E & H_{AB} - ES_{AB} \\ H_{AB} - ES_{AB} & E_B - E \end{vmatrix} = (E_A - E)(E_B - E) - (H_{AB} - ES_{AB})^2 = 0. \tag{12.18}$$

This can be rearranged to

$$E - E_A = \frac{(H_{AB} - ES_{AB})^2}{E - E_B}. \tag{12.19}$$

To estimate the root closest to E_A, we can replace E by E_A on the right hand side of the equation. This leads to

$$E^{-} \approx E_A + \frac{(H_{AB} - E_A S_{AB})^2}{E_A - E_B}, \tag{12.20}$$

and analogously for the other root,

$$E^{+} \approx E_B - \frac{(H_{AB} - E_B S_{AB})^2}{E_A - E_B}. \tag{12.21}$$

The relative energies of these AO's and MO's are represented by a correlation diagram, as shown in Fig. 12.7.

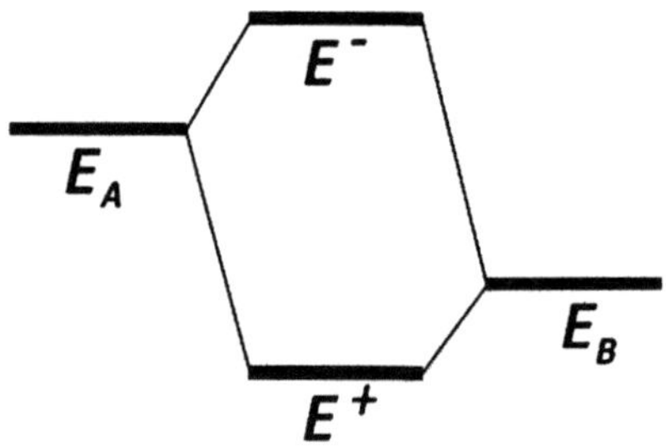

FIGURE 12.7 Correlation diagram for bonding and antibonding molecular orbitals in heteronuclear case.

An analysis of Eqs. (12.20)-(12.21) implies that, in order for two atomic orbitals ψ_A and ψ_B to form effective molecular orbitals the following conditions must be met:

1. The AO's must have compatible symmetry.

For example, ψ_A and ψ_B can be either s or $p\sigma$ orbitals to form a σ-bond or both can be $p\pi$ (with the same orientation) to form a π-bond.

2. The charge clouds of ψ_A and ψ_B should overlap as much as possible.

This was the rationale for hybridizing the s and p orbitals in carbon. A larger value of S_{AB} implies a larger value for H_{AB}.

3. The energies E_A and E_B must be of comparable magnitude.

Otherwise, the denominator in (12.20) and (12.21) will be too large and the MO's will not differ significantly from the original AO's. A rough criterion is that E_A and E_B should be within about 0.2 hartree or 5 eV. For example, the chlorine $3p$ orbital has an energy of -13.0 eV, comfortably within range of the hydrogen $1s$, with energy -13.6 eV. Thus these can interact to form a strong bonding (plus an antibonding) MO in HCl. The chlorine $3s$ with an energy of -24.5 eV could *not* form an effective bond with hydrogen, even if it were available. Generally, the greater the AO-energy difference, the more polar will be the bond. The limiting case is an ionic bond, in which no effective MO is formed but two electrons occupy the lower AO.

12.6 Electronegativity

The character of a chemical bond is dependent on the electronegativity difference between the bonded atoms. Electronegativity was first introduced by Pauling as a measure of an atom's power to attract electron charge. He proposed the following formula for the difference in electronegativity χ (chi) between two atoms X and Y:

$$\left|\chi_{\mathrm{X}}-\chi_{\mathrm{Y}}\right|=0.102\left[D_{\mathrm{XY}}-\frac{1}{2}(D_{\mathrm{XX}}-D_{\mathrm{YY}})\right]^{1/2}, \tag{12.22}$$

where the D's are the dissociation energies of the diatomic molecules expressed in kJ/mol (1 eV=96.485 kJ/mol). Following are conventional electronegativity values for a few common elements:

	Na	Si	B	P	H	C	S	N	Cl	O	F
χ	0.93	1.90	2.04	2.19	2.20	2.55	2.58	3.04	3.16	3.44	3.98

An empirical relation between the electronegativity difference $\Delta\chi$ and the fractional ionic character of a covalent bond was suggested by Hannay and Smythe (1946):

$$\%\text{ionic} \approx 16\,\Delta\chi + 3.5(\Delta\chi)^2. \tag{12.23}$$

For a diatomic molecule, the fractional ionic character can be defined by

$$\text{fraction ionic} = \frac{\mu(XY)}{\mu(X^+Y^-)} = \frac{\mu}{eR_e}, \tag{12.24}$$

where μ is the observed dipole moment and R_e, the equilibrium internuclear distance. For a 100% ionic bond, the positive and negative charges would be separated by a distance R_e. As an example, HF has a dipole moment of 1.82 D. One debye unit (D) equals 3.336×10^{-30} C m. For charges $\pm e$ separated by R Å, a simple numerical relation is $\mu = 4.80R$ debye. For HF, $R_e = 0.916$ Å, so the bond is 41% ionic. Eq. (12.23) with $\Delta\chi = 1.78$ predicts 40% ionic character.

Mulliken gave an alternative definition of electronegativity in terms of the ionization energy I and electron affinity A of an atom. The ionization energy equals ΔE for the reaction $\mathrm{X} \rightarrow \mathrm{X}^+ + \mathrm{e}$, while the electron affinity equals ΔE for $\mathrm{X}^- \rightarrow \mathrm{X} + \mathrm{e}$. To a rough approximation,

$$I \approx -E_{\mathrm{HOMO}} \qquad \text{and} \qquad A \approx -E_{\mathrm{LUMO}} \tag{12.25}$$

Strictly speaking, these energies should pertain to the appropriate *valence states* of the atoms. Consider now the hypothetical alternative atomic ionization reactions:

$$\mathrm{X}+\mathrm{Y} \longrightarrow \begin{cases} \mathrm{X}^+ + \mathrm{Y}^- \\ \mathrm{X}^- + \mathrm{Y}^+ \end{cases}$$

For the first reaction, $\Delta E_1 = I_X - A_Y$, while for the second $\Delta E_2 = I_Y - A_X$. If the electronegativities of X and Y were equal then we would have $\Delta E_1 = \Delta E_2$ and so $I_X + A_X = I_Y + A_Y$. This suggests Mulliken's definition of the electronegativity of an element as the average of I and A (in eV):

$$\chi = \frac{I + A}{2}. \tag{12.26}$$

The Mulliken and Pauling scales are related approximately by

$$\chi_{\mathrm{P}} \approx 1.35\,(\chi_{\mathrm{M}})^{1/2} - 1.37. \tag{12.27}$$

A parameter akin to electronegativity is *hardness*, introduced by Pearson and Parr (1983):

$$\eta = \frac{I - A}{2} \approx \frac{1}{2}(E_{\mathrm{LUMO}} - E_{\mathrm{HOMO}}). \tag{12.28}$$

This is equal to ΔE for the disproportionation reaction

$$\mathrm{X} \rightarrow \frac{1}{2}\mathrm{X}^{+} + \frac{1}{2}\mathrm{X}^{-}.$$

Such reactions become increasingly difficult as the hardness of X increases. Hardness is thus a measure of *resistance* to separation of charge; more generally, resistance to chemical reactivity, consistent with its relation to the HOMO-LUMO energy gap.

Chapter 12. Exercises

12.1. After separation of variables in the H_2^+ problem, the function $\Xi(\xi)$ is found to obey the differential equation

$$\frac{d}{d\xi}\left[(\xi^2 - 1)\frac{d\Xi}{d\xi}\right] + \left(A + 2R\,\xi + \tfrac{1}{4}R^2 E\,\xi^2 - \frac{\lambda^2}{\xi^2 - 1}\right)\Xi(\xi) = 0$$

where A is a constant, R is the internuclear distance, λ is the angular momentum quantum number, an integer, and E is the energy, a negative number for bound states. Find the asymptotic solution of the above equation as $\xi \rightarrow \infty$.

12.2. Consider the LCAO functions

$$\psi = N(\psi_A \pm \psi_B)$$

with ψ_A and ψ_B each normalized and their overlap integral equal to S. Show that ψ is normalized when $N = [2(1 \pm S)]^{-1/2}$.

12.3. Write out the two-electron orbital function for the H_2 molecule

$$\psi(1,2) = \psi_{1\sigma g}(1)\,\psi_{1\sigma g}(2)$$

assuming the LCAO approximation for each MO. Expand this result and show how it relates to the corresponding valence-bond wavefunction. What is the meaning of the left-over terms?

12.4. Give the electron configuration, term symbol and bond order for the ground state of each of the following species: N_2^+, N_2 and N_2^-.

12.5. Predict the electronic configuration and term symbol for the ground state of the superoxide ion O_2^- and of the peroxide ion O_2^{2-}.

12.6. Propose electron configurations and term symbols for the two lowest singlet excited states of O_2.

12.7. Rationalize why the Be_2 molecule, unlike He_2, is weakly bound ($D_2 \approx 0.1$ eV). Hint: 2s-2p hybridization is involved.

12.8. The overlap integral between a $1s$ and a $2p\sigma$ orbital on nuclei separated by a distance R (in bohr) is given by

$$S = \left(R + R^2 + \frac{R^3}{3}\right) e^{-R}$$

Determine the value of R which gives the maximum overlap. (It may be of interest that the internuclear distance in HF equals 0.916 Å.)

Chapter 13

Polyatomic molecules and solids

13.1 Hückel molecular orbital theory

Molecular orbital theory has been very successfully applied to large conjugated systems, especially those containing chains of carbon atoms with alternating single and double bonds. An approximation introduced by Hückel in 1931 considers only the delocalized p-electrons moving in a framework of σ-bonds. This is, in fact, a more sophisticated version of the free-electron model introduced in Chap. 3. We again illustrate the model using butadiene $CH_2{=}CH{-}CH{=}CH_2$. From four p atomic orbitals, designated p_1, p_2, p_3, p_4, with nodes in the plane of the carbon skeleton, one can construct four π molecular orbitals by an extension of the LCAO approach:

$$\psi = c_1 p_1 + c_2 p_2 + c_3 p_3 + c_4 p_4. \tag{13.1}$$

Applying the linear variational method, the energies of the MO's are the roots of the 4×4 secular equation

$$\begin{vmatrix} H_{11} - \epsilon & H_{12} - \epsilon S_{12} & \dots \\ H_{12} - \epsilon S_{12} & H_{22} - \epsilon & \dots \\ \dots & \dots & \dots \end{vmatrix} = 0. \tag{13.2}$$

Four simplifying assumptions are now made:

(1) All overlap integrals S_{ij} are sufficiently small that they can be set equal to zero.

This is quite reasonable since the p-orbitals are directed perpendicular to the direction of their bonds.

(2) All resonance integrals H_{ij} between non-neighboring atoms are set equal to zero.

(3) All resonance integrals H_{ij} between neighboring atoms are set equal to β.

(4) All coulomb integrals H_{ii} are set equal to α.

Introduction to Quantum Mechanics. https://doi.org/10.1016/B978-0-12-822310-9.00021-5

The secular equation thus reduces to

$$\begin{vmatrix} \alpha-\epsilon & \beta & 0 & 0 \\ \beta & \alpha-\epsilon & \beta & 0 \\ 0 & \beta & \alpha-\epsilon & \beta \\ 0 & 0 & \beta & \alpha-\epsilon \end{vmatrix} = 0. \tag{13.3}$$

Dividing by β^4 and defining

$$x = \frac{\alpha-\epsilon}{\beta}, \tag{13.4}$$

the equation simplifies further to

$$\begin{vmatrix} x & 1 & 0 & 0 \\ 1 & x & 1 & 0 \\ 0 & 1 & x & 1 \\ 0 & 0 & 1 & x \end{vmatrix} = 0. \tag{13.5}$$

This is essentially the connection matrix for the molecule. Each pair of connected atoms is represented by 1, each non-connected pair by 0 and each diagonal element by x. Expansion of the determinant gives the 4th order polynomial equation

$$x^4 - 3x^2 + 1 = 0. \tag{13.6}$$

Noting that this is a quadratic equation in x^2, the roots are found to be $x^2 = \left(3 \pm \sqrt{5}\right)/2$, so that $x = \pm 0.618, \pm 1.618$. This corresponds to the four MO energy levels

$$\epsilon = \alpha \pm 1.618\beta, \qquad \alpha \pm 0.618\beta. \tag{13.7}$$

Since α and β are negative, the lowest MO's have

$$\epsilon_{1\pi} = \alpha + 1.618\beta \quad \text{and} \quad \epsilon_{2\pi} = \alpha + 0.618\beta,$$

and the total π-electron energy of the $1\pi^2 2\pi^2$ configuration equals

$$E_\pi = 2(\alpha + 1.618\beta) + 2(\alpha + 0.618\beta) = 4\alpha + 4.472\beta. \tag{13.8}$$

The coefficients c_i in (13.1) can now be found by solving four simultaneous equations. For the lowest energy orbital $\epsilon_{1\pi}$:

$$\sum_{i=1}^{4} (H_{ij} - \epsilon_{1\pi}\delta_{ij})c_j^{1\pi} = 0 \qquad j = 1, \ldots, 4, \tag{13.9}$$

and analogously for each of the higher MO's. The normalized Hückel MO's are given by

$$\begin{aligned}\psi_{1\pi} &= 0.372p_1 + 0.602p_2 + 0.602p_3 + 0.372p_4,\\ \psi_{2\pi} &= 0.602p_1 + 0.372p_2 - 0.372p_3 - 0.602p_4,\\ \psi_{3\pi} &= 0.602p_1 - 0.372p_2 - 0.372p_3 + 0.602p_4,\\ \psi_{4\pi} &= 0.372p_1 - 0.602p_2 + 0.602p_3 - 0.372p_4.\end{aligned} \tag{13.10}$$

A schematic representation of these four orbitals is given in Fig. 13.1, with the scale of each p-orbitals proportional to its coefficient in (13.10). Note the topological resemblance to free-electron model wavefunctions in Sect. 3.3.

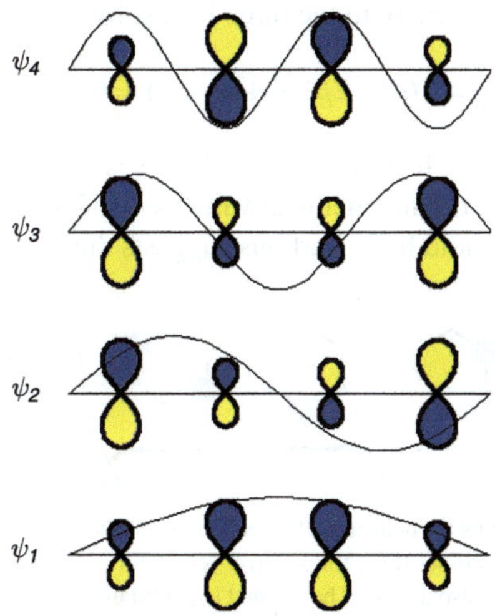

FIGURE 13.1 Hückel molecular orbitals for butadiene. Positive regions are shown as blue, negative as yellow. The corresponding free-electron model wavefunctions are drawn in the background.

The simplest application of Hückel theory, the ethylene molecule $CH_2{=}CH_2$, gives the secular equation

$$\begin{vmatrix} x & 1 \\ 1 & x \end{vmatrix} = 0. \tag{13.11}$$

This is easily solved for the energies $\epsilon = \alpha \pm \beta$. [Compare to Eqs. (12.16) and (12.17) with $S = 0$.] The lowest orbital has $\epsilon_{1\pi} = \alpha + \beta$ and the $1\pi^2$ ground state has $E_\pi = 2(\alpha + \beta)$. If butadiene had two localized double bonds, as in its dominant valence-bond structure, its π-electron energy would be given by $E_\pi = 4(\alpha + \beta)$. Comparing this with the Hückel result (13.8), we see that the energy lies lower than the that of two double bonds by 0.48β. The thermochemical value is approximately -17 kJmol^{-1}. This stabilization of a conjugated system is known as the *delocalization energy*. It corresponds to the resonance-stabilization energy in valence-bond theory.

Aromatic systems provide the most significant applications of Hückel theory. For benzene, we find the secular equation

$$\begin{vmatrix} x & 1 & 0 & 0 & 0 & 1 \\ 1 & x & 1 & 0 & 0 & 0 \\ 0 & 1 & x & 1 & 0 & 0 \\ 0 & 0 & 1 & x & 1 & 0 \\ 0 & 0 & 0 & 1 & x & 1 \\ 1 & 0 & 0 & 0 & 1 & x \end{vmatrix} = 0, \tag{13.12}$$

with the six roots $x = \pm 2, \pm 1, \pm 1$. The energy levels are $\epsilon = \alpha \pm 2\beta$ and two-fold degenerate $\epsilon = \alpha \pm \beta$. With the three lowest MO's occupied, we have

$$E_\pi = 2(\alpha + 2\beta) + 4(\alpha + \beta) = 6\alpha + 8\beta. \tag{13.13}$$

Since the energy of three localized double bonds is $6\alpha + 6\beta$, the delocalization energy equals 2β. The thermochemical value is $-152\ \text{kJmol}^{-1}$. A least-squares fit of a series of benzenoid hydrocarbons suggests the value $|\beta| \approx 2.72$ eV.

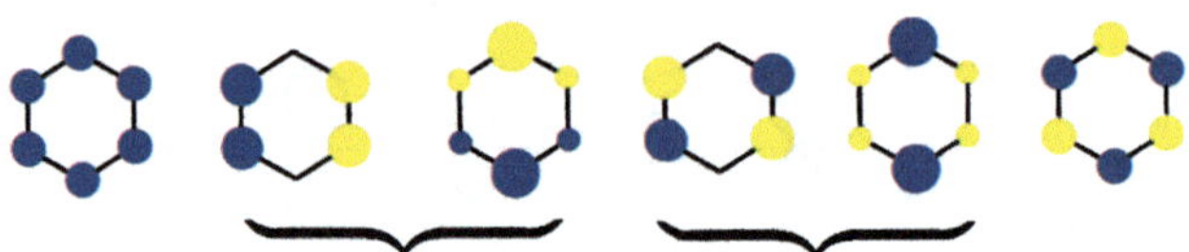

a_{2u} e_{1g} e_{2u} b_{2g}

FIGURE 13.2 Hückel MO's for benzene. The coefficients of the carbon p atomic orbitals are proportional to the sizes of the circles (blue for positive, yellow for negative). Since these are p orbitals, the plane of the molecule is also a node. The e_{2u} and b_{2g} orbitals are unoccupied in the ground state.

The six π-electron molecular orbitals for benzene are diagrammed in Fig. 13.2. The group-theoretical notation for the orbitals will be explained in Chap. 15. It should be evident from the figure that the number of nodes increases with energy. The ground-state electron configuration is $a_{2u}^2 e_{1g}^4$.

13.2 Conservation of orbital symmetry; Woodward-Hoffmann rules

In the course of their synthesis of Vitamin B_{12}, R. B. Woodward and coworkers were puzzled by the failure of certain cyclic products to form from apparently appropriate starting materials—in particular, the stereochemistry of interconversions of cyclohexadienes with conjugated trienes in thermal and photochemical reactions. Woodward, in collaboration with Roald Hoffmann (ca 1965), discovered that the course of such reactions depended on identifiable symmetries of the participating molecular orbitals. The principle of *conservation of orbital symmetry* can be stated thus:

R. B. Woodward

In the course of a concerted reaction, the MO's of the reactant molecules are transformed into the MO's of the products by a continuous pathway.

A *concerted reaction* is one which takes place in a single step, through a *transition state*, but *without* the formation of reactive intermediates. Breaking of bonds in the reactants and formation of new bonds in the products takes place in one continuous process. Often this involves interconversion of σ- and π-bonds. Such reactions are generally insensitive to such factors as solvent polarity and catalysis but are characterized by a high degree of stereospecificity.

As emphasized by Fukui, the mechanism of chemical reactions can often be understood in terms of *frontier orbitals*—the HOMO's and LUMO's of reacting molecules. Ideally, the frontier orbitals of the reactants interact to form the MO's of the products. And it is in such transformations that orbital symmetry is conserved. We will consider two relevant examples from organic chemistry: electrocyclic reactions and cycloadditions.

Simple examples of *electrocyclic reactions* are the formation of cyclobutene from butadiene and cyclohexadiene from hexatriene:

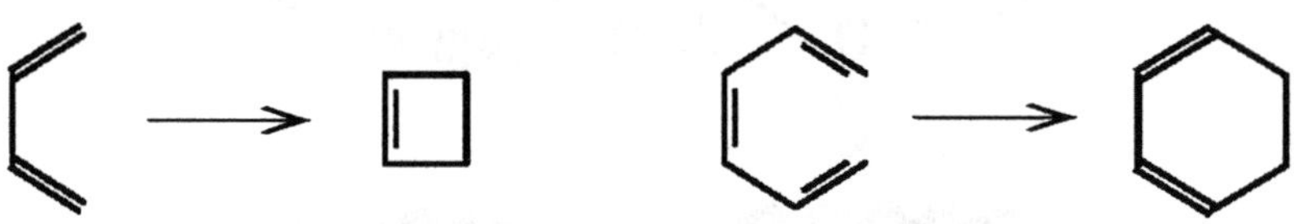

To appreciate the conformational implications of orbital-symmetry conservation we consider these two reactions with groups R_1 and R_2 replacing two of the terminal hydrogens. (As per convention, the remaining hydrogens are not drawn.) If the reactions are carried out under thermal conditions, they proceed as follows:

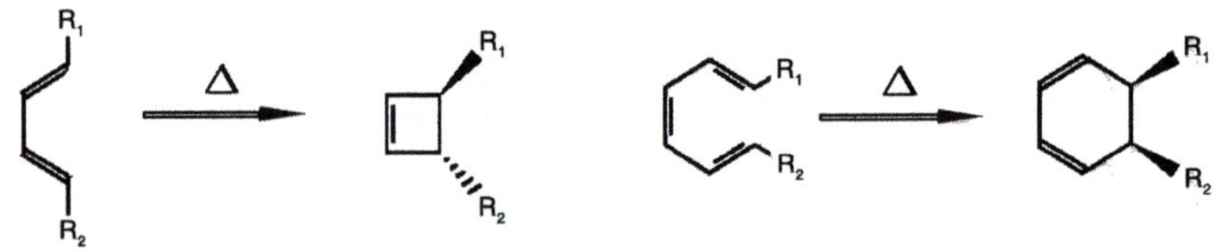

The butadiene reaction gives a *trans*-configuration of substituents R_1 and R_2 while the hexatriene gives a *cis*-configuration. If, on the other hand, the reactions are photochemically induced, the opposite configurations are produced:

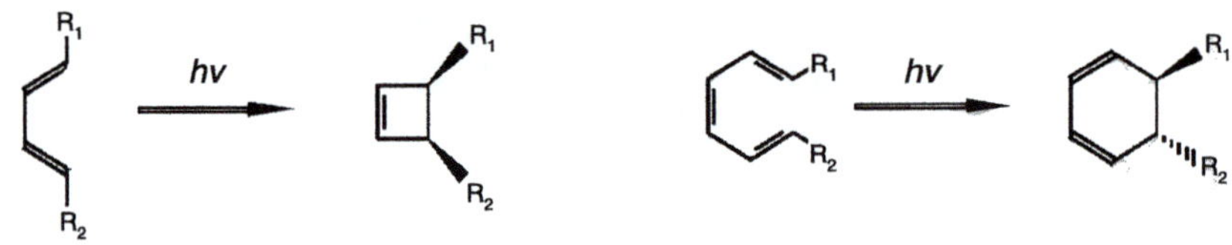

The stereoselectivity of the above reactions can be explained by the geometry of the highest-occupied molecular orbitals (HOMO's). For butadiene, the HOMO for the ground electronic state is the orbital ψ_2 in Fig. 13.1. Ring closure occurs when the two terminal p-orbitals reorient themselves to create a σ-bond, as shown in Fig. 13.3. Since the p_1 and p_4 lobes are *out of phase*, the orbitals rotate in the *same* direction—*conrotatory*—to give a positive overlap necessary for bonding. This is accompanied by the substituents R_1 and R_2 moving to opposite sides of the ring, into a *trans* configuration. This accounts for the geometry of the *thermal* electrocyclic reaction of butadiene. The reaction can be alternatively induced *photochemically* by irradiation with ultraviolet. What happens then is electrons in the ψ_2 orbital are excited to ψ_3. As is evident from Fig. 13.1, the p_1 and p_4 lobes for this new HOMO are now *in phase*. They must now rotate in *opposite* directions—*disrotatory*—to give a bonding overlap. Thus R_1 and R_2 wind up on the same side of the ring, the *cis* configuration.

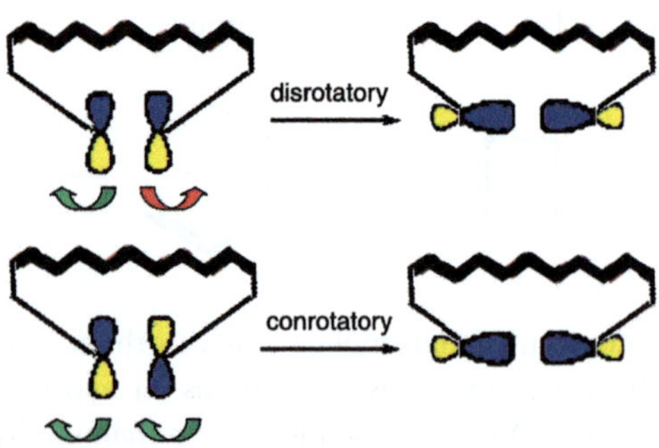

FIGURE 13.3 Woodward-Hoffmann rules for conservation of orbital symmetry in electrocyclic ring closure, as determined by relative phases in HOMO.

For ring closure in hexatriene, you can easily show that the HOMO for the ground state has its terminal lobes in phase, while the photochemically induced state has its terminal lobes out of phase. Thus the cis-trans stereospecificity is exactly the opposite of that for butadiene. The general result can be formalized by the *Woodward-Hoffmann rule* for concerted electrocyclic reactions: If the total number of electrons in the transition state equals $4n$ $[4n+2]$, the thermal reaction will produce the conrotarory [disrotatory] configuration while the photochemical reaction will produce the disrotatory [conrotarory] configuration.

The prototype of a cycloaddition is the Diels-Alder reaction between a diene and a dienophile. Two examples:

Fig. 13.4 shows two possible ways for this to happen: the HOMO of the diene can combine with the LUMO of the dienophile or the LUMO of the diene with the HOMO of the dienophile. The thermal reaction with this 6-electron transition state is allowed but the corresponding photochemical mechanism is forbidden. More generally, the Woodward-Hoffmann rule for concerted cycloaddition reactions can be stated: If the number of electrons in the transition state equals $4n$ $[4n+2]$, then the photochemical [thermal] reaction will be allowed but the thermal [photochemical] reaction will be forbidden.

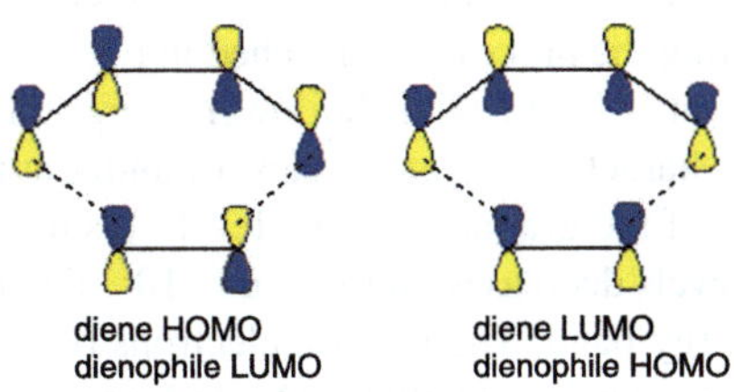

FIGURE 13.4 Two possible orbital-symmetry combinations in Diels-Alder reaction.

The hydrogen-iodine reaction

$$H_2 + I_2 \rightarrow 2HI$$

was one of the first whose kinetics were studied in detail. It had long been assumed that the reaction proceeded through a square intermediate, followed by breaking of H–H and I–I bonds and simultaneous formation of H–I bonds. Application of Woodward-Hoffmann orbital-symmetry concepts shows, however, that such a mechanism could not possibly explain the course of the reaction. According to frontier-orbital picture, the formation and dissociation of this intermediate must involve electron flow either from the hydrogen HOMO to the iodine LUMO or from the iodine HOMO to the hydrogen LUMO. The symmetries of these valence-shell MO's are the same as those illustrated in Fig. 11.3.

The hydrogen HOMO is a σ_g bonding orbital while the LUMO is a σ_u antibonding orbital. For iodine, the HOMO is a π_g antibonding orbital while the LUMO is a σ_u antibonding orbital from $p\sigma$-$p\sigma$ overlap. The hydrogen HOMO and iodine LUMO are symmetry-incompatible. The hydrogen LUMO-iodine HOMO interaction would be symmetry-compatible but further analysis indicates that this intermediate would not lead to the desired products. As electrons flow into the hydrogen LUMO the H–H bond would indeed weaken, but the I–I bond would strengthen as electrons vacated the π antibonding orbital. Thus no H–I bonds are likely to be formed. Subsequent calculations and experiments confirmed that this was *not* a concerted reaction but rather proceeded through a sequence of steps beginning with the dissociation of I_2 into iodine atoms.

13.3 Band theory of metals and semiconductors

The importance of metals and semiconductors to modern technology is difficult to overestimate. We consider in this Section the band theory of solids, which can account for many of the characteristic properties of these materials.

The LCAO approximation, including the Hückel model, exhibits a "conservation law" for orbitals in which the number of molecular orbitals is equal to the number of constituent atomic orbitals. Consider, for example, a 3-dimensional array of n sodium atoms, each contributing one $3s$ valence electron. Two overlapping AOs will interact to form one bonding plus one antibonding orbital. Three AOs will give, in addition, an orbital of intermediate energy, essentially nonbonding. Continuing the process, as sketched in Fig. 13.5, n interacting AOs will produce a stack of n MOs, with the lower energy orbitals being of predominantly bonding character, the upper ones, of antibonding character. The n electrons will fill half of the available energy levels. As n increases, the spacing between successive levels decreases, until for $n \sim 10^{23}$ the discrete levels merge into a continuous energy band. The valence electrons become delocalized over the entire crystal lattice, consistent with the Drude-Sommerfeld model of a metal as an electron gas surrounding cores of positive ions. This simple model accounts for many of the familiar attributes of metals. High electrical and thermal conductivity are obvious consequences of the large number of mobile electrons. Metals are usually malleable and ductile because metallic bonding, although strong, is nondirectional and tolerant of lattice deformation.

Metals can be usually recognized by their shiny appearance or "metallic luster," their ability of reflect light. The high-frequency electromagnetic fields of light induces oscillations of the loosely-bound electrons near the metal surface. These vibrating charges, in turn, reemit radiation, equivalent to a reflection of the incident light. The closely-spaced energy levels in the conduction bands allow metals (with notable exceptions of copper and gold) to absorb all wavelengths across the visible range.

Elements with a valence-shell configuration ns^2, such as beryllium and magnesium, might be expected to have completely filled bands and thus behave as

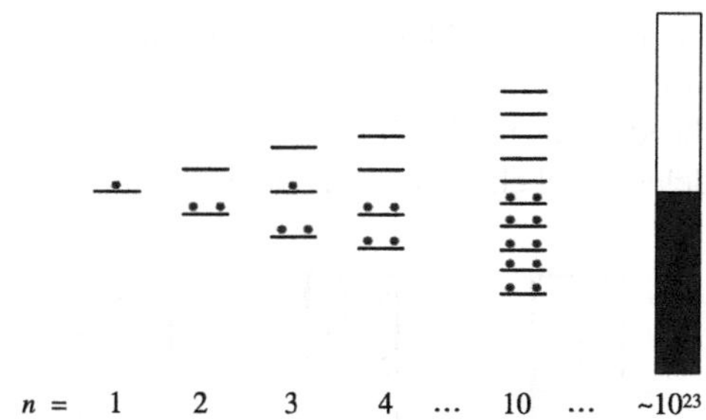

FIGURE 13.5 Approach to continuous energy bands in a crystal with increasing number of interacting atoms n.

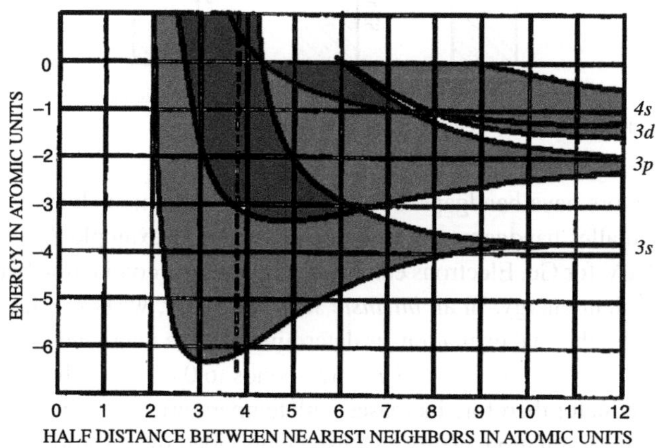

FIGURE 13.6 Energy bands in sodium as function of atomic spacing. After J. C. Slater, *Phys. Rev.* 45 794 (1934). Dotted line represents equilibrium spacing.

nonmetals. However the nearby p-orbitals likewise form a band which overlaps the upper part of the s-band to give a continuous conduction band with an abundance of unoccupied orbitals. Transition metals can also contribute their d-orbitals to the conduction bands. Fig. 13.6 is a detailed plot of the band structure of metallic sodium, which shows how combinations of s, p and d energy bands can overlap. Only the outermost atomic orbitals are involved in band formation. Inner atomic orbitals remain localized and are not involved in bonding or electrical conduction.

The band theory of solids very succinctly describes the essential differences between conductors, insulators and semiconductors, as shown in Fig. 13.7. A metallic conductor possesses either a partially filled valence band or overlapping valence and conduction bands so that electrons can be excited into the empty levels by an external electric field. Energy bands in crystalline solids can be separated by forbidden zones or bandgaps. When the orbitals below a sufficiently-large bandgap are completely filled, the element or compound becomes an insulator. In sodium chloride, for example, the Cl $3s$, $3p$ band is completely filled by the valence electrons and separated by a large gap from the empty Na $3s$ band. Thus the NaCl is an ionic crystal made up of Na^+ and Cl^- units but no delocalized electrons at moderate temperatures.

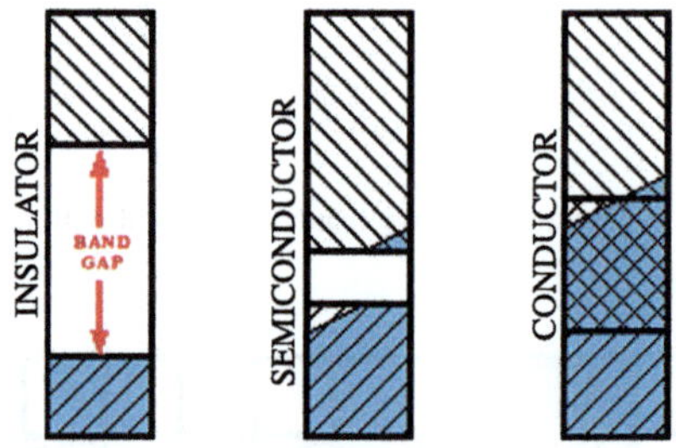

FIGURE 13.7 General features of band structure for insulators, semiconductors and conductors. Blue regions show levels filled by electrons at room temperature.

Good insulators have bandgaps E_g of at least 5 eV. Semiconductors are materials with smaller bandgaps, of the order of 1 eV. For example $E_g = 1.12$ eV for Si, 0.66 eV for Ge. Electrons can be excited into the conduction band if they absorb sufficient energy. In an *intrinsic semiconductor*, weak conductivity can be achieved by thermal excitation, as determined by the magnitude of the Boltzmann factor $e^{-E_g/kT}$. At 300 K, kT corresponds to 0.026 eV. The conductivity of a semiconductor therefore increases with temperature (opposite to the behavior of a metal). Excitation energy can also be provided by absorption of light, so many semiconductors are also photoconductors.

The spectacular success of the semiconductor industry is based on the production of materials selectively designed for specialized applications in electronic and optical devices. By carefully-controlled doping of semiconductors with selected impurities—electron donors or electron acceptors—the conductivity and other properties can be modulated with great precision. Fig. 13.8 shows schematically how doped semiconductors work. In an intrinsic semiconductor (a), conducting electron-hole pairs can only by produced by thermal or photoexcitation across the band gap. In (b), addition of a small concentration of an electron donor creates an impurity band just below the conduction band. Electrons can then jump across a much-reduced gap to the conduction band and act as negatively-charged current carriers. This produces a *n-type semiconductor*. In (c), an electron acceptor creates an empty impurity band just above the valence band. In this case electrons can jump from the valence band to leave positive holes. These can also conduct electricity since electrons falling into positive holes create new holes, a sequence which can propagate across the crystal, in the direction *opposite* to the electron flow. The result is a *p-type semiconductor*.

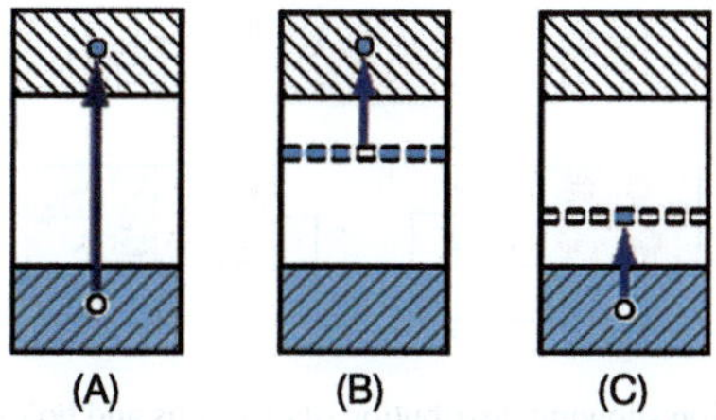

FIGURE 13.8 Band structures for (A) intrinsic semiconductor, (B) n-type, (C), p-type. Filled bands are shown in blue. Arrows show excitations creating electrons and holes.

The most popular semiconductor material is silicon (hence Silicon Valley). Fig. 13.9A is a schematic representation of a pure Si crystal. Each Si atom has 4 valence electrons and bonds to 4 other atoms to form Lewis octets. The crystal can become a conductor if some of the valence electrons are shaken loose. This produces both negative and positive charge carriers—electrons and holes. Much more important are *extrinsic semiconductors* in which the Si crystal is doped with impurity atoms, usually at concentrations of several parts per million (ppm). For example, Si can be doped with P (or As or Sb) atoms, which has 5 valence electrons. As shown in Fig. 13.9B, a P atom can replace a Si atom in the lattice. The fifth electron on the P is not needed for bonding and becomes available as a current carrier. Thus Si doped with P is a n-type semiconductor, as electrons can be excited from the donor band to the conduction band. If Si is instead doped with B (or Ga or Al), which has only 3 valence electrons, as shown in Fig. 13.9C, a B atom replacing a Si atom leaves an electron vacancy in one of its 4 bonds. Such positive holes can likewise become current carriers, making Si doped with B an p-type semiconductor, with electrons excited from the valence band to leave positive holes.

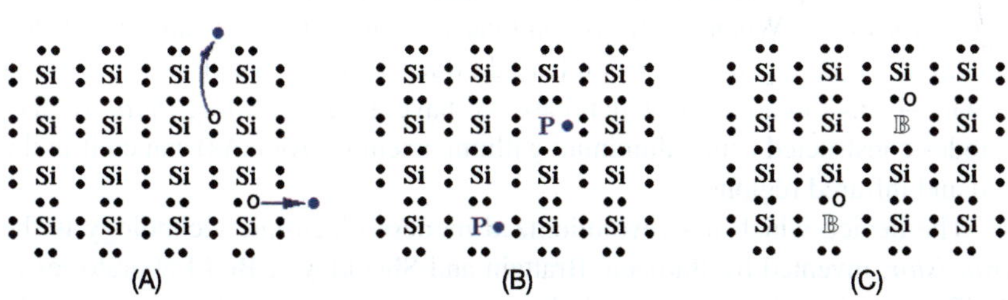

FIGURE 13.9 Lewis structures for pure and doped silicon crystals. (A) Pure silicon showing excitation of two electron-hole pairs. (B) Si doped with P, an electron donor. (C) Si doped with B, an electron acceptor.

Besides the semiconducting elements Si and Ge from Group IV of the periodic table, there exist III-V semiconducting compounds including GaAs, InP and GaN and II-VI compounds such as ZnS and CdTe. GaAs (Gallium arsenide), with a band gap of 1.43 eV, has been especially useful for solar cells, light-emitting diodes, lasers, and other optoelectronic devices.

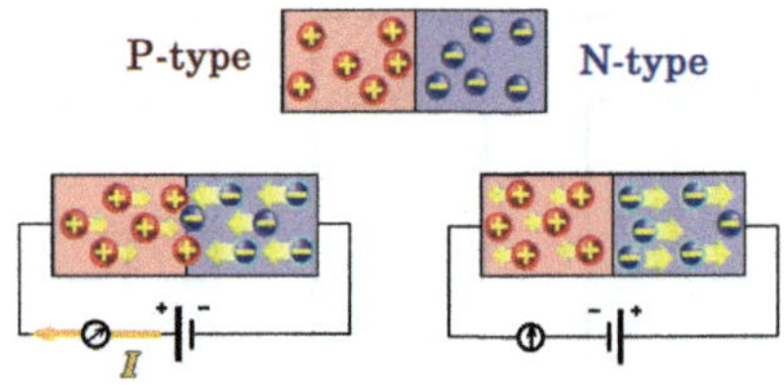

FIGURE 13.10 P-N junction showing distribution of electrons and holes. Current can flow (bottom left) only if the diode is forward biased.

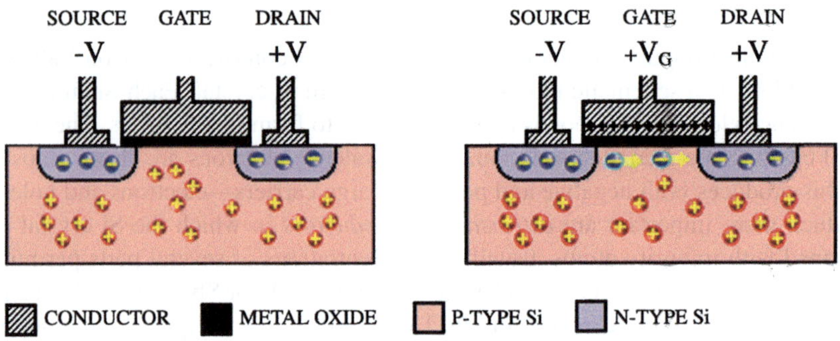

FIGURE 13.11 Principle of metal-oxide-semiconductor field-effect transistor (MOSFET). A narrow conducting channel is created when the gate is at positive voltage, allowing electrons to pass from the source to the drain.

A key element in solid state electronics is the P-N junction, formed when p-type and n-type semiconductors are placed in contact, shown in Fig. 13.10. Electric current can flow through the junction in one direction (forward biased) but not in the opposite direction (reverse biased). The junction can act as a *semiconductor diode*. When electrons combine with holes in a forward biased P-N junction, the bandgap energy is released either as heat—which is usual for Si or Ge—or as radiation, in which case we have a *light-emitting diode* (LED). Diodes constructed using aluminum gallium arsenide (AlGaAs) can emit in the red and infrared regions.

The device which has revolutionized modern electronic technology is the *transistor*, invented by Bardeen, Brattain and Shockley at Bell Laboratories in 1947. In digital circuits, particularly computers, transistors function as high-speed electronic switches. Transistors are building blocks for logic gates, RAM memory and other components in integrated circuits. In analog circuits, transistors are used as amplifiers and oscillators, having replaced vacuum tubes since the 1960s. A transistor is based on a three-layer assembly of doped semiconductors, either NPN or PNP. Most of the transistors manufactured today are *metal-oxide-semiconductor field-effect transistors* (MOSFETs). These have largely supplanted the original bipolar transistors. The leading developers of semiconductor technology have been Bell Laboratories, Fairchild Semiconductor, Texas Instruments and many other companies based in Silicon Valley and Japan. A simplified representation of the operation of a MOSFET is shown in Fig. 13.11. The gate is a metal electrode with a very thin insulating coating of its oxide. When the gate is uncharged, very little current flows between the *source* and the *drain* through the NPN sequence of semiconductor layers. When the gate becomes positively charged, its electric field causes a concentration of electrons to build up in a narrow channel opposite the gate. This closes the circuit between the source and the drain. A small change in the gate voltage can produce a large and rapid variation in the current. The transister can thereby act as a signal amplifier or as a high-speed electronic switch which can open and close several million times a second.

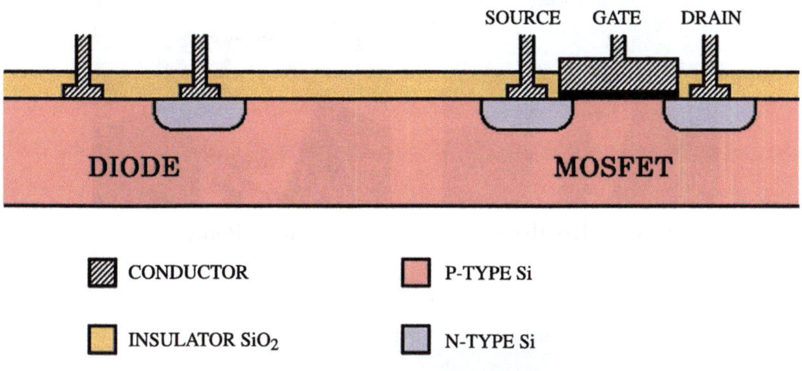

FIGURE 13.12 MOSFET and diode in an integrated circuit. The substrate here is p-type silicon.

Today, most microelectronic technology is based on *integrated circuits* (ICs). These are assemblies containing thousands or millions of microscopic-size electronic components—resistors, capacitors, diodes, transistors—and their connections built into a chip of semiconductor (in most cases, silicon) called the substrate. Fig. 13.12 shows a simplified representation of a MOSFET and a diode as components of an integrated circuit.

13.4 Computational chemistry

There is no question that quantum mechanics provides the correct mathematical framework for description of all chemical phenomena. However its fundamental equation can be solved exactly for only a single problem—the hydrogenlike system. Every other application of quantum mechanics to atoms or molecules involves approximations of varying levels of sophistication. Some of these, although rather rudimentary, can provide useful models for some aspect of chemical behavior. For many purposes, however, it is necessary to seek more accurate solutions to the Schrödinger equation, the goal being "chemical accuracy," which is usually understood to mean an error less than about 1 kJ/mol (approximately .001 hartree). The application of electronic computers to chemical problems during the last third of the Twentieth Century is now considered a specialty in its own right—computational chemistry—taking its place alongside the traditional modes of experimental and theoretical research. Computational chemistry now routinely contributes to the design and synthesis of materials with novel properties, the understanding of complex biological processes and the rational design of therapeutic drugs.

C. C. J. Roothaan

John Pople

Roald Hoffmann

Computational chemistry had its roots in the early attempts to solve the Schrödinger equation for two-electron systems, notably the work of Hylleraas in 1929 and of Coolidge and James in 1933. At its most advanced level, this work was performed using hand-cranked calculating machines. In the 1950s, with the advent of digital computers, accurate computations became possible for molecular systems with up to 5 atoms. By the 1970s, these could be extended to as many as 20 atoms. Spectacular advances in computing technology and algorithms in subsequent years has made it possible now to do computations, to varying levels of rigor, even on biological molecules containing thousands of atoms.

Formulation of the methods of computational chemistry is reasonably straightforward. The hard work comes in its implementation. We begin with the electronic Hamiltonian for a molecule, a generalization of that for a many-electron atom given in Eq. (10.2):

$$H = \sum_i \left\{ -\frac{1}{2}\nabla_i^2 \right\} + \sum_{i\,A} \left\{ -\frac{Z_A}{r_{iA}} \right\} + \sum_{i<j} \frac{1}{r_{ij}} + \sum_{A<B} \frac{Z_A Z_B}{R_{AB}}, \tag{13.14}$$

where

$$r_{ij} = |\mathbf{r}_i - \mathbf{r}_j| \qquad r_{i\alpha} = |\mathbf{r}_i - \mathbf{R}_A| \qquad R_{AB} = |\mathbf{R}_A - \mathbf{R}_B|. \tag{13.15}$$

Here the indices $i,\ j$ run over electrons while $A,\ B$ run over nuclei. The second sum must now account for the attraction of each electron to *multiple* nuclei. The last sum represents repulsion between each pair of nuclei. It acts as a constant term added to the electronic energy. The Born-Oppenheimer approximation is assumed, so that the positions of the nuclei $\mathbf{R}_A$ are fixed with no nuclear kinetic-energy contributions. The Schrödinger equation for the N-electron problem is written symbolically as

$$H\psi(\mathbf{r}_1, \mathbf{r}_2, \ldots \mathbf{r}_N; \mathbf{R}_A, \ldots) = E(\mathbf{R}_A, \ldots)\psi(\mathbf{r}_1, \mathbf{r}_2, \ldots \mathbf{r}_N; \mathbf{R}_A, \ldots) \tag{13.16}$$

In concept, the equation is to be solved for a multitude of different nuclear configurations.

The starting point for most high-level molecular computations is the *Hartree-Fock method.* This is based on a Slater determinant of molecular orbital functions, analogous to Eq. (10.1) for atoms:

$$\psi(1, 2 \ldots N) = \frac{1}{\sqrt{N!}} \begin{vmatrix} \phi_a(1) & \phi_b(1) & \ldots & \phi_n(1) \\ \phi_a(2) & \phi_b(2) & \ldots & \phi_n(2) \\ & & \vdots & \\ \phi_a(N) & \phi_b(N) & \ldots & \phi_n(N) \end{vmatrix}. \tag{13.17}$$

The variational-principle energy for a system with Hamiltonian (13.14) works out to a sum analogous to (10.4) for atoms:

$$E = \sum_a I_a + \frac{1}{2}\sum_{a,b}(J_{ab} - K_{ab}). \tag{13.18}$$

The one-electron, Coulomb and exchange integrals are analogous to (10.5)–(10.7), but in terms of molecular rather than atomic orbitals. (The I_a must now contain contributions from *all* the nuclei in the molecule.) The optimized wavefunction of the form (13.17) involves, in principle, the solution of N simultaneous integrodifferential *Hartree-Fock equations*. It is much more computationally efficient to transform these into a set of N linear algebraic equations. To do this, each of the MO's is expressed in terms of a set of n *basis functions*:

$$\phi_a = c_{a1} f_1 + c_{a2} f_2 + \cdots + c_{an} f_n = \sum_{i=1}^{n} c_{ai} f_i. \tag{13.19}$$

This represents a generalization of the LCAO approximation. In its simplest realization—called a *minimum basis set*—the functions f_i are in fact just AO's centered on the component atoms. More accurate computations require a larger basis set in which a simple correspondence between basis functions and AO's no longer applies. Strictly speaking, a MO and its constituent basis functions are divided into subsets belonging to different symmetry species, which provides a significant reduction in computational effort. We will take up molecular symmetry and group theory in Chap. 15. In terms of the basis functions f_i, the molecular integrals can be expressed as follows:

$$I_a = \sum_{ij} c_{ai}^* c_{aj} [i|j], \qquad J_{ab} = \sum_{ijkl} c_{ai}^* c_{aj} c_{bk}^* c_{bl} [ij|kl],$$

$$K_{ab} = \sum_{ijkl} c_{ai}^* c_{aj} c_{bk}^* c_{bl} [il|kj] \quad (\text{spin}\, a = \text{spin}\, b), \tag{13.20}$$

in terms of *one-electron integrals*

$$[i|j] = \int f_i^*(\mathbf{r}) \{-\frac{1}{2}\nabla^2 - \sum_\alpha \frac{Z_A}{r_A}\} f_j(\mathbf{r})\, d\tau \tag{13.21}$$

and *two-electron integrals*

$$[ij|kl] = \int\int f_i^*(\mathbf{r}_1) f_k^*(\mathbf{r}_2) \frac{1}{r_{12}} f_j(\mathbf{r}_1) f_l(\mathbf{r}_2)\, d\tau_1\, d\tau_2. \tag{13.22}$$

The one-electron integrals are reasonably straightforward (remember a computer is doing the work), as are two-electron integrals in which the four basis functions f_i, f_j, f_k, f_l are centered about no more than *two* different atoms. But it is a real challenge to evaluate *three-* and *four-center* integrals.

In terms of the above integrals, we can define Hamiltonian matrix elements

$$H_{ij} = [i|j] + \sum_b \sum_{kl} c_{bk}^* c_{bl} \Big([ij|kl] - [il|kj] \Big). \tag{13.23}$$

We also need the overlap integrals

$$S_{ij} = \int f_i^* f_j \, d\tau. \tag{13.24}$$

The optimized coefficients c_{ai} for the molecular orbitals ϕ_a are then found by the linear variational method introduced in Sect. 12.4. Specifically this leads to the n simultaneous equations for the coefficients c_{aj} for *each* MO ϕ_a:

$$\sum_i (H_{ij} - \epsilon_a S_{ij}) c_{aj} = 0. \tag{13.25}$$

These are known as the *Roothaan equations*. They represent an algebraic equivalent to the Hartree-Fock equations. The approximate eigenvalues ϵ_a represent *orbital* energies. By Koopmans' theorem, $-\epsilon_a$ approximates the ionization energy for an electron occupying orbital a. The orbital energies can be determined directly from the n roots of the secular equation

$$|H_{ij} - \epsilon S_{ij}| = 0, \tag{13.26}$$

where we have used a common abbreviation for the $n \times n$ determinant. The Roothaan equations are complicated by the fact that the Hamiltonian matrix elements (13.23) contain the coefficients c_{ai}, which depend on solutions of the same equations. This suggests using an iterative procedure. From an initial "guess" of the coefficients, the H_{ij} are computed. The Roothaan equations are then solved to give an "improved" set of coefficients. From these, a new Hamiltonian matrix is computed, followed by a new set of coefficients. And the process is continued until self-consistency of the input and output results is obtained. For this reason, the Hartree-Fock method is also known as the *self-consistent field* (SCF).

Most of the earlier SCF computations, on both atoms and molecules, used basis functions in the form of *Slater-type orbitals* (STO's):

$$f_{n\ell m}(r, \theta, \phi) = r^{n-1} e^{-\zeta_{n\ell} r} Y_{\ell m}(\theta, \phi), \tag{13.27}$$

which have the same general form as hydrogenic atomic orbitals. Experiments with alternative types of basis functions were stimulated largely because of the difficulties in computing 3- and 4-center integrals among the $[ij|kl]$, of which there might be several thousand for a medium-sized molecule. The most successful alternative has turned out to be *gaussian-type orbitals* (GTO's), first suggested by S. F. Boys in the 1950's. One variant of the GTO's are the cartesian gaussian functions

$$g_{nmk}(\mathbf{r}) = (x^n + y^m + z^k) e^{-\alpha r^2}. \tag{13.28}$$

The advantage of GTO's can be seen by considering the following possible 4-center integral involving gaussian functions

$$[ij|kl] = \int\int e^{-\alpha_i r_{1A}^2}\, e^{-\alpha_k r_{2B}^2}\, \frac{1}{r_{12}}\, e^{-\alpha_j r_{1C}^2}\, e^{-\alpha_l r_{2D}^2}\, d\tau_1\, d\tau_2, \tag{13.29}$$

where A, B, C, D represent four different nuclei. The key simplification is that the product of two gaussians around different centers can be reduced to a single gaussian around a new center. Note that

$$e^{-\alpha_i r_{1A}^2}\, e^{-\alpha_j r_{1C}^2} = \exp[-\alpha_i |\mathbf{r}_1 - \mathbf{R}_A|^2 - \alpha_j |\mathbf{r}_1 - \mathbf{R}_C|^2] = \exp[-\alpha |\mathbf{r}_1 - \mathbf{R}|^2 + \beta], \tag{13.30}$$

where α, β and $\mathbf{R}$ are constants. Thus a 4-center integral is, in effect, reduced to a 2-center integral. A gaussian function cannot represent the cusp at the position of a nucleus, which an STO containing $e^{-\zeta r}$ does naturally. However the behavior near the nucleus can be *simulated* by a superposition of n gaussian functions, designated by STO-nG. Fig. 13.13 shows radial distribution functions for contracted gaussians representing the $1s$ STO. The overlap integral is shown for each of the STO-nG functions.

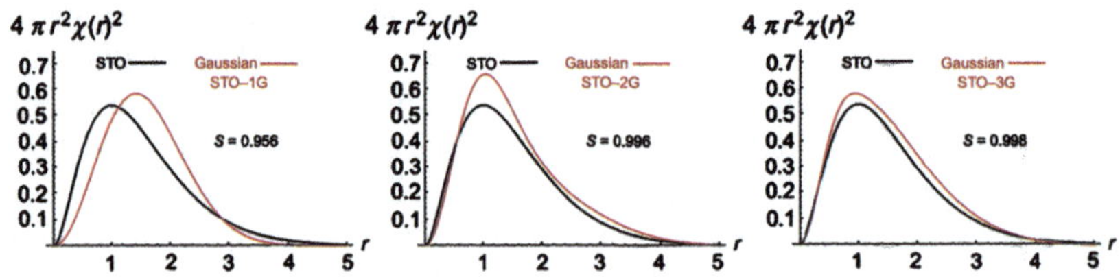

FIGURE 13.13 Approximation of $1s$ STO by gaussian functions.

J. A. Pople and coworkers, in the 1970s, developed suites of software programs called GAUSSIAN (latest version: *Gaussian 16*), which has made molecular computation available to a whole community of chemists, physicists and biologists without special expertise in computational chemistry (Pople shared the 1998 Nobel Prize in Chemistry). Determination of the structure of complex molecules, for example, the anticancer agent taxol (Fig. 13.14) has been one of the more spectacular accomplishments of *Gaussian*.

Several extensions of the HF method, referred to as *post Hartree-Fock methods*, have been developed to enable more accurate computations on atoms and molecules, in particular, to account for the effects of electron correlation. Such improved accuracy is necessary to treat chemically significant properties of molecules, including bond energies and geometric parameters—bond lengths and angles. In addition, there are spectroscopic frequencies, dipole moments, magnetic properties, and NMR coupling constants. These post-HF

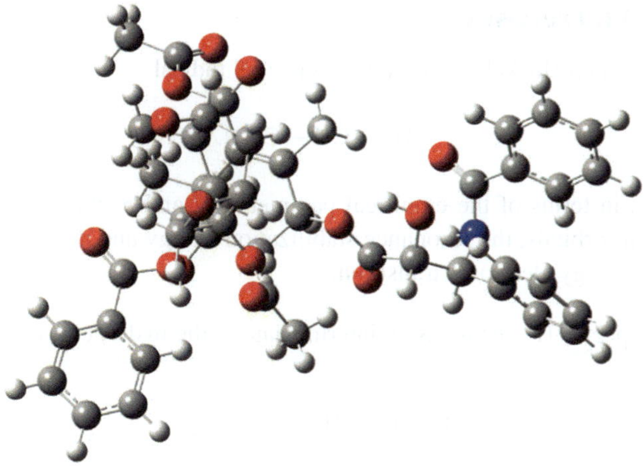

FIGURE 13.14 Optimized structure of taxol, computed using Gaussian 94.

methods are beyond the scope of this book; we simply mention configuration interaction (CI), Møller-Plesset perturbation theory and the coupled-cluster method.

The types of computation discussed thus far can be categorized as *ab initio*, since no empirical information is used other than the identities and connectivities of the constituent atoms. Very useful results can be obtained with far less extensive computational effort by means of *semi-empirical* methods. These are formally based on the Hartree-Fock approach, but most of the requisite integrals are estimated using empirical formulas rather than being calculated. Representative of this category is *extended Hückel theory* (EHT), exploited by Roald Hoffmann (Nobel Prize 1981) in the 1960's to predict the structures of hydrocarbons, boron hydrides and organometallic compounds. EHT generally considers only the valence shells of molecules. Only the overlap integrals S_{ij} are explicitly computed, usually with Slater-type atomic orbitals. Everything else is empirical. Diagonal matrix elements H_{ii} are estimated from atomic ionization energies. Off-diagonal elements of the Hamiltonian are approximated by

$$H_{ij} = \frac{1}{2} K (H_{ii} + H_{jj}), \tag{13.31}$$

where K is an empirical constant, usually taken as 1.75. EHT represents an 'extension' of the simple Hückel method in the sense that valence-shell σ- as well as π-orbitals are considered and overlap integrals are not neglected.

There are a number of packages of *ab initio* and semi-empirical quantum chemistry programs which can be downloaded to personal computers. A very versatile software suite called GAMESS (General Atomic and Molecular Electronic Structure System) can be downloaded at no cost from:
https://www.msg.chem.iastate.edu/GAMESS/download.html

Chapter 13. Exercises

13.1. Carry out a Hückel calculation on the allyl radical

$$CH_2 = CH - CH_2{}^{\circ}.$$

Determine, in terms of the empirical parameters α and β, the energies of the π-molecular orbitals, the resonance stabilization energy and the wavelength of the lowest-energy electronic transition.

13.2. The species H_3 occurs as an intermediate in the hydrogen exchange reaction

$$H_2 + H \rightleftharpoons H_3 \rightleftharpoons H + H_2.$$

Is H_3 a linear or a triangular molecule? For both the linear and equilateral triangular configurations, apply a variant of the Hückel theory based on hydrogen 1s orbitals (rather than carbon $2p$) to predict which has the lower energy. Also predict the shapes of the ions H_3^+ and H_3^-.

13.3. Do a Hückel calculation on the conjugated 4-carbon ring cyclobutadiene. Calculate the π-electron delocalization energy. Comment on the applicability of Hückel's $4N + 2$ rule for aromaticity.

13.4. Do a Hückel computation on the naphthalene molecule:

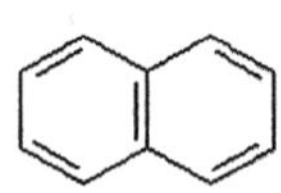

Estimate the resonance energy of naphthalene.

13.5. An isomer of naphthalene is azulene:

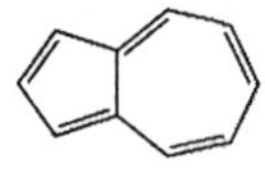

an aromatic system of fused 5- and 7-member rings. Do a Hückel computation on azulene and estimate the wavelength of the lowest-energy electronic excitation. Can you guess the color of the compound?

13.6. Apply the Woodward-Hoffmann rules to the electrocyclic reaction of hexatriene to cyclohexadiene considering the appropriate Hückel MO's. Determine whether the mechanism is conrotatory or disrotatory for both thermal and photochemical reactions.

13.7. If you have access to *ab initio* software such as GAMESS, you might find it amusing to compute which of following possible structures for diborane B_2H_6 is the more stable:

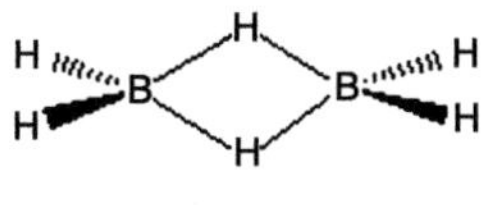

$\mathcal{D}_{2h}$

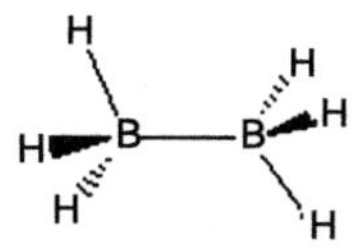

$\mathcal{D}_{3d}$

Chapter 14

Density functional theory

Approximations based on manipulation of wavefunctions have long been the dominant techniques of computational chemistry. In the past few years, however, these are being overtaken by the newly-developed methods of *density functional theory* (DFT). In fact, DFT computations are rapidly becoming the preferred technique for molecular computations in chemistry, physics, materials science and engineering. DFT is based on the total electronic charge density $\rho(\mathbf{r})$, which is more directly related to observable quantities than is the N-electron wavefunction $\psi(\mathbf{r}_1, \mathbf{r}_2 \ldots \mathbf{r}_N)$. The density function can always be determined from the wavefunction using

$$\rho(\mathbf{r}) = N \int \int \cdots \int |\psi(\mathbf{r}, \mathbf{r}_2 \ldots \mathbf{r}_N)|^2 \, ds \, dx_2 \ldots dx_N, \tag{14.1}$$

where $\mathbf{r}_1$ has been singled out for replacement by the unlabeled three-dimensional variable $\mathbf{r}$. Note that integration runs over space *and* spin coordinates of electrons $2, 3 \ldots N$, but only over the spin of the first electron.

14.1 Thomas-Fermi model

This statistical model for the electronic structure of many-body systems, proposed shortly after the introduction of the Schrödinger equation (in 1927), is a precursor to modern density functional theory. The TF model is formulated in terms of the electronic density alone. It is quantitatively correct only in the limit of infinite nuclear charge but does introduce some components of density functional theory, in particular, an expression for the kinetic energy.

The fundamental requisite of the Thomas-Fermi model is that the potential energy $V(r)$ varies slowly enough over an electron wavelength $\lambda = h/p$ that many electrons can be contained in a small element of volume in which the potential changes by a small fraction of its value. Let the element of volume be given by $\Delta V = \lambda^3$ and an associated element of volume in momentum space for a single electron be given by ΔP. Then the volume in phase space is equal to

$$\Delta V \Delta P = h^3. \tag{14.2}$$

The n electrons occupying the element of volume ΔV are then assumed to fill an energy band, similar to that of a solid, with momenta up to a *Fermi momentum* p_F, with two electrons of spins α and β in each level. The total momentum

Introduction to Quantum Mechanics. https://doi.org/10.1016/B978-0-12-822310-9.00022-7

space volume ΔP occupied by n electrons, with momentum from 0 to p_F is given by $2 \times \frac{4\pi}{3} p_F^3$. Thus

$$\Delta V \frac{8\pi}{3} p_F^3 = nh^3. \tag{14.3}$$

The electron density per unit volume, $\rho(r) = n/\Delta V$, is then given by

$$\rho(r) = \frac{8\pi}{3h^3} p_F^3 = \frac{p_F^3}{3\pi^2\hbar^3}. \tag{14.4}$$

The most energetic electrons occupying the volume element ΔV have the Fermi energy $E_F = p_F^2/2m$. This can be equated to $-V(r)$, assuming maximum occupancy of the available energy band. This leads to a relation between electron density and potential energy:

$$\rho(r) = \frac{1}{3\pi^2\hbar^3}[-2mV(r)]^{3/2}. \tag{14.5}$$

The electrostatic potential is also related to the charge density by Poisson's equation (see Eq. (1.23)):

$$\nabla^2 V(r) = \frac{1}{r^2}\frac{d}{dr}\left(r^2 \frac{dV(r)}{dr}\right) = -4\pi\rho(r). \tag{14.6}$$

Eqs. (14.5) and (14.6) provide simultaneous equations for $\rho(r)$ and $V(r)$. For a neutral atom with atomic number Z, we have $V(r) \approx -Ze^2/r$ as $r \to 0$. The behavior of the solution as $r \to \infty$ is not physically accurate.

The differential equation can be reduced to dimensionless form with the definitions

$$r = \left(\frac{3\pi}{4}\right)^{2/3} \frac{\hbar^2}{2me^2 Z^{1/3}} x, \quad V(r) = -\frac{Ze^2}{r}\chi(x). \tag{14.7}$$

With these substitutions, we obtain

$$x^{1/2}\frac{d^2\chi(x)}{dx^2} = \chi(x)^{3/2}, \tag{14.8}$$

with boundary conditions

$$\chi(0) = 1, \quad \text{and} \quad \chi(\infty) = 0. \tag{14.9}$$

The equation cannot be easily solved by any standard method. Bush and Caldwell in 1931 obtained a numerical solution using the then state-of-the-art Differential Analyzer at MIT. The result is plotted in Fig. 14.1. A good approximation was derived by Sommerfeld:

$$\chi(x) \approx \left[1 + \left(\frac{x}{x_0}\right)^{3/\lambda}\right]^{-\lambda}, \quad x_0 = 144^{1/3}, \ \lambda \approx 3.886. \tag{14.10}$$

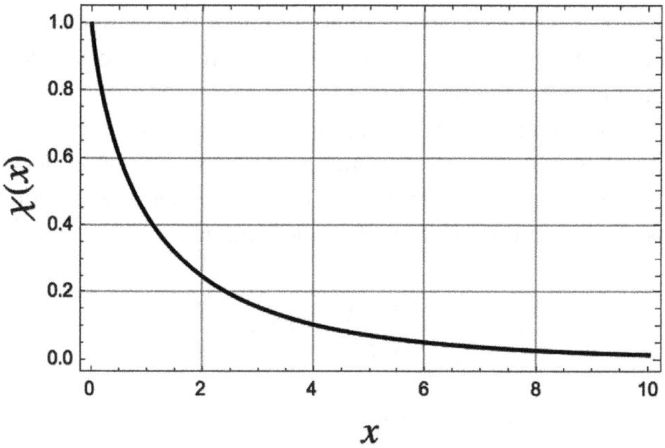

FIGURE 14.1 Solution of the Thomas-Fermi differential equation.

For compactness of subsequent formulas, we will now be using atomic units $\hbar = m = e = 1$. For example, we now write $x = bZ^{1/3}r$, with $b = 2(4/3\pi)^{2/3} \approx 1.13$.

The energy of the system can be determined in terms of the density function $\rho(r)$. The potential energy terms are simple transcriptions of the classical results:

$$E^V = -Z \int \frac{\rho(r)}{r} d^3\mathbf{r} \tag{14.11}$$

for the nuclear attractive energy and

$$E^J = \frac{1}{2} \int\int \frac{\rho(r_1)\rho(r_2)}{r_{12}} d^3\mathbf{r}_1 d^3\mathbf{r}_2 \tag{14.12}$$

for the interelectronic repulsion energy. The kinetic energy is represented by application of the *local density approximation* (LDA) from the theory of a free electron gas. The kinetic energy of electrons filling a unit of phase space $d^3\mathbf{p}\, d^3\mathbf{r}/h^3$ to the Fermi level p_F is given by

$$E^T = \int \left(\frac{2}{h^3} \int_0^{p_F} \frac{p^2}{2m} 4\pi p^2 \, dp \right) d^3\mathbf{r}. \tag{14.13}$$

Note the factor of 2 for the double occupancy. With atomic units $m = 1$ and $h = 2\pi\hbar = 2\pi$, we evaluate the integral over p:

$$\frac{1}{(2\pi)^3} \int_0^{p_F} p^2 \, 4\pi p^2 \, dp = \frac{1}{10\pi^2} p_F^5. \tag{14.14}$$

Using (14.4), the relation of the density $\rho(r)$ to p_f, we obtain

$$E^T = \frac{3}{10}(3\pi^2)^{2/3} \int \rho(r)^{5/3} \, d^3\mathbf{r}. \tag{14.15}$$

This result is necessarily an approximation within the LDA, since it is known that the kinetic energy in quantum mechanics actually involves derivatives of functions, and is thus not strictly local.

The Coulomb integral (14.12) for the interelectronic interaction does not take account of the exchange contribution and, in particular, contains spurious electron self interactions. A term for the exchange energy was added by Dirac in 1928, also based on the LDA in a free electron gas:

$$E^X = -\frac{9}{8}\left(\frac{3}{\pi}\right)^{1/3} \int \rho(r)^{4/3}\, d^3\mathbf{r}. \tag{14.16}$$

This elaboration of the theory is known as the *Thomas-Fermi-Dirac model.*

A rough approximation to the density function, from solution of (14.8), is a simple exponential function $\rho(r) = \text{const}\, e^{-cZ^{1/3}r}$. This function needs to be normalized such that

$$\int \rho(r)\, d^3\mathbf{r} = N, \tag{14.17}$$

where N is the total number of electrons, equal to Z for a neutral atom. The normalized approximate function we cited can then be written

$$\rho(r) \approx \frac{c^3 Z^2}{8\pi} e^{-cZ^{1/3}r}. \tag{14.18}$$

The maximum of the radial distribution function $4\pi r^2 \rho(r)$ occur at $r \approx 1.8\, Z^{-1/3}$, giving a rough approximation of the radius of a neutral atom of atomic number Z. As a very rough approximation to the total energy of an atom, consider the potential energy of Z electrons, at an average distance of $Z^{-1/3}$ from a nucleus of charge Z. The potential energy is approximately $-Z^2/Z^{-1/3} = -Z^{7/3}$. A more accurate calculation give a numerical result $E \approx -0.76\, Z^{7/3}$ hartrees. Fig. 14.2 shows the experimental energies of atoms from $Z = 1$ to 54, along with a plot of the TFD predictions (red curve). (A more accurate fit of experimental atomic energies gives $E \approx -0.59\, Z^{7/3}$ hartrees.)

Although the Thomas-Fermi method can give decent energy results for atoms, it was found by Teller that no molecules based on the model can ever be stable, the dissociated energy of the atoms always having a lower energy.

14.2 The Hohenberg-Kohn theorems

The conceptual basis of DFT is a remarkable pair of theorems due to Hohenberg and Kohn (1964) which is, in a sense the converse relation to (14.1). (Walter Kohn shared the 1998 Nobel Prize with John Pople.) We give a simplified statement of the first Hohenberg-Kohn theorem:

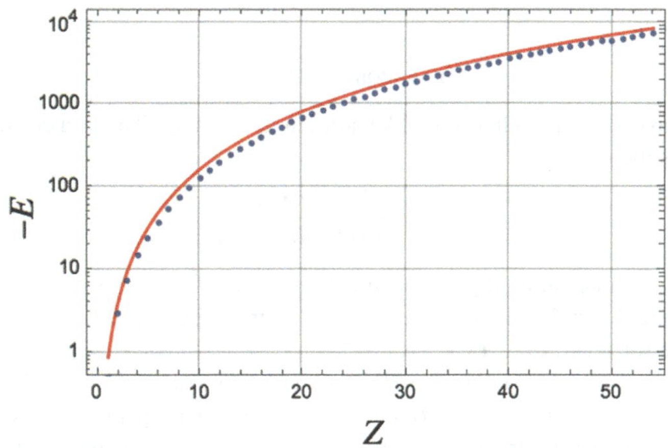

FIGURE 14.2 Energies of atoms Z = 1 to 54, with red curve showing TFD results.

Pierre Hohenberg

Walter Kohn

If the density function $\rho(\mathbf{r})$ for the ground state of a quantum system is known, then the N-electron wavefunction $\psi(\mathbf{r}_1, \mathbf{r}_2 \ldots \mathbf{r}_N)$ is, in principle, determined.

Since the wavefunction leads to *all* other observable quantities (energy, etc.), knowledge of the density is tantamount to a complete description of the quantum system. To prove the HK theorem we note first that the number of electrons in the system is determined by integrating the density, since

$$\int \rho(\mathbf{r})\, d^3\mathbf{r} = N \tag{14.19}$$

by completing the integration of the normalized wavefunction in (14.1). At the position of every nucleus, the density exhibits a *cusp* which has the functional dependence

$$\rho(\mathbf{r}) \approx \text{const}\, e^{-2Z_A|\mathbf{r}-\mathbf{R}_A|} \tag{14.20}$$

(this is easy to see for a hydrogenlike atom in its $1s$ state). This is a consequence of Kato's theorem:

$$Z_A = -\frac{1}{2\rho(\mathbf{r})} \frac{d\rho(\mathbf{r})}{dr}\bigg|_{r\to\mathbf{R}_A}. \tag{14.21}$$

Knowing the nuclear configuration and the number of electrons, we can, in principle, write down the Hamiltonian and the Schrödinger equation. And then we can solve for $\psi(\mathbf{r}_1, \mathbf{r}_2 \ldots \mathbf{r}_N)$.

The cautionary phrase in the HK theorem is *in principle*. It is also true in principle that a solution of the N-electron Schrödinger equation exists, but just try to find it! But the real breakthrough in DFT is the fact that we now need deal with a function of only one 3-dimensional variable, rather than N. The electronic energy is a *functional* of the density, which we write

$$E = E[\rho] \tag{14.22}$$

By *functional* we understand a function which depends on the form of another function—loosely, "a function of a function." In the present context, a functional can be considered a recipe for extracting a single number from a function. For example the variational principle involves a functional of the wavefunction, $E = E[\psi]$. A more elegant formulation of the first Hohenberg-Kohn theorem is the statement: *the wavefunction is a unique functional of the density*.

The second Hohenberg-Kohn theorem is a variational principle for the density functional, requiring that

$$E_0 \leq E[\tilde{\rho}] \tag{14.23}$$

for trial density functions $\tilde{\rho}$ which satisfy the condition (14.2).

14.3 Density functional theory

The fundamental quantity is the energy considered as a functional of the electron density: $E[\rho]$. This is conveniently divided into four contributions:

$$E[\rho] = E^K[\rho] + E^V[\rho] + E^J[\rho] + E^{XC}[\rho]. \tag{14.24}$$

An exact form for the kinetic-energy functional is not known but, as a first approximation,

$$E^K[\rho] = \frac{3}{10}(3\pi^2)^{2/3} \int \rho(\mathbf{r})^{5/3}\, d^3\mathbf{r}, \tag{14.25}$$

which is suggested by the local density approximation used in the Thomas-Fermi atomic model. A correction to $E^K[\rho]$ which considers also the *gradient* of the density is the *Weizsacker correction*:

$$\Delta E_W^K[\rho] = \frac{\lambda}{8} \int \frac{|\nabla \rho(\mathbf{r})|^2}{\rho(\mathbf{r})} d^3\mathbf{r}, \tag{14.26}$$

where λ is an empirical constant. The potential energy of nuclear-electronic and internuclear interactions in a molecule is given by

$$E^V[\rho] = -\sum_A Z_A \int \frac{\rho(\mathbf{r})}{|\mathbf{r} - \mathbf{R}_A|} d^3\mathbf{r} + \sum_{A<B} \frac{Z_A Z_B}{R_{AB}}, \tag{14.27}$$

while the electron-electron potential energy is

$$E^J[\rho] = \frac{1}{2} \int\int \frac{\rho(\mathbf{r}_1)\rho(\mathbf{r}_2)}{r_{12}} d^3\mathbf{r}_1 d^3\mathbf{r}_2. \tag{14.28}$$

This has to be corrected by subtracting out electron self-interactions, which is done by the exchange potential. Both potential-energy parts have analogs in classical electromagnetic theory. The exchange-correlation functional is the most challenging. It is entirely of quantum-mechanical origin—the Pauli exclusion principle and electron correlation. An early approximation to the exchange part of this contribution, introduced by Slater, is known as the X-alpha functional, to be discussed in the following Section.

$$E^X[\rho] = -\frac{3\alpha}{2} \left(\frac{3}{\pi}\right)^{1/3} \int \rho(\mathbf{r})^{4/3} d^3\mathbf{r} \tag{14.29}$$

where α is another empirical constant. The Lieb-Oxford bound provides an upper limit to the exact exchange-correlation energy, namely

$$E^{XC}[\rho] \geq -1.68 \int \rho(\mathbf{r})^{4/3} d^3\mathbf{r}. \tag{14.30}$$

By further manipulation of both functional forms and empirical parameters, which we will not describe in detail, a very successful formulation of DFT has been realized. The currently favored version is designated B3LYP (after Becke 3-parameter, Lee, Yang and Parr).

14.4 Slater's X-alpha method

A precursor of the Kohn-Sham equations, to be considered in the next Section, was Slater's X-alpha method (1951) for simplifying the Hartree-Fock equations. Recall from Sects. 10.2 and 13.4 that the Hartree-Fock method is based on a variational optimization of the atomic or molecular orbitals in a Slater determinant which approximates an N-electron wavefunction. The computational problem reduces to the solution of a coupled set of N integrodifferential equations. For an atom, we have the Hartree-Fock equations

$$-\frac{1}{2}\nabla^2\phi_a(x) - \frac{Z}{r}\phi_a(x) + \Big[\sum_b \int \frac{|\phi_b(x')|^2}{|\mathbf{r}-\mathbf{r}'|}dx'\Big]\phi_a(x)$$
$$-\sum_b \Big[\int \frac{\phi_b^*(x')\phi_a(x')}{|\mathbf{r}-\mathbf{r}'|}dx'\Big]\phi_b(x) = \epsilon_a\phi_a(x), \qquad (14.31)$$

for each of the N spinorbitals $a, b, \ldots, n$. A major complication comes from the exchange terms in the last sum which keeps the equation from being just an effective eigenvalue equation for the spinorbital $\phi_a(x)$. This is spoiled by the terms multiplying the $\phi_b(x)$, with $b \neq a$. The X-alpha method seeks to remedy this by replacing the last sum by an effective term which acts directly on $\phi_a(x)$ alone.

Note that the summations over b include the contribution $b = a$. Thus the first summation, representing the mutual interelectronic Coulomb interactions, if not corrected, would include a self-interaction of spinorbital ϕ_a with itself. This is corrected by the $b = a$ term in the second summation, which cancels the spurious self-interaction. The rest of the second sum represents the exchange interactions of ϕ_a with all the other orbitals of the same spin. The reduction of the spurious electron density can be approximated by imagining the existence of a *Fermi hole* around each electron, from which electrons of the *same* spin are partially excluded. The total charge excluded by the Fermi hole should equal that of one electron, so that the first sum in the Hartree-Fock equation is reduced by one term, thus eliminating the self-interaction. Assuming that the Fermi hole is spherical, with radius r_F, the excluded charge is given by $\frac{4\pi}{3}r_F^3|\rho(r)|$, where $\rho(r) = -|\phi(x)|^2$. Setting the excluded charge equal to 1 (electron charge), we find

$$r_F = (3/4\pi)^{1/3}|\rho(r)|^{-1/3}. \qquad (14.32)$$

We assume that within the Fermi hole that the charge density of the appropriate spin, which can be approximated as $\frac{1}{2}\rho(r)$, is spherical about the electron. The exchange potential energy on electrons within the Fermi hole can be approximated as

$$V^X \approx -\int_0^{r_F} \frac{\rho(r)/2}{|\mathbf{r}-\mathbf{r}'|}4\pi r'^2\,dr' \approx -\text{const}\,\rho(r)r_F^2 \approx -\text{const}\,\rho(r)^{1/3}. \quad (14.33)$$

Slater found const$=\frac{2}{3}\left(\frac{3}{\pi}\right)^{1/3}$. Alternative derivations, including that of Kohn and Sham, found a factor 1 rather than $\frac{2}{3}$. Since the result is an approximation in any case, an empirical factor α is included in the constant. Thus the X-alpha method results in Hartree-Fock equations for the spinorbitals ϕ_a in the form:

$$\left[-\frac{1}{2}\nabla^2 - \frac{Z}{r} + \int \frac{\rho(x')}{|\mathbf{r}-\mathbf{r}'|}dx' - \frac{3\alpha}{2}\left(\frac{3}{\pi}\right)^{1/3} \rho(x)^{1/3}\right]\phi_a(x) = \epsilon_a\phi_a(x). \tag{14.34}$$

14.5 The Kohn-Sham equations

Kohn and Sham (1965) developed an alternative computational scheme which is a hybrid between DFT and the Hartree-Fock method. In particular, the complicated many-electron potential in the Hartree-Fock equation for a spinorbital is replaced by an effective potential $V_{\text{KS}}(\mathbf{r})$ incorporating Coulomb, exchange and correlation effects. The actual many-electron system is thereby represented by an equivalent system with independent particles, such that a single Slater determinant gives the exact solution. A major advantage of the KS method is that it is no longer necessary to approximate the kinetic energy by local density type approximations. The kinetic energy of a spinorbital is now represented by the correct quantum mechanical operator $-\frac{\hbar^2}{2m}\nabla^2$. The Hartree-Fock-Kohn-Sham equation for a spinorbital $\phi_n(\mathbf{r})$ (in atomic units) can be written:

$$\left\{-\frac{1}{2}\nabla^2 + V_{\text{KS}}(\mathbf{r})\right\}\phi_n(\mathbf{r}) = \epsilon_n\phi_n(\mathbf{r}), \tag{14.35}$$

where

$$V_{\text{KS}}(\mathbf{r}) = -\sum_A \frac{Z_A}{|\mathbf{r}-\mathbf{R}_A|} + V^J[\rho] + V^{XC}[\rho]. \tag{14.36}$$

The last functional includes correlation (absent in HF) as well as exchange.

A quite successful empirical approximation to the exchange-correlation potential was found by Gunnarsson and Lundqvist (1976):

$$V^{XC}[\rho] = -0.458/r_s - 0.0666\, G(r_s/11.4), \tag{14.37}$$

where

$$G(x) = \frac{1}{2}\left[(1+x^3)\log(1+x^{-1}) - x^2 + \frac{x}{2} - \frac{1}{3}\right]. \tag{14.38}$$

Here, r_s is the Wigner-Seitz radius,

$$r_s = \left(\frac{3}{4\pi|\rho(\mathbf{r})|}\right)^{1/3}, \tag{14.39}$$

the radius of a spherical volume containing an average of one electron.

At each stage of an iterative computation, the density is computed using

$$\rho(\mathbf{r}) = \sum_a |\psi_a(\mathbf{r})|^2 \tag{14.40}$$

The exchange-correlation potential is formally related to the exchange-correlation energy in by a *functional derivative*

$$V^{XC}[\rho] = \frac{\delta E^{XC}[\rho]}{\delta\rho} \equiv \lim_{\Delta\rho\to 0} \frac{E^{XC}[\rho+\Delta\rho] - E^{XC}[\rho]}{\Delta\rho} \tag{14.41}$$

An intensive search for a “divine functional” to accurately represent exchange and correlation continues to be actively being pursued by computational chemists.

14.6 Chemical potential

Some quantities in density-functional theory suggest analogies with chemical and thermodynamic concepts. For example, a *chemical potential* can be defined as the derivative of energy with respect to electron number

$$\mu = \left(\frac{\partial E}{\partial N}\right)_V \tag{14.42}$$

with constant potential energy V, meaning no change in nuclear configuration. Since N actually can change only by integer steps, the physical significance of μ resides in its finite-difference analog. Thus, applied to a chemical species X

$$\mu \approx \frac{E(X^+) - E(X^-)}{2} = -\frac{I + A}{2} = -\chi \tag{14.43}$$

which gives a intuitively reasonable identification of electronegativity with the negative of chemical potential. Analogously, the derivative of chemical potential gives

$$\left(\frac{\partial\mu}{\partial N}\right)_V = \left(\frac{\partial^2 E}{\partial N^2}\right)_V \approx E(X^+) - 2E(X) + E(X^-) = I - A = 2\eta \tag{14.44}$$

where η is known as *hardness*. This is a measure of energy *curvature*, and thus describes resistance to chemical change.

Chapter 14. Exercises

14.1. Apply the Slater X-alpha method to compute the ionization energy of the helium atom. Determine the value of α which gives the experimental value $\epsilon_{1s} = -0.9037$, corresponding to $I_1 = 24.59$ eV. Assume that the optimized orbital function is given by $\phi(r) = \frac{\zeta^{3/2}}{\sqrt{\pi}} e^{-\zeta r}$, with $\zeta = Z - \frac{5}{16} = 1.6875$. Make use of the results in Sect. 9.2.

Chapter 15

Molecular symmetry

In many cases, the symmetry of a molecule provides a great deal of information about its quantum states and allowed transitions, even without explicit solution of the Schrödinger equation. A geometrical transformation which turns a molecule into an indistinguishable copy of itself is called a *symmetry operation*. A symmetry operation can consist of a rotation about an axis, a reflection in a plane, an inversion through a point, or some combination of these. In this Chapter, we will consider in detail the symmetry groups of ammonia and water, C_{3v} and C_{2v}, respectively.

15.1 The ammonia molecule

We shall introduce the concepts of symmetry and group theory by considering a concrete example–the ammonia molecule NH_3. In any symmetry operation on NH_3, the nitrogen atom remains fixed but the hydrogen atoms can be permuted in 3!=6 different ways. The axis of the molecule is called a C_3 axis, since the molecule can be rotated about it into 3 equivalent orientations, 120° apart. More generally, a C_n axis has n equivalent orientations, separated by $2\pi/n$ radians. The axis of highest symmetry in a molecule is called the *principal axis*. Three mirror planes, designated $\sigma_1, \sigma_2, \sigma_3$, run through the principal axis in ammonia. These are designated as σ_v or *vertical* planes of symmetry. Ammonia belongs to the symmetry group designated C_{3v}, characterized by a three-fold axis with three vertical planes of symmetry.

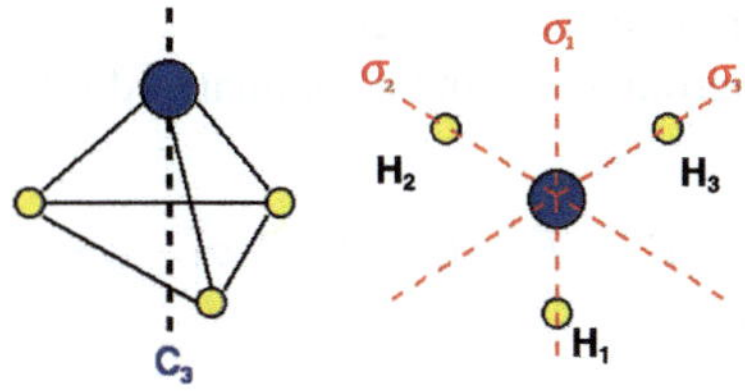

FIGURE 15.1 Two views of the ammonia molecule.

Let us designate the orientation of the three hydrogen atoms in Fig. 15.1 as $\{1, 2, 3\}$, reading in clockwise order from the bottom. A counterclockwise rotation by 120°, designated by the operator C_3, produces the orientation $\{2, 3, 1\}$. A second counterclockwise rotation, designated C_3^2, produces gives $\{3, 1, 2\}$. Note that two successive counterclockwise rotations by 120° is equivalent to

Introduction to Quantum Mechanics. https://doi.org/10.1016/B978-0-12-822310-9.00023-9

one clockwise rotation by 120°, so the last operation could also be designated C_3^{-1}. The three reflection operations $\sigma_1, \sigma_2, \sigma_3$ applied to the original configuration $\{1,2,3\}$ produces $\{1,3,2\}$, $\{3,2,1\}$ and $\{2,1,3\}$, respectively. Finally, we must include the identity operation, designated E, which leaves an orientation unchanged. The effects of the six possible operations of the symmetry group C_{3v} can be summarized as follows:

$$\begin{aligned} E\{1,2,3\} &= \{1,2,3\}, & C_3\{1,2,3\} &= \{2,3,1\}, \\ C_3^2\{1,2,3\} &= \{3,1,2\}, & \sigma_1\{1,2,3\} &= \{1,3,2\}, \\ \sigma_2\{1,2,3\} &= \{3,2,1\}, & \sigma_3\{1,2,3\} &= \{2,1,3\}. \end{aligned} \tag{15.1}$$

This accounts for all 6 possible permutations of the three hydrogen atoms.

The successive application of two symmetry operations is equivalent to some single symmetry operation. For example, applying C_3, then σ_1 to our starting orientation, we have

$$\sigma_1\, C_3\{1,2,3\} = \sigma_1\{2,3,1\} = \{2,1,3\}. \tag{15.2}$$

But this is equivalent to the single operation σ_3. This can be represented as an algebraic relation among symmetry operators

$$\sigma_1\, C_3 = \sigma_3 \tag{15.3}$$

Note that successive operations are applied in the order *right to left* when represented algebraically. For the same two operations in reversed order, we find

$$C_3\,\sigma_1\{1,2,3\} = C_3\{1,3,2\} = \{3,2,1\} = \sigma_2\{1,2,3\}. \tag{15.4}$$

Thus symmetry operations do *not*, in general commute

$$A\,B \not\equiv B\,A, \tag{15.5}$$

although they *may* commute, as do C_3 and C_3^2.

The algebra of the group C_{3v} can be summarized by the following multiplication table.

2nd \ 1st	E	C_3	C_3^2	σ_1	σ_2	σ_3
E	E	C_3	C_3^2	σ_1	σ_2	σ_3
C_3	C_3	C_3^2	E	σ_3	σ_1	σ_2
C_3^2	C_3^2	E	C_3	σ_2	σ_3	σ_1
σ_1	σ_1	σ_2	σ_3	E	C_3	C_3^2
σ_2	σ_2	σ_3	σ_1	C_3^2	E	C_3
σ_3	σ_3	σ_1	σ_2	C_3	C_3^2	E

(15.6)

Notice that each operation appears exactly once, and only once, in each row and in each column.

15.2 Mathematical theory of groups

In mathematics, a *group* is defined as a set of h elements $\mathcal{G} \equiv \{G_1, G_2 \ldots G_h\}$ together with a rule for combination of elements, which we usually refer to as a *product*. The elements of a group must fulfill the following four conditions:

1. The product of any two elements of the group gives another element of the group. That is, $G_i G_j = G_k$ with $G_k \in \mathcal{G}$.
2. Group multiplication obeys an associative law, $G_i(G_j G_k) = (G_i G_j)G_k \equiv G_i G_j G_k$.
3. There exists an *identity element* E such that $EG_i = G_i E = G_i$ for all G_i.
4. Every element G_i has a unique inverse G_i^{-1}, such that $G_i G_i^{-1} = G_i^{-1} G_i = E$ with $G_i^{-1} \in \mathcal{G}$.

The number of elements h is called the *order* of the group. Thus C_{3v} is a group of order 6.

A set of quantities which obeys the group multiplication table is called a *representation* of the group. Because of the possible noncommutativity of group elements, simple numbers are not always adequate to represent groups; we must often use matrices. The group C_{3v} has three *irreducible representations*, or IR's, which cannot be broken down into simpler representations. A trivial, but nonetheless important, representation of every group is the *totally symmetric representation*, in which each group element is represented by 1. The multiplication table then simply reiterates that $1 \times 1 = 1$. For C_{3v} this is called the A_1 representation:

$$A_1 : E = 1,\ C_3 = 1,\ C_3^2 = 1,\ \sigma_1 = 1,\ \sigma_2 = 1,\ \sigma_3 = 1. \tag{15.7}$$

A slightly less trivial representation is A_2:

$$A_2 : E = 1,\ C_3 = 1,\ C_3^2 = 1,\ \sigma_1 = -1,\ \sigma_2 = -1,\ \sigma_3 = -1. \tag{15.8}$$

Much more exciting is the E representation, which requires 2×2 matrices:

$$\begin{aligned}
E &= \begin{pmatrix} 1 & 0 \\ 0 & 1 \end{pmatrix}, \quad C_3 = \begin{pmatrix} -1/2 & -\sqrt{3}/2 \\ \sqrt{3}/2 & -1/2 \end{pmatrix}, \\
C_3^2 &= \begin{pmatrix} -1/2 & \sqrt{3}/2 \\ -\sqrt{3}/2 & -1/2 \end{pmatrix}, \quad \sigma_1 = \begin{pmatrix} -1 & 0 \\ 0 & 1 \end{pmatrix}, \\
\sigma_2 &= \begin{pmatrix} 1/2 & -\sqrt{3}/2 \\ -\sqrt{3}/2 & -1/2 \end{pmatrix}, \quad \sigma_3 = \begin{pmatrix} 1/2 & \sqrt{3}/2 \\ \sqrt{3}/2 & -1/2 \end{pmatrix}.
\end{aligned} \tag{15.9}$$

The operations C_3 and C_3^2 are said to belong to the same *class* since they perform the same geometric operation, but oriented differently in space. Analogously, σ_1, σ_2 and σ_3 are obviously in the same class. E is in a class by itself. The class structure of the group is designated by $\{E, 2C_3, 3\sigma_v\}$. We state without proof that the number of irreducible representations of a group is equal to the number of classes. Another important theorem states that the sum of the squares of the dimensionalities of the irreducible representations of a group adds up to the order of the group. Thus, for C_{3v}, we find $1^2 + 1^2 + 2^2 = 6$.

The *trace* or *character* of a matrix is defined as the sum of the elements along the main diagonal:

$$\chi(M) \equiv \sum_k M_{kk}. \tag{15.10}$$

For many purposes, it suffices to know just the characters of a matrix representation of a group, rather than the complete matrices. For example, the characters for the E representation of C_{3v} in Eq. (15.9) are given by

$$\begin{aligned} \chi(E) = 2, \quad \chi(C_3) = -1, \quad \chi(C_3^2) = -1, \\ \chi(\sigma_1) = 0, \quad \chi(\sigma_2) = 0, \quad \chi(\sigma_3) = 0. \end{aligned} \tag{15.11}$$

It is true in general that the characters for all operations in the same class are equal. Thus, in abbreviated form

$$\chi(E) = 2, \quad \chi(C_3) = -1, \quad \chi(\sigma_v) = 0. \tag{15.12}$$

For one-dimensional representations, such as A_1 and A_2, the characters are equal to the matrices themselves, so Eqs. (15.7) and (15.8) can be directly read as character tables.

The essential information about a symmetry group is summarized in its *character table*. The character table for C_{3v} is shown here:

	E	2 C_3	3 σ_v		
A_1	1	1	1	z	x^2+y^2, z^2
A_2	1	1	−1	R_z	
E	2	−1	0	$(R_x, R_y), (x, y)$	$(x^2-y^2, xy), (xz, yz)$

The last two columns show how the cartesian coordinates x, y, z, combinations of cartesian coordinates and rotations R_x, R_y, R_z transform under the operations of the group.

15.3 Group theory in quantum mechanics

When a molecule has the symmetry of a group $\mathcal{G}$, this means that each member of the group commutes with the molecular Hamiltonian

$$[G_i, H] = 0, \qquad i = 1 \ldots h, \tag{15.13}$$

where we now explicitly designate the group elements G_i as operators on wavefunctions. As was shown in Chap. 4, commuting operators can have simultaneous eigenfunctions. A representation of the group of dimension d means that there must exist a set of d degenerate eigenfunctions of H that transform among themselves in accord with the corresponding matrix representation. For example, if the eigenvalue E_n is d-fold degenerate, the commutation conditions imply that, for $i = 1 \ldots h$,

$$G_i\, H\, \psi_{nk} = H\, G_i\, \psi_{nk} = E_n\, G_i\, \psi_{nk} \quad \text{for} \quad k = 1 \ldots d. \tag{15.14}$$

Thus each $G_i\, \psi_{nk}$ is also an eigenfunction of H with the same eigenvalue E_n, and must therefore be representable as a linear combination of the eigenfunctions ψ_{nk}. More precisely, the eigenfunctions transform among themselves according to

$$G_i\, \psi_{nk} = \sum_{m=1}^{d} D(G_i)_{km} \psi_{nm}, \tag{15.15}$$

where $D(G_i)_{km}$ means the $\{k, m\}$ element of the matrix representing the operator G_i.

The character of the identity operation E immediately shows the degeneracy of the eigenvalues of that symmetry. The C_{3v} character table reveals that NH_3, and other molecules of the same symmetry, can have only nondegenerate and two-fold degenerate energy levels. The following notation for symmetry species was introduced by Mulliken:

1. One dimensional representations are designated either A or B. Those symmetric wrt rotation by $2\pi/n$ about the C_n principal axis are labeled A, while those antisymmetric are labeled B.
2. Two dimensional representations are designated E; 3, 4 and 5 dimensional representations are designated T, F and G, respectively. These latter cases occur only in groups of high symmetry: cubic, octahedral and icosahedral.
3. In groups with a center of inversion, the subscripts g and u indicate even and odd parity, respectively.
4. Subscripts 1 and 2 indicate symmetry and antisymmetry, respectively, wrt a C_2 axis perpendicular to C_n, or to a σ_v plane.
5. Primes and double primes indicate symmetry and antisymmetry to a σ_h plane.

For individual orbitals, the lower case analogs of the symmetry designations are used. For example, MOs in ammonia are classified a_1, a_2 and e.

15.4 Molecular orbitals for ammonia

For ammonia and other C_{3v} molecules, there exist three species of eigenfunctions. Those belonging to the classification A_1 are transformed into themselves by all symmetry operations of the group. The $1s$, $2s$ and $2p_z$ AOs on nitrogen are in this category. The z-axis is taken as the 3-fold axis. There are no low-lying orbitals belonging to A_2. The nitrogen $2p_x$ and $2p_y$ AOs form a two-dimensional representation of the group C_{3v}. That is to say, any of the six operations of the group transforms either one of these AOs into a linear combination of the two, with coefficients given by the matrices (13.8). The three hydrogen $1s$ orbitals transform like a 3×3 representation of the group. If we represent the hydrogens by a column vector $\{H_1, H_2, H_3\}$, then the six group operations generate the following algebra

$$E = \begin{pmatrix} 1 & 0 & 0 \\ 0 & 1 & 0 \\ 0 & 0 & 1 \end{pmatrix}, \quad C_3 = \begin{pmatrix} 0 & 1 & 0 \\ 0 & 0 & 1 \\ 1 & 0 & 0 \end{pmatrix}, \quad C_3^2 = \begin{pmatrix} 0 & 0 & 1 \\ 1 & 0 & 0 \\ 0 & 1 & 0 \end{pmatrix},$$

$$\sigma_1 = \begin{pmatrix} 1 & 0 & 0 \\ 0 & 0 & 1 \\ 0 & 1 & 0 \end{pmatrix}, \quad \sigma_2 = \begin{pmatrix} 0 & 0 & 1 \\ 0 & 1 & 0 \\ 1 & 0 & 0 \end{pmatrix}, \quad \sigma_3 = \begin{pmatrix} 0 & 1 & 0 \\ 1 & 0 & 0 \\ 0 & 0 & 1 \end{pmatrix}. \tag{15.16}$$

Let us denote this representation by Γ. It can be shown that Γ is a *reducible* representation, meaning that by some unitary transformation the 3×3 matrices can be factorized into block-diagonal form with 2×2 plus 1×1 submatrices. The reducibility of Γ can be deduced from the character table. The characters of these matrices are

$$\Gamma: \quad \chi(E) = 3, \quad \chi(C_3) = 0, \quad \chi(\sigma_v) = 1. \tag{15.17}$$

The character of each of these permutation operations is equal to the number of H atoms left untouched: 3 for the identity, 1 for a reflection and 0 for a rotation. The characters of Γ are seen to equal the sum of the characters of A_1 plus E. This reducibility relation is expressed by writing

$$\Gamma = A_1 \oplus E. \tag{15.18}$$

The three H atom $1s$ functions can be combined into LCAO functions which transform according to the IR's of the group. Clearly the sum

$$\psi = \psi_{1s}(1) + \psi_{1s}(2) + \psi_{1s}(3) \tag{15.19}$$

transforms like A_1. The two remaining linear combinations which transform like E must be orthogonal to this and to one another. One possible choice is

$$\psi' = \psi_{1s}(2) - \psi_{1s}(3), \quad \psi'' = 2\psi_{1s}(1) - \psi_{1s}(2) - \psi_{1s}(3). \tag{15.20}$$

A pictorial representation of these *symmetry-adapted* AO's is given in Fig. 15.2.

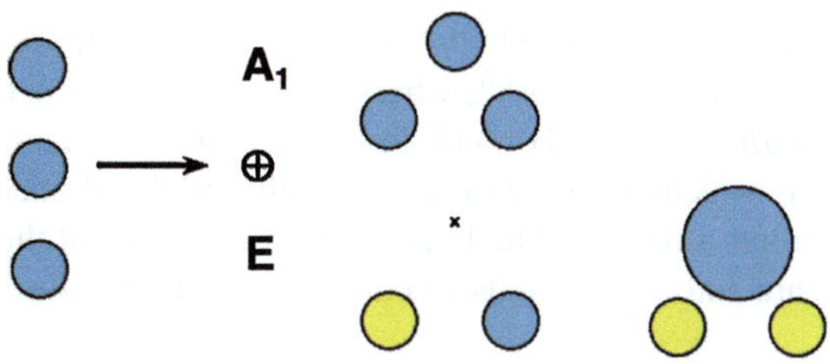

FIGURE 15.2 Symmetry adapted AOs for the three hydrogens in ammonia.

Now (15.19) can be combined with the N $1s$, $2s$ and $2p_z$ to form MO's of A_1 symmetry, while (15.20) can be combined with the N $2p_x$ and $2p_y$ to form MO's of E symmetry. In the electronic ground state of ammonia, the N $1s$, $2s$ and $2p_z$ AO's, together with the H $1s$ combination (13.18), interact to form the three MO's, $1a_1$, $2a_1$ and $3a_1$. Similarly the N $2p_x$ and $2p_y$ combine with the two H $1s$ combinations (15.20) to give the doubly-degenerate $1e$ MO. The explicit coefficients in these linear combinations of atomic orbitals are determined by a detailed variational calculation, most likely done using one of the computational chemistry programs. The resulting molecular electron configuration can then be written $1a_1^2 2a_1^2 3a_1^2 1e^4\ {}^1A_1$. A closed shell configuration will have the totally-symmetric term symbol, in this case A_1.

The molecular orbitals computed above are delocalized to conform to the symmetry of the molecule. They are not obviously associated with individual atoms or chemical bonds. A puzzled chemist might ask, "Where are the N-H bonds; where is the nitrogen lone pair; what has happened to hybridized atomic orbitals?" To answer these questions we note that it is possible to transform the spinorbitals of a Slater determinant in such a way that the value of the determinant is not changed. Consider, for example, the 2×2 determinant

$$\psi(1,2) = \frac{1}{\sqrt{2}}\begin{vmatrix} a(1) & b(1) \\ a(2) & b(2) \end{vmatrix} \tag{15.21}$$

and let the spinorbitals a and b be transformed into two new spinorbitals c and d such that

$$a = \frac{1}{\sqrt{2}}(c+d), \qquad b = \frac{1}{\sqrt{2}}(c-d) \tag{15.22}$$

It is easily shown that the determinant (13.20) transforms to

$$\psi(1,2) = \frac{1}{\sqrt{2}}\begin{vmatrix} c(1) & d(1) \\ c(2) & d(2) \end{vmatrix} \tag{15.23}$$

Although the orbitals are now different, the total molecular wavefunction has not changed. In a similar way, the 10×10 Slater determinant representing the configuration $1a_1^2 2a_1^2 3a_1^2 1e^4$ can be transformed into an equivalent configuration we might designate $is^2 lp^2 nh^6$, where is stands for N inner shell, lp, for N lone pair and nh, for an N–H bond. In this way we have recovered a set of localized orbitals which are more in accord with traditional chemical ideas of molecular electronic structure. The hybridization of AO's and the formation of bonding MO's are automatic results of computation and need not be explicitly introduced.

15.5 Selection rules

The strongest interaction of an atom or molecule with an electromagnetic field is usually through dipole coupling with the electric field of the radiation. This has the form $-\boldsymbol{\mu}\cdot\mathbf{E}$, where the electric dipole operator is given by

$$\boldsymbol{\mu} = \sum_i q_i \mathbf{r}_i \tag{15.24}$$

summed over all charges q_i, both electrons and nuclei. As shown in Supplement 4B, the probability of a radiative transition between two states ψ_n and ψ_m is proportional to the square of the matrix element of one of the components of $\mathbf{r}$

$$\int \psi_n^* \, x \, \psi_m \, d^3\mathbf{r}, \quad \int \psi_n^* \, y \, \psi_m \, d^3\mathbf{r}, \quad \int \psi_n^* \, z \, \psi_m \, d^3\mathbf{r}. \tag{15.25}$$

The particular component determines the direction of polarization for the radiation absorbed or emitted. A selection rule is a condition on the symmetries of ψ_n and ψ_m such that one of the above integrals is *not* identically equal to zero. If all three of the above integrals equal zero, the transition is said to be *dipole forbidden*. It might still be weakly allowed by some other mechanism.

An integral of the form

$$\int f(\mathbf{r})\, d^3\mathbf{r}, \tag{15.26}$$

which could be an overlap integral or one of the above matrix elements, will not change in value if the coordinate system is rotated, reflected or otherwise transformed. This is shown schematically in Fig. 13.3. In the notation of group theory,

$$G_i \int f(\mathbf{r})\, d^3\mathbf{r} = \int f(\mathbf{r})\, d^3\mathbf{r}. \tag{15.27}$$

The C_3 rotation of the coordinates, at the left in Fig. 15.3, clearly leaves the value of the integral unchanged. However, the σ_v reflection at the right changes the sign of the integral. But since the integral should have the same value after any group operation, this implies that the second integral must equal zero. In this case, the vanishing of the integral is obvious from its equal positive (blue)

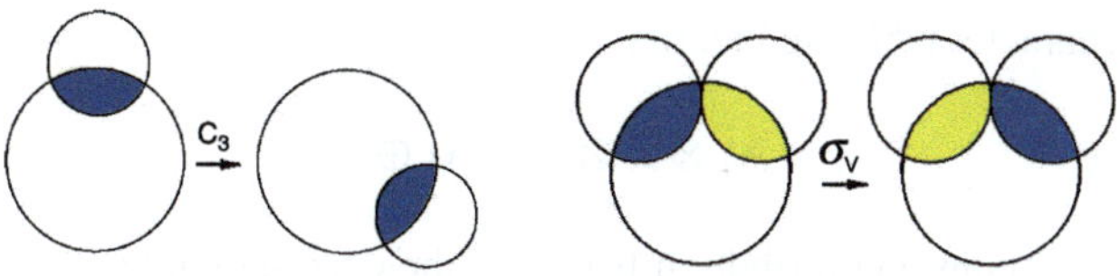

FIGURE 15.3 Transformations on two overlap integrals.

and negative (yellow) contributions, but in other cases it might be necessary to go through the group theoretical argument.

The condition for an integral *not* to be identically equal to zero is that its integrand belongs to the totally symmetric representation of the group. More precisely, the integrand must contain at least a *part* that is totally symmetrical. This condition can be deduced from the group character table. Suppose the integrand has the general form $f_1 f_2 f_3$, where each factor f_1, f_2, f_3 transforms according to one of the IR's of the group, say Γ_1, Γ_2, Γ_3. The product of the three functions will transform as the *direct product* of the three representations, which is written

$$\Gamma = \Gamma_1 \otimes \Gamma_2 \otimes \Gamma_3. \tag{15.28}$$

Only if the reduction of Γ includes the totally symmetric representation will the integral *not* be identically equal to zero. As an example, let us determine whether dipole transitions can occur between A_1 and E states of ammonia. For z-polarized transitions, the integral to be considered is

$$\int \psi(\mathrm{A}_1)\, z\, \psi(\mathrm{E})\, d^3\mathbf{r}. \tag{15.29}$$

According to the C_{3v} character table, the coordinate z transforms as A_1. For the representation Γ_z, we find

$$\chi(\mathrm{E}) = 1 \times 1 \times 2 = 2,\ \chi(\mathrm{C}_3) = 1 \times 1 \times -1 = -1,\ \chi(\sigma_\mathrm{v}) = 1 \times 1 \times 0 = 0, \tag{15.30}$$

so Γ_z has the same character as the E representation and transforms in the same way. Since the product gives no contribution containing A_1, the integrals must vanish and therefore z-polarized transitions between A_1 and E states are forbidden. For x- or y-polarized transitions we consider the integral

$$\int \psi(\mathrm{A}_1)\, x\, \psi(\mathrm{E})\, d^3\mathbf{r} \tag{15.31}$$

Since x or y transforms as E, we have $\Gamma_{x,y} = \mathrm{A}_1 \otimes E \otimes E$ and

$$\chi(\mathrm{E}) = 1 \times 2 \times 2 = 4,\ \chi(\mathrm{C}_3) = 1 \times -1 \times -1 = 1,\ \chi(\sigma_\mathrm{v}) = 1 \times 0 \times 0 = 0. \tag{15.32}$$

From the character table we identify

$$\Gamma_{x,y} = \mathrm{A}_1 \otimes E \otimes E = \mathrm{A}_1 \oplus \mathrm{A}_2 \oplus E. \tag{15.33}$$

Since $\Gamma_{x,y}$ contains a contribution from A_1, the corresponding x- and y-polarized transitions are allowed.

15.6 The water molecule

The symmetry elements of H_2O are shown in Fig. 15.4. The molecule belongs to the symmetry group $\mathrm{C}_{2\mathrm{v}}$ with a two-fold axis of rotation C_2 and two vertical mirror planes σ_v and σ'_v. In contrast to the case of ammonia, these two reflections are in *different* classes since one lies in the plane of the molecule while the other bisects the plane.

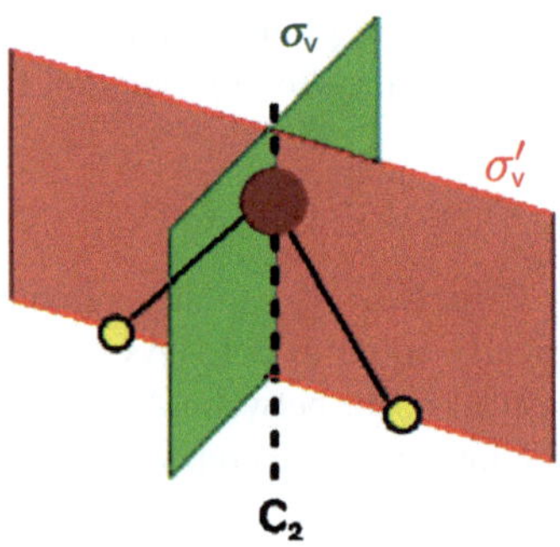

FIGURE 15.4 Symmetry elements for water molecule.

The $\mathrm{C}_{2\mathrm{v}}$ character table is shown here:

	E	C_2	σ_v	$\sigma_v{}'$		
A_1	1	1	1	1	z	x^2, y^2, z^2
A_2	1	1	−1	−1	R_z	xy
B_1	1	−1	1	−1	R_y, x	xz
B_2	1	−1	−1	1	R_x, y	yz

The group has four one-dimensional representations, all nondegenerate. The z-axis coincides with C_2, while the y and x axes are in the molecular plane and perpendicular to it, respectively. Oxygen and hydrogen AOs can be classified according to $\mathrm{C}_{2\mathrm{v}}$ symmetry as follows:

A_1: $\mathrm{O}1s$, $\mathrm{O}2s$, $\mathrm{O}2p_z$, $\mathrm{H}1s_\mathrm{A}$+$\mathrm{H}1s_\mathrm{B}$

B_1: $\mathrm{O}2p_x$

B_2: $\mathrm{O}2p_y$, $\mathrm{H}1s_\mathrm{A}$-$\mathrm{H}1s_\mathrm{B}$

The ground-state electronic configuration, in the localized valence-bond description, can be written $is^2lp^4oh^4$. It can be deduced that the corresponding delocalized MO configuration is $1a_1^2 2a_1^2 3a_1^2 1b_1^2 1b_2^2\ {}^1A_1$. The $3a_1$ and $1b_2$ are predominantly O-H bonding orbitals, while the $1b_1$ and $2a_1$ are predominantly oxygen lone pairs.

15.7 Walsh diagrams

If a C_{2v} molecule were straightened out into a linear molecule, its symmetry would turn into $D_{\infty h}$, which is the group for homonuclear diatomic and symmetric triatomic molecules. The orbital classification we studied in Chap. 12 would then apply and we can reclassify the AO's in the triatomic molecule XH_2 as follows

σ_g: X1s, X2s, H1s_A+H1s_B

σ_u: X2p_x, H1s_A-H1s_B

π_u: X2p_y, X2p_z

The correlation between C_{2v} and $D_{\infty h}$ orbitals can be represented in a Walsh diagram, shown in Fig. 15.5. The inner shell $1a_1$ or $1\sigma_g$ is not shown. The $2s$ will tend to hybridize with $2p$ as the molecule is bent. The largest effect of decreasing the bond angle from 180° is to convert the nonbonding $2p_y$ on the central atom to a bonding combination with H1s_A+H1s_B. The 5th electron in the valence shell should go into this orbital and cause a transition from a linear to a bent molecule. Indeed, it is found that BeH_2 is linear while BH_2 is bent, as are all the subsequent dihydrides, CH_2, NH_2 and H_2O.

15.8 Molecular symmetry groups

Our studies of NH_3 and H_2O have shown how much information can be deduced just from knowledge of a molecule's symmetry group, even without solving the Schrödinger equation. It is therefore useful to develop a general strategy for determining the symmetry group of any molecule. Specifically, we are dealing with *point groups*, so called because at least one point in the molecule remains fixed under all the symmetry operations of the group. We will work within the Born-Oppenheimer approximation, treating electronic states in a fixed configuration of nuclei. Following are the types of symmetry operations we need to consider:

E: The identity transformation meaning *do nothing* (from the German *Einheit*, meaning unity).
C_n: Clockwise rotation by an angle of $2\pi/n$ radians (n is an integer). The axis for which n is maximum is called the *principal axis*.
σ: Reflection through a plane (*Spiegel* is German for mirror).
σ_h: Horizontal reflection plane, one passing through the origin and *perpendicular* to the principal axis.

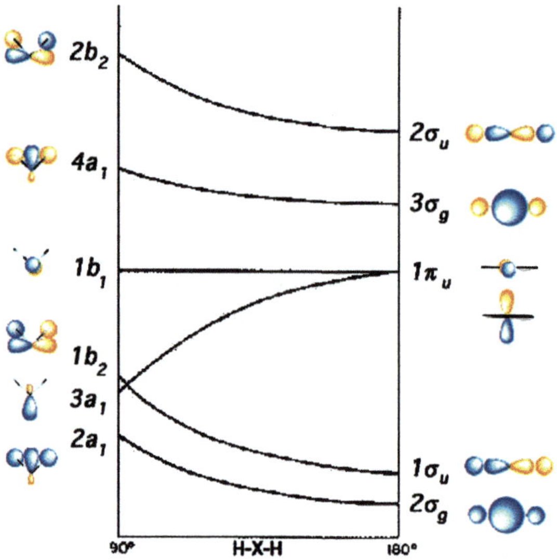

FIGURE 15.5 Walsh diagram for XH_2 triatomic molecules.

σ_v: Vertical reflection plane, one passing through the origin and *containing* the principal axis.

σ_d: Diagonal or dihedral reflection in a plane through the origin and containing the principal axis. Similar to σ_v, except that it also bisects the angles between two C_2 axes perpendicular to the principal axis.

i: Inversion through the origin. In Cartesian coordinates, the transformation $(x, y, z) \rightarrow (-x, -y, -z)$. Irreducible representations under this symmetry operation are classified as g (even) or u (odd).

S_n: An improper rotation or rotation-reflection axis. Clockwise rotation through an angle of $2\pi/n$ radians followed by a reflection in the plane perpendicular to the axis of rotation. Also known as an alternating axis of symmetry. Note that S_1 is equivalent to σ_h and S_2 is equivalent to i. See Fig. 15.6.

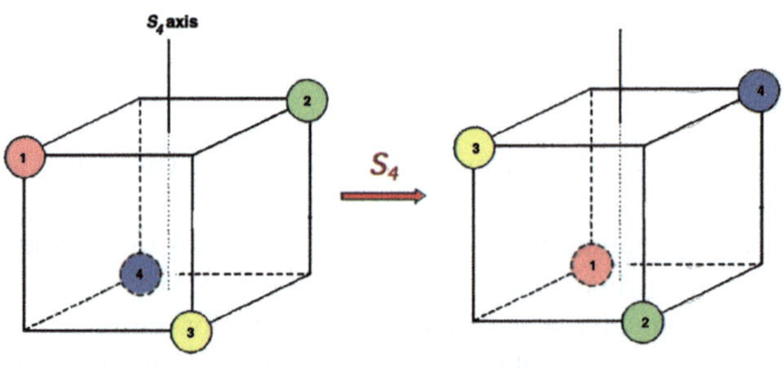

FIGURE 15.6 Pictorial representation of the operation S_4.

Every molecule can be characterized as belonging to some *group* of symmetry operations from the above list, under which it can be transformed into indistinguishable copies of itself. We cannot, however, have arbitrary combinations of symmetry operations. For example, a molecule with a C_n axis of rotation can only have mirror planes which either contain the axis or are perpendicular to it. We will use the *Schonflies* classification scheme, most favored by chemists. (Crystallographers generally use the International or Hermann-Mauguin classification.)

A symbol such as C_n actually has a triple meaning in group theory. It represents a *symmetry element*, namely an n-fold axis of rotation. It also designates the *symmetry operation* of rotation by $2\pi/n$ radians about this axis. Finally, it is used to characterize the *symmetry group* containing the elements $\{E, C_n, C_n^2 \ldots C_n^{n-1}\}$. In this Section, we will use calligraphic symbols such as $\mathcal{C}_n$ when referring to a symmetry group. We will begin with point groups that have the *lowest* symmetry and work up to those of high symmetry.

Low-symmetry groups

We consider first groups lacking a C_n axis. A molecule with no elements of symmetry (other than the identity E) belongs to the point group $\mathcal{C}_1$. The identity element is equivalent to C_1, a rotation by $2\pi/1$ radians or 360°. Very few small molecules are so unsymmetrical, one example being the trisubstituted methane:

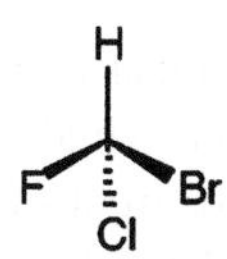

$\mathcal{C}_1$

Molecules designated $\mathcal{C}_s$ have a plane of symmetry but no other symmetry elements. Their group consists of just the two elements: $\{E, \sigma\}$. An example is a monosubstituted naphthalene:

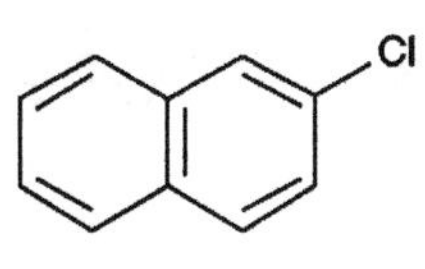

$\mathcal{C}_s$

Molecules belonging to $\mathcal{C}_i$ have only a center of inversion, their group consisting of $\{E, i\}$. An example is the staggered configuration of an exotic substituted ethane:

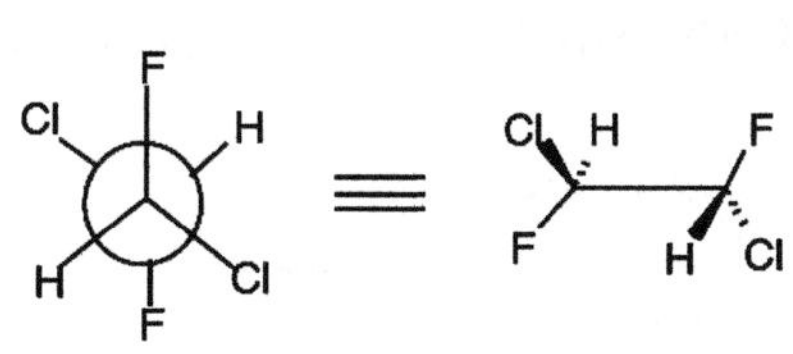

$\mathcal{C}_i$

Rotational groups

Next we consider groups with a single C_n axis. Molecules belonging to the groups C_n have only the n-fold rotational axis. Their groups are of dimension n with elements $\{E, C_n, C_n^2 \dots C_n^{n-1}\}$. Some examples of C_2 molecules are

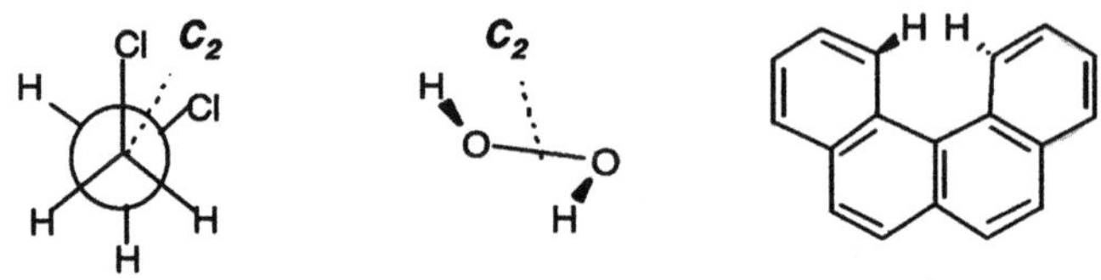

C_2

This group is not very common since most molecules will have additional elements of symmetry. Molecules belonging to the symmetry groups C_{nv} are much more numerous. These possess n vertical planes of symmetry σ_v in addition to the n-fold axis of rotation. We have already studied in detail C_{2v} and C_{3v}, the symmetry groups of H_2O and NH_3. Other examples are

C_{2v}

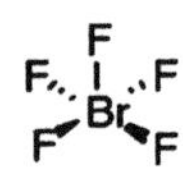

C_{4v}

Heteronuclear diatomic molecules, as well as nonsymmetrical linear molecules, such as HCN, are classified as $C_{\infty v}$, since cylindrical symmetry can be regarded as a rotational axis of infinite order. The groups C_{nh} contain a horizontal plane of symmetry σ_h in addition to the n-fold rotation axis. Many of these molecules are planar, such as *trans*-dichloroethylene and boric acid:

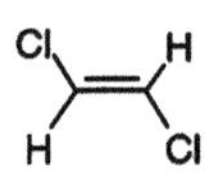

C_{2h}

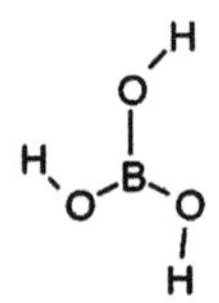

C_{3h}

The groups S_n $(n = 4, 6, 8 \dots)$ involve the symmetry operations associated with an S_n-axis. For odd n the group S_n is identical to C_{nh}, while S_1 and S_2 are the same as C_s and C_i, respectively. A 1,3-disubstituted allene and the cobalt thiocyanate complex ion $[Co(NCS)_4]^{2+}$ belong to this group S_4:

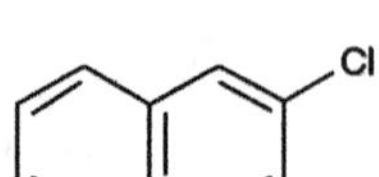

C_s

Dihedral groups

Next we come to the dihedral groups $\mathcal{D}_n$, $\mathcal{D}_{nh}$ and $\mathcal{D}_{nd}$, which are often the trickiest to identify. These all have a C_n axis with n C_2 axes perpendicular to it. The group $\mathcal{D}_2$ has mutually three perpendicular C_2 axes and no other symmetry elements. An example is twisted biphenyl (by some angle between 0° and 90°):

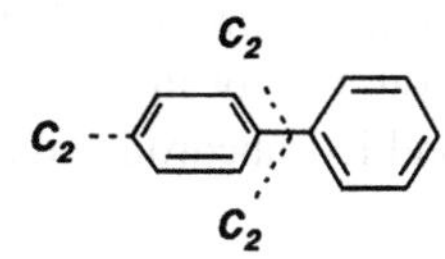

$\mathcal{D}_2$

A horizontal plane σ_h in addition to the C_n principal axis with n perpendicular 2-fold axes gives the symmetry groups $\mathcal{D}_{nh}$. Following is a sampling of molecules belonging to $\mathcal{D}_{nh}$ groups:

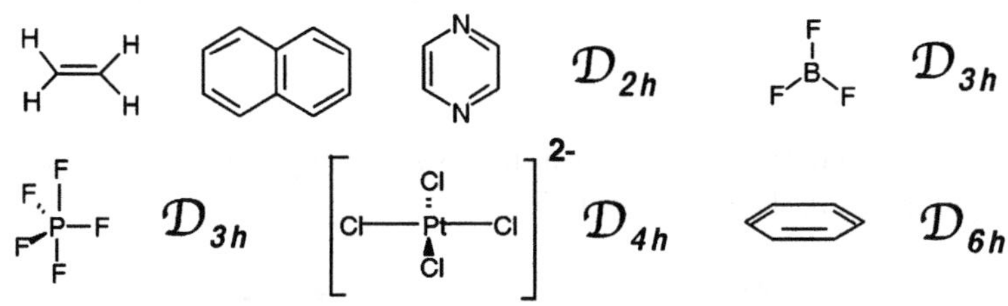

Homonuclear diatomic molecules and other symmetric linear molecules such as H–C≡C–H are classified as $\mathcal{D}_{\infty h}$.

The point groups $\mathcal{D}_{nd}$ have, in addition to the axes defining $\mathcal{D}_n$, n diagonal planes σ_d which bisect the angles between successive 2-fold axes. The σ_d and C_2 axes imply that there is also an S_{2n} and, if n is odd, a center of symmetry i. Molecules of $\mathcal{D}_{2d}$ symmetry, have the shape of two equivalent halves twisted by 90°, for example allene and spiran shown here:

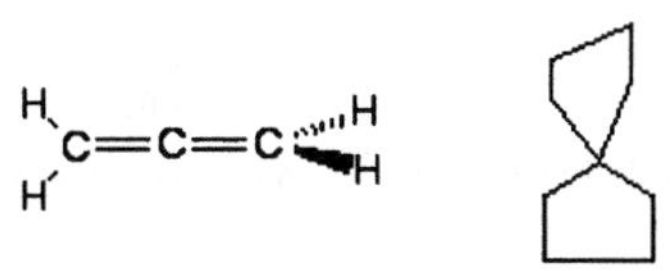

$\mathcal{D}_{2d}$

The staggered configuration of ethane has $\mathcal{D}_{3d}$ symmetry, while the eclipsed configuration is $\mathcal{D}_{3h}$:

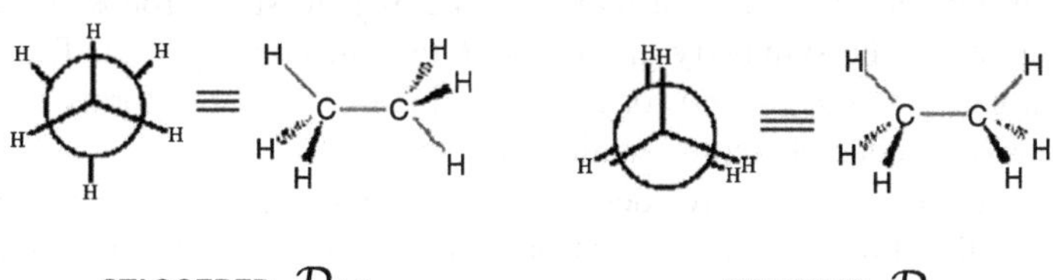

STAGGERED $\mathcal{D}_{3d}$ ECLIPSED $\mathcal{D}_{3h}$

Note that in going from $\mathcal{D}_{3h}$ to $\mathcal{D}_{3d}$, the σ_v planes become σ_d, while the σ_h operation is replaced by i.

Groups of higher symmetry

These are groups which contain more than one 3-fold or higher axis. We will limit our consideration to the symmetry groups which describe the Platonic solids: $\mathcal{T}_d$ for the regular tetrahedron, $\mathcal{O}_h$ for the cube and regular octahedron, $\mathcal{I}_h$ for the regular dodecahedron and icosahedron and $\mathcal{K}_h$ for the sphere. Some molecules in the cubic groups:

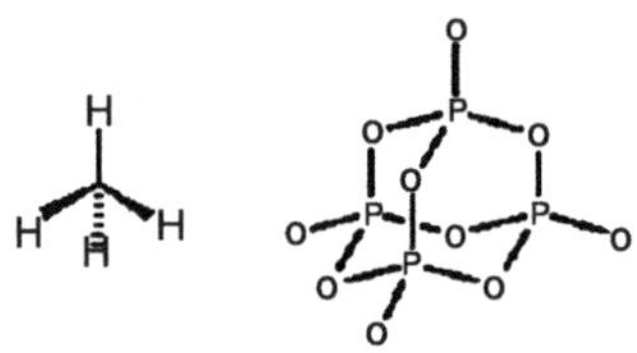

$\mathcal{T}_d$

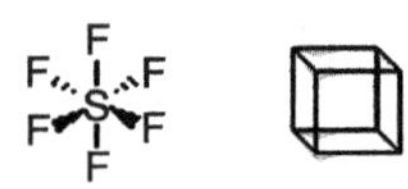

$\mathcal{O}_h$

Ball and stick models of two chemical species belonging to the icosahedral symmetry group $\mathcal{I}_h$ are shown in Fig. 15.7.

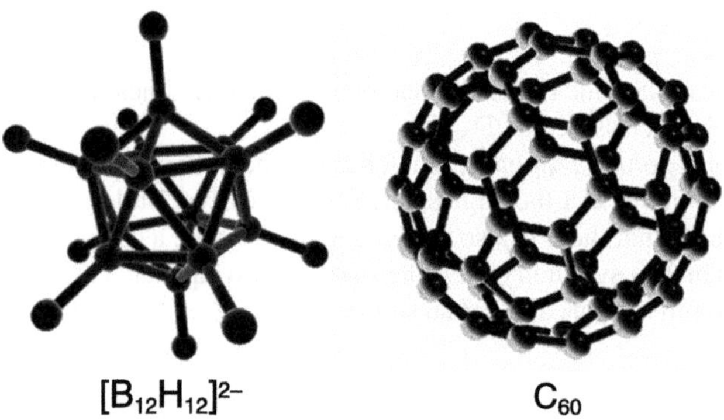

FIGURE 15.7 Icosahedral boron hydride ion $\left[B_{12}H_{12}\right]^{2-}$ and soccer-ball shaped buckminsterfullerene C_{60}.

The unadorned symmetry groups $\mathcal{T}$, $\mathcal{O}$ and $\mathcal{I}$ contain only the rotational axes but none of the σ planes. This can be accomplished by exotic arrays of stripes, making it unlikely that any real molecules belongs to such groups. Spherical atoms belong to the symmetry group $\mathcal{K}_h$ (*Kugel* meaning sphere). The set of spherical harmonics $Y_{\ell m}(\theta, \phi)$ for a given value of $\ell = 0$ transforms as $(2\ell + 1)$-dimension irreducible representation of $\mathcal{K}_h$.

A systematic procedure for determining the point group of a molecule is outlined in the flowchart on the next page. Follow the path by answering the question in each diamond box until you arrive at one of the group designations.

Note that if you answer "NO" to every question, you will wind up at C_1.

Flowchart for Molecular Symmetry Groups

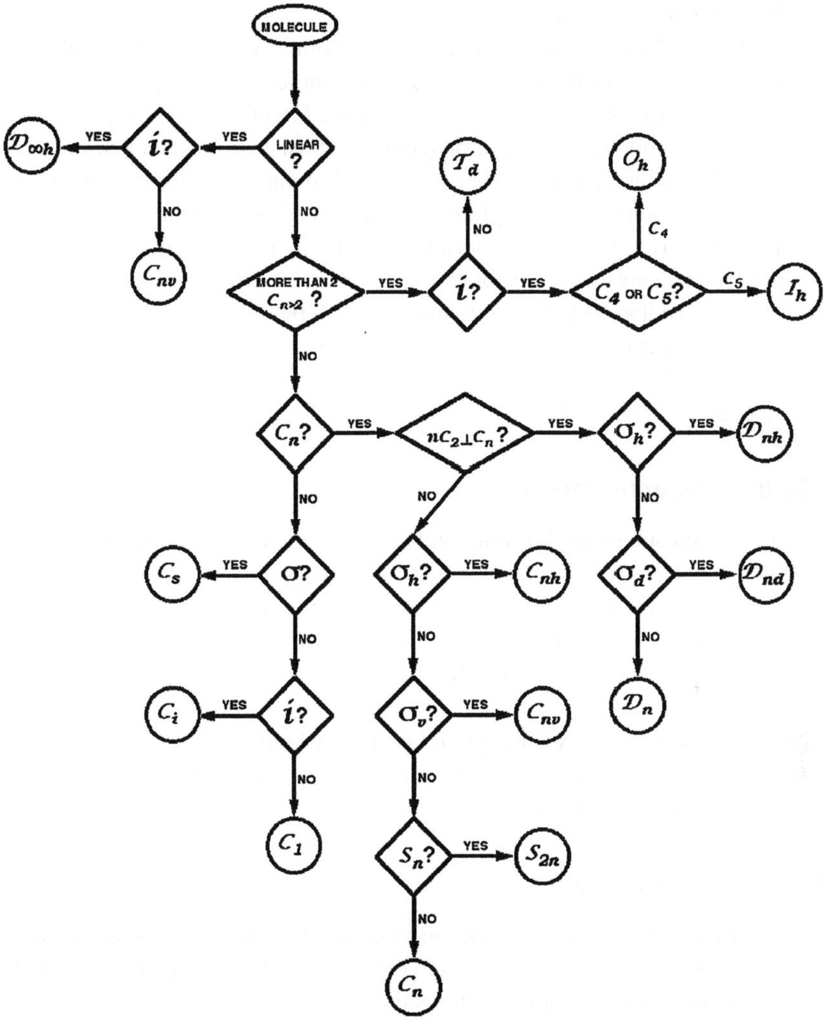

15.9 Dipole moments and optical activity

A molecule has a permanent electric dipole moment if and only if the centroids of nuclear and electronic charge do not coincide. If there is a C_n axis, the dipole must lie along this axis. If there is a plane of symmetry, the dipole must lie in that plane. Both NH_3 and H_2O have dipole moments coincident with their rotational axes which are also the intersections of all their σ_v planes. Any molecule with more than one C_n axis must have a zero dipole moment, since the dipole cannot have two different directions. A center of inversion i also implies a zero dipole moment, since both the positive and negative charges must be symmetrically distributed. Thus only molecules belonging to C_s, C_n and C_{nv}, including $C_{\infty v}$, can have dipole moments.

A *chiral* molecule (from the Greek *cheir* meaning "hand") is one which *cannot* be superposed on its mirror image. Such molecules are *optically active*, meaning that they can rotate the plane of polarized light. A molecule possessing any plane of symmetry or a center of symmetry must be *achiral* and lack optical activity. Most achiral molecules belong to the C_1 point group, for example those containing asymmetric carbon atoms. But it is possible for other C_n species to be optically active. An elegant criterion for achirality is that the molecule's symmetry group contains an S_n operation of any order, including $S_1 = \sigma$ and $S_2 = i$. It is well known that proteins, DNA and almost all biologically-active molecules are chiral. Molecules with the wrong chirality are either ineffective or actually harmful (e.g., thalidomide). The occurrence of optical activity is often regarded as evidence for some lifeform.

15.10 Character tables

In our detailed studies of the water and ammonia molecules, we introduced the C_{2v} and C_{3v} character tables and showed how to use them. Rather than listing character tables for all chemically-important point groups, we refer to websites which give convenient tabulations:
http://symmetry.jacobs-university.de
also
https://www.webqc.org/symmetrypointgroup-c2v.html
These can also be found as appendices in innumerable physical chemistry and quantum chemistry textbooks.

Chapter 15. Exercises

15.1. For the H_2O molecule, work out selection rules for all possible electric-dipole transitions among states of symmetry A_1, A_2, B_1, B_2. Identify the polarization (x, y or z) of each allowed transition.

15.2. Construct a set of C_{2v} symmetry orbitals for H_2O starting with a minimum basis set of hydrogen and oxygen AO's. [Optionally, consider also "polarization functions," d-orbitals on O and p-orbitals on H.]

15.3. Enumerate all the possible symmetry groups for hypothetical molecules with the formula X_2Y_2, with a sketch of each structure. These will include (*i*) linear structures, (*ii*) planar nonlinear structures and (*iii*) nonplanar structures.

15.4. The geometry of each of the following molecules can be predicted by the valence-shell repulsion model. Determine the symmetry classification for each molecule.

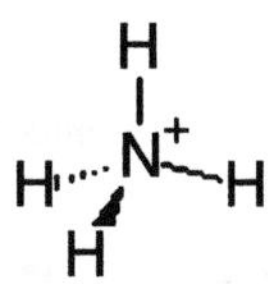

NH_4^+

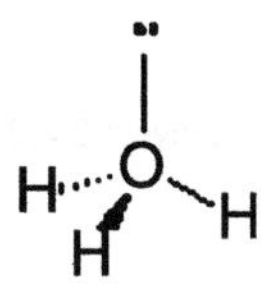

H_3O^+

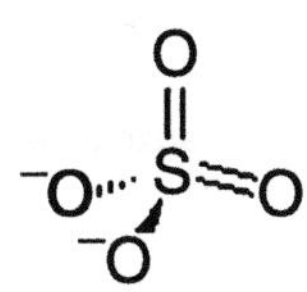

SO_4^{2-}

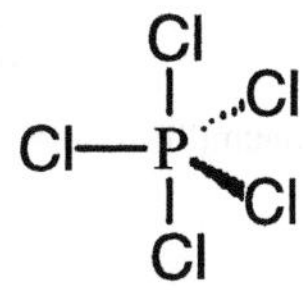

PCl_5

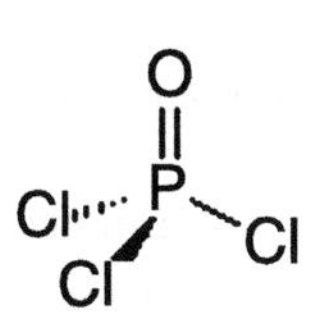

$POCl_3$

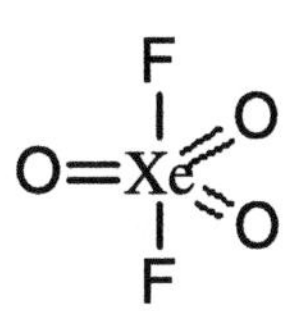

XeO_3F_2

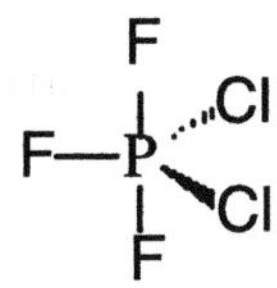

PF_3Cl_2

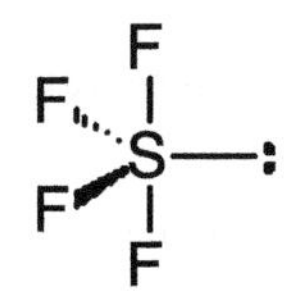

SF_4

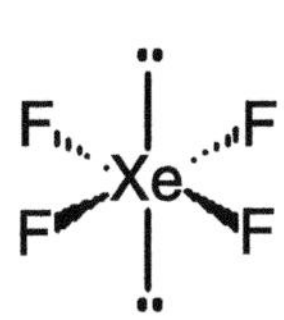

XeF_4

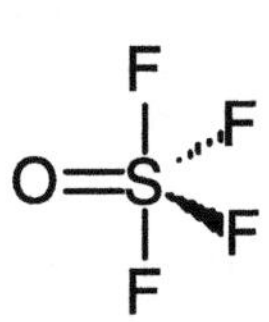

SOF_4

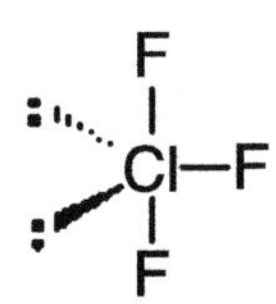

ClF_3

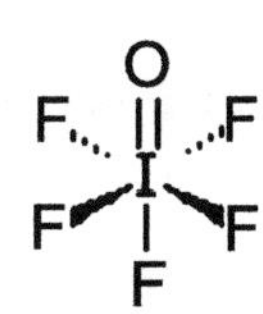

IOF_5

15.5. Determine the point group for each of the following molecules:

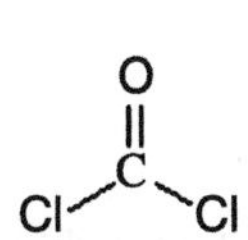

(A)

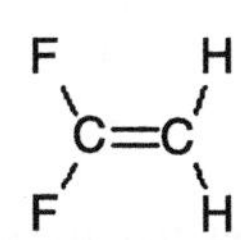

(B)

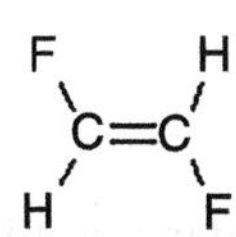

(C)

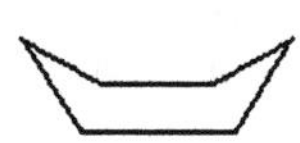

(D)

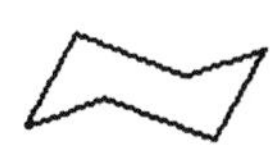

(F)

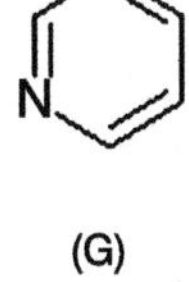

(G)

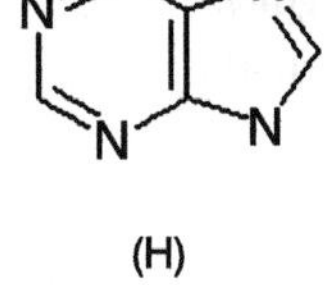

(H)

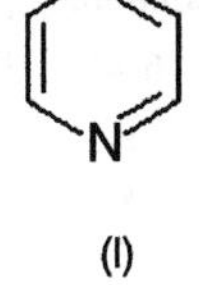

(I)

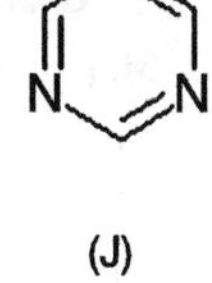

(J)

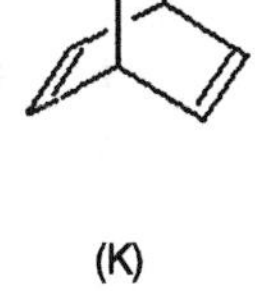

(K)

15.6. Determine the point group for each of the following structures:

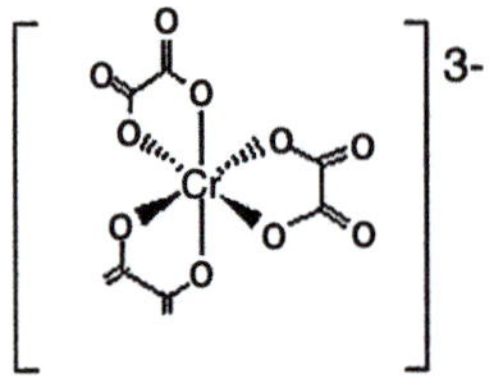

chromium(III)oxylate

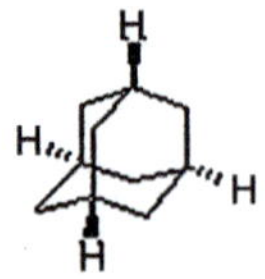

adamantane

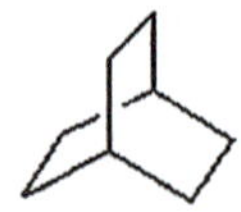

bicyclo[2,2,2]octane

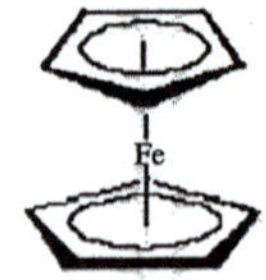

ferrocene

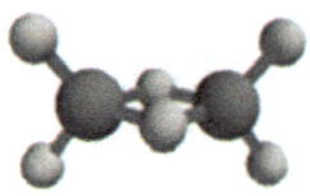

diborane B_2H_6

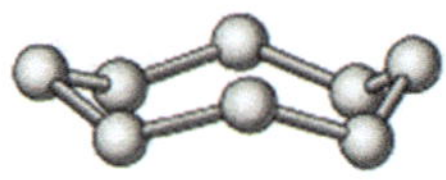

sulfur S_8

baseball (front and back)

We'll give you an intentional pass on the last one since it's not a molecule. It is easy to see two σ planes. But less obvious are two C_2 axes bisecting these planes. Therefore the symmetry group is $\mathcal{D}_{2d}$.

15.7. Following is the character table for $\mathcal{D}_{6h}$:

D_{6h}	E	$2C_6$	$2C_3$	C_2	$3C_2'$	$3C_2''$	i	$2S_3$	$2S_6$	σ_h	$3\sigma_d$	$3\sigma_v$		
A_{1g}	1	1	1	1	1	1	1	1	1	1	1	1		x^2+y^2, z^2
A_{2g}	1	1	1	1	−1	−1	1	1	1	1	−1	−1	R_z	
B_{1g}	1	−1	1	−1	1	−1	1	−1	1	−1	1	−1		
B_{2g}	1	−1	1	−1	−1	1	1	−1	1	−1	−1	1		
E_{1g}	2	1	−1	−2	0	0	2	1	−1	−2	0	0	(R_x, R_y)	(xz, yz)
E_{2g}	2	−1	−1	2	0	0	2	−1	−1	2	0	0		(x^2-y^2, xy)
A_{1u}	1	1	1	1	1	1	−1	−1	−1	−1	−1	−1		
A_{2u}	1	1	1	1	−1	−1	−1	−1	−1	−1	1	1	z	
B_{1u}	1	−1	1	−1	1	−1	−1	1	−1	1	−1	1		
B_{2u}	1	−1	1	−1	−1	1	−1	1	−1	1	1	−1		
E_{1u}	2	1	−1	−2	0	0	−2	−1	1	2	0	0	(x, y)	
E_{2u}	2	−1	−1	2	0	0	−2	1	1	−2	0	0		

Refer to the benzene MO's pictured in Fig. 13.2. Show that the a_{2u}, e_{1g}, e_{2u} and b_{2g} orbitals transform according to the corresponding IR's of $\mathcal{D}_{6h}$.

Character table for T_d point group							
	E	$8C_3$	$3C_2$	$6S_4$	$6\sigma_d$	linear, rotations	quadratic
A_1	1	1	1	1	1		$x^2+y^2+z^2$
A_2	1	1	1	-1	-1		
E	2	-1	2	0	0		$(2z^2-x^2-y^2, x^2-y^2)$
T_1	3	0	-1	1	-1	(R_x, R_y, R_z)	
T_2	3	0	-1	-1	1	(x, y, z)	(xy, xz, yz)

15.8. The character table for the tetrahedral group $\mathcal{T}_d$ is shown above. Referring to Fig. 11.9, verify that the five d-orbitals in a tetrahedral crystal field transform according to the E and T_2 representations.

15.9. Do an analogous classification of the five d-orbitals in a octahedral crystal field.

Chapter 16

Molecular spectroscopy

Our most detailed knowledge of atomic and molecular structure has been obtained from spectroscopy—study of the emission, absorption and scattering of electromagnetic radiation accompanying transitions among atomic or molecular energy levels. Whereas atomic spectra involve only electronic transitions, the spectroscopy of molecules is more intricate because vibrational and rotational degrees of freedom come into play as well. Early observations of absorption or emission by molecules were characterized as *band spectra*—in contrast to the line spectra exhibited by atoms. It is now understood that these bands reflect closely-spaced vibrational and rotational energies augmenting the electronic states of a molecule. With improvements in spectroscopic techniques over the years, it has become possible to resolve individual vibrational and rotational transitions. This has provided a rich source of information on molecular geometry, energetics and dynamics. Molecular spectroscopy has also contributed significantly to analytical chemistry, environmental science, astrophysics, biophysics and biochemistry. In this Chapter, we will focus on quantum-mechanical principles useful in spectroscopy. We will not take up any of the experimental techniques.

16.1 Vibration of diatomic molecules

A diatomic molecule with nuclear masses m_A, m_B has a reduced mass

$$\mu = \frac{m_A m_B}{m_A + m_B}, \tag{16.1}$$

recalling the definition of reduced mass in Sect. 8.8. Solution of the electronic Schrödinger equation gives the energy as a function of internuclear distance $E_{\text{elec}}(R)$. This plays the role of a potential energy function for motion of the nuclei $V(R)$, as sketched in Fig. 16.1. We can thus write the Schrödinger equation for vibration, in analogy with Eq. (7.63),

$$\left\{-\frac{\hbar^2}{2\mu}\frac{d^2}{dR^2} + V(R)\right\}\chi(R) = E\chi(R). \tag{16.2}$$

If the potential energy is expanded in a Taylor series about $R = R_e$

$$V(R) = V(R_e) + (R - R_e)V'(R_e) + \tfrac{1}{2}(R - R_e)^2 V''(R_e) + \ldots. \tag{16.3}$$

Introduction to Quantum Mechanics. https://doi.org/10.1016/B978-0-12-822310-9.00024-0

An approximation for this expansion has the form of a harmonic oscillator with

$$V(R) \approx \tfrac{1}{2}k(R - R_e)^2 \tag{16.4}$$

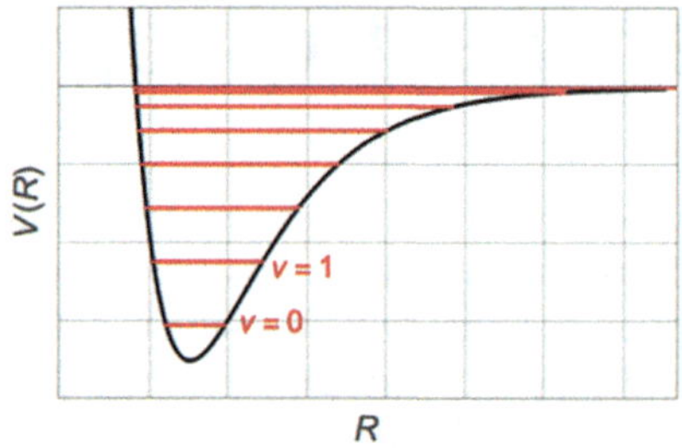

FIGURE 16.1 Vibrational energies of a diatomic molecule, as approximated by a Morse oscillator.

The energy origin can be chosen so $V(R_e) = 0$. At the minimum of the potential, $V'(R_e) = 0$. The best fit to this parabola is obtained with a force constant set equal to

$$k \approx \left. \frac{d^2V(R)}{dR^2} \right|_{R=R_e}. \tag{16.5}$$

The force constant is a measure of the stiffness of a chemical bond. Larger values of k are generally associated with deep and sharply-curved potential energy functions. Molecular force constants are typically in the range of 200–2000 N/m, remarkably, not very different from those for bedsprings. From the solution for the harmonic oscillator, we identify the ground state vibrational energy, with quantum number $v = 0$

$$E_0 = \hbar\omega = \hbar\sqrt{\frac{k}{\mu}}. \tag{16.6}$$

The actual dissociation energy from ground vibrational state is then approximated by

$$D_0 \approx D_e - \tfrac{1}{2}\hbar\omega, \tag{16.7}$$

or, in wavenumber units

$$hcD_0 \approx hcD_e - \tfrac{1}{2}\tilde{\nu} \quad \text{cm}^{-1}. \tag{16.8}$$

An improved treatment of molecular vibration must account for *anharmonicity*, deviation from a harmonic oscillator. Anharmonicity results in a *finite* number of vibrational energy levels and the possibility of dissociation of the molecule at sufficiently high energy. A very successful approximation for the energy of a diatomic molecule is the Morse potential (see Fig. 16.1):

$$V(R) = hcD_e\left(1 - e^{-a(R-R_e)}\right)^2, \qquad a = \left(\frac{\mu\omega^2}{2hcD_e}\right)^{1/2}. \tag{16.9}$$

Note that $V(R_e) = 0$ at the minimum of the potential well. The Schrödinger equation for a Morse oscillator can be solved to give the energy levels

$$E_v = (v + \tfrac{1}{2})\hbar\omega_e - (v + \tfrac{1}{2})^2\,\hbar\omega x_e, \tag{16.10}$$

or, expressed in wavenumber units,

$$\frac{E_v}{hc} = (v + \tfrac{1}{2})\omega_e - (v + \tfrac{1}{2})^2\omega_e x_e. \tag{16.11}$$

Spectroscopists generally write ω_e in place of $\tilde{\nu}$ for diatomic molecules. The anharmonicity $\omega_e x_e$ is usually tabulated as a single parameter. Higher vibrational energy levels are spaced closer together, just as in real molecules. The anharmonicity for a Morse oscillator is determined by

$$x_e = \frac{a^2\hbar}{2\mu\,\omega} = \frac{\omega_e}{4D_e}. \tag{16.12}$$

More accurate characterization of the energy levels—beyond the Morse potential—might include additional anharmonic contribution of the form $(v + \frac{1}{2})^3\,\omega_e y_e$. Vibrational transitions of diatomic molecules occur in the infrared, in the broad range of 50–15,000 cm^{-1}. A molecule will absorb or emit radiation only if it has a non-zero dipole moment. Thus heteronuclear diatomic molecules such as HCl are *infrared active* while homonuclear diatomics such as H_2 and Cl_2 are not.

The quantum number of the highest bound vibrational level in a diatomic molecule is estimated by

$$v_{\max} \approx \left\lfloor \frac{\omega_e}{2\,\omega_e x_e} - \frac{1}{2} \right\rfloor, \tag{16.13}$$

where the brackets identify the *floor*, or integer part of the expression. This can be found by equating the energy (14.11) to D_e or by setting $dE_v/dv \approx 0$.

The selection rule for vibrational transitions for a harmonic oscillator is $\Delta v = \pm 1$ since the integral $\int_{-\infty}^{\infty} \psi_v\, x\, \psi_{v'}\, dx \neq 0$ only when $v' = v \pm 1$ (see Exercise 6.6). The transition $v = 1 \leftarrow v = 0$ determines the *fundamental vibrational frequency*. For an anharmonic oscillator, *overtone* transitions such as

$2 \leftarrow 0$, $3 \leftarrow 0$, etc. are also possible, usually with much weaker intensities than the fundamental.

16.2 Vibration of polyatomic molecules

A molecule with N atoms has a total of $3N$ degrees of freedom for its nuclear motions, since each nucleus can be independently displaced in three perpendicular directions. Three of these degrees of freedom correspond to translational motion of the center of mass. For a nonlinear molecule, three more degrees of freedom determine the orientation of the molecule in space, and thus its rotational motion. This leaves $3N - 6$ vibrational modes. For a linear molecule, there are just two rotational degrees of freedom, which leaves $3N - 5$ vibrational modes. For example, the nonlinear molecule H_2O has three vibrational modes while the linear molecule CO_2 has four vibrational modes. The vibrations consist of coordinated motions of several atoms in such a way as to keep the center of mass stationary and nonrotating. These are called the *normal modes*. Each normal mode has a characteristic resonance frequency ν_i (expressed in cm^{-1}), which is usually determined experimentally. To a reasonable approximation, each normal mode behaves as an independent harmonic oscillator of frequency ν_i. The normal modes of H_2O and CO_2 are shown in Figs. 16.2 and 16.3. A normal mode will be infrared active only if it involves an oscillation of the dipole moment. All three modes of H_2O are active. The symmetric stretch of CO_2 is inactive because the two C–O bonds, each of which is polar, exactly compensate. Note that the bending mode of CO_2 is doubly degenerate. Bending of adjacent bonds in a molecule generally involves less energy than bond stretching, thus bending modes generally have lower wavenumbers than stretching modes. Each normal mode transforms as one of the irreducible representations of the symmetry group of the molecule. Thus water molecule, with symmetry $\mathcal{C}_{2v}$, has vibrations belonging to A_1 and B_1. Carbon dioxide, with symmetry $\mathcal{D}_{\infty h}$, has vibrations classified as Σ_g^+, Σ_u^+, and Π_g.

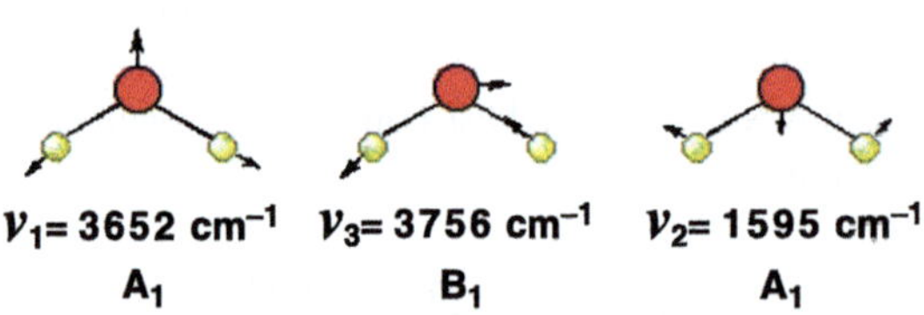

FIGURE 16.2 Normal modes of H_2O: symmetric stretch, asymmetric stretch and bend, respectively. All are IR active.

The normal modes of ammonia, a $\mathcal{C}_{3v}$ molecule, are shown in Fig. 16.4. All modes for NH_3 are IR active. For molecules of any symmetry, the IR active modes of vibration can be determined from the character table. Modes which are infrared-active belong to the same irreducible representation as one of the cartesian coordinates x, y or z, shown in one of the right-hand columns of the

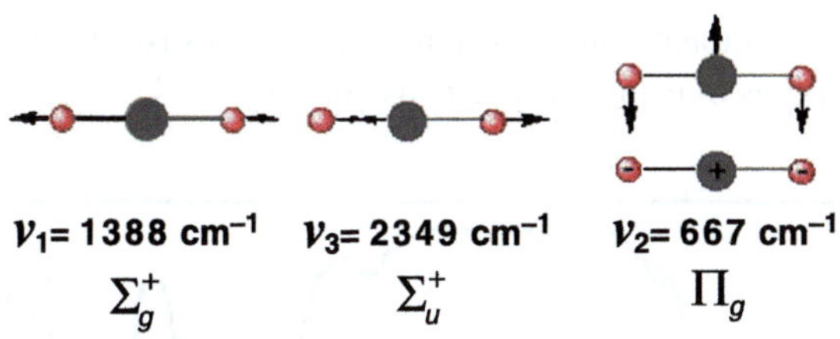

FIGURE 16.3 Normal modes of CO_2: all except ν_1 are IR active.

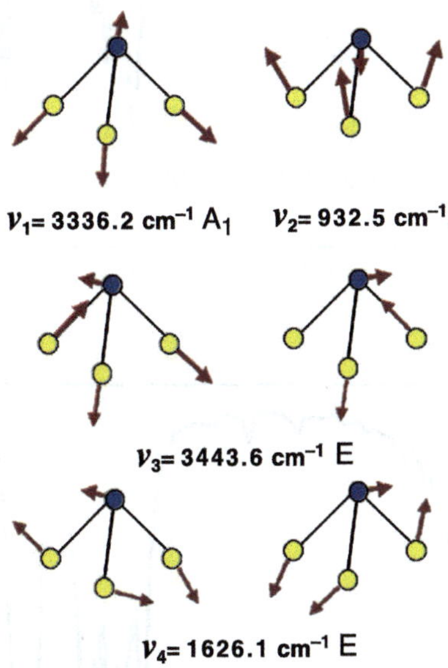

FIGURE 16.4 Normal modes of NH_3: ν_1 is a symmetric stretch, ν_2, a symmetric bend, ν_3 and ν_4 are doubly degenerate asymmetric stretches and bends. The symmetry species A_1 and E are indicated.

character table. Check this for the $\mathcal{C}_{2v}$, $\mathcal{C}_{3v}$ and $\mathcal{D}_{\infty h}$ molecules we have considered.

Infrared spectra of solutions are extensively used for identification and characterization of organic compounds. The features between 200 and 1400 cm^{-1}, known as the *fingerprint region*, are especially useful. These are absorptions due mainly to bending modes. Although it is difficult to assign the individual bonds producing this spectrum, each compound produces a characteristic pattern which can serve to identify the molecule. Fig. 16.5 shows the IR spectra of 1-propanol and 2-propanol. The broad absorption around 3200–3600 cm^{-1} is characteristic of the O–H bond stretch in alcohols, while the peaks around 2900 cm^{-1} are from C–H bond stretches. A very distinctive feature of carbonyl compounds is the C=O stretch around 1700 cm^{-1}. In condensed phases, rotational transitions can not be individually resolved. Instead, they contribute to the broadening of vibrational transitions into "bands." This is actually a simplifying feature for purposes of using IR spectra as "fingerprints" to identify molecules.

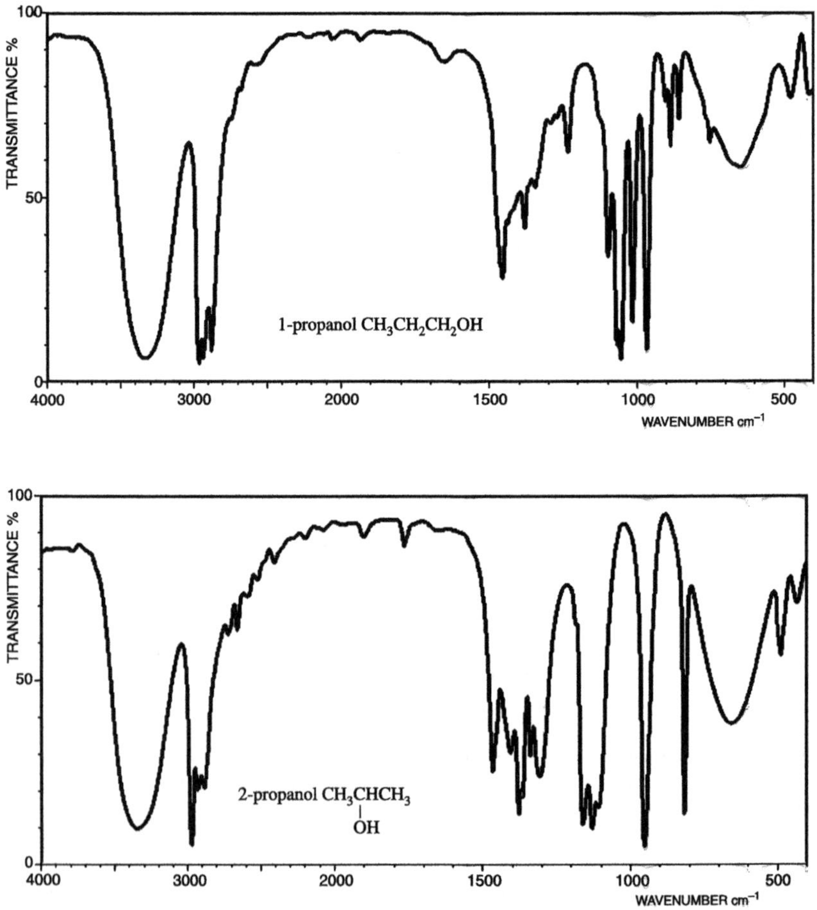

FIGURE 16.5 IR spectra of 1-propanol and 2-propanol. From the spectroscopic database for organic compounds on the website of the National Institute of Advanced Industrial Science and Technology (Tsukuba, Ibaraki, Japan).

16.3 Rotation of diatomic molecules

The rigid rotor model assumes that the internuclear distance R is a constant. This is not a bad approximation since the amplitude of vibration is generally of the order of 1% of R. The Schrödinger equation for nuclear motion then involves the three-dimensional angular momentum operator, written J rather than L when it refers to molecular rotation. The solutions to this equation are already known and we can write

$$\frac{J^2}{2\mu R^2} Y_{JM}(\theta,\phi) = E_J\, Y_{JM}(\theta,\phi),$$
$$J = 0, 1, 2 \ldots, \qquad M = 0, \pm 1 \cdots \pm J, \tag{16.14}$$

where $Y_{JM}(\theta,\phi)$ are spherical harmonics in terms of the quantum numbers J and M, rather than ℓ and m. Since the eigenvalues of J^2 are $J(J+1)\hbar^2$, the rotational energy levels are

$$E_J = \frac{\hbar^2}{2I} J(J+1). \tag{16.15}$$

The moment of inertia is given by

$$I = \mu R_e^2 = m_A R_{\mathrm{A}}^2 + m_B R_{\mathrm{B}}^2, \tag{16.16}$$

where R_e is the equilibrium internuclear separation, while R_{A} and R_{B} are the distances from nuclei A and B, respectively, to the center of mass. In wavenumber units, the rotational energy is expressed

$$\frac{E_J}{hc} = B_e J(J+1)\ \mathrm{cm}^{-1}, \tag{16.17}$$

where B_e is the rotational constant. A rotational energy-level diagram is shown in Fig. 16.6. Each level is $(2J+1)$-fold degenerate. For small molecules, the rotational constant B_e is generally in the range 0.1 to 10 cm^{-1}, thus pure rotational transitions are observed in microwave spectroscopy. Again, only polar molecules can absorb or emit radiation in the course of rotational transitions. The radiation is in the microwave or far infrared region. The selection rules for rotational transitions are $\Delta J = \pm 1$, $\Delta M = 0, \pm 1$. If the molecule has nonzero electronic orbital angular momentum (term symbol other than Σ) such as NO, $\Delta J = 0$ transitions are also allowed. These selection rules are consistent with photons carrying one unit of angular momentum, so that the vector sum $\mathbf{J}_{\text{initial}} + \mathbf{J}_{\text{photon}}$ can equal $\mathbf{J}_{\text{final}}$. It must be possible for the quantum numbers J_{initial}, $J_{\text{photon}} = 1$ and J_{final} to form a triangle.

At temperatures around 300 K, the great majority of diatomic molecules will occupy their $v = 0$ vibrational ground state. The thermal energy kT at 300 K corresponds to about $kT/hc \approx 200\ \mathrm{cm}^{-1}$. Thus for a molecule with vibrational constant $\omega_e \approx 2000\ \mathrm{cm}^{-1}$, the Boltzmann factor is of the order of

$$e^{-(E_1-E_0)/kT} \approx 5 \times 10^{-5}.$$

By contrast, a significant number of *rotational* levels are occupied at room temperature. Taking account of the degeneracy $g_J = 2J+1$, the Boltzmann distribution for rotational levels has the form

$$N_J = \mathrm{const}\,(2J+1)\, e^{-B_e J(J+1)hc/kT}. \tag{16.18}$$

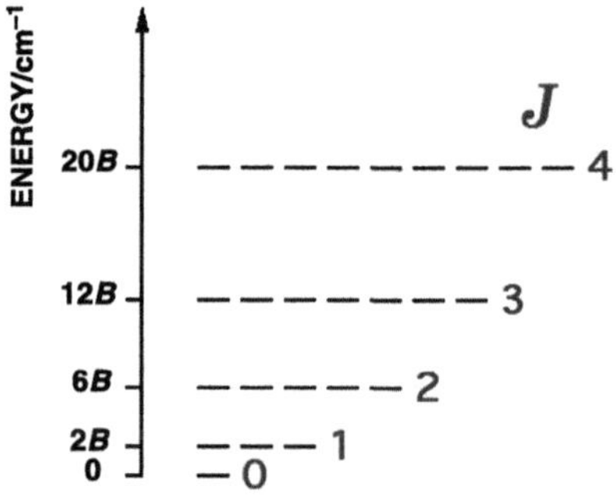

FIGURE 16.6 Rotational energies on wavenumber scale: $E_J/hc = BJ(J+1)$. Note $(2J+1)$-fold degeneracy.

The maximum population occurs for the level

$$J_{\max} = \left[\left(\frac{kT}{2hcB_e}\right)^{1/2} - \frac{1}{2}\right], \tag{16.19}$$

where the square brackets mean rounding to the nearest integer.

Just as deviations from the harmonic model for molecular vibration become significant for larger values of v, the rigid rotor model for rotation requires correction for larger values of J. The physical effect is *centrifugal distortion*, which causes a rapidly rotating molecule to stretch and thereby increase its moment of inertia. Taking account of centrifugal distortion, the rotational energy can be approximated by

$$\frac{E_J}{hc} = B_e J(J+1) - D_J J^2(J+1)^2. \tag{16.20}$$

Note that the increase in moment of inertia causes the rotational levels to be closer together, hence the minus sign. Centrifugal distortion constants D_J are typically of the order of $10^{-4} B_e$ and approximated by

$$D_J \approx 4B_e^3/\omega_e^2. \tag{16.21}$$

(Spectroscopists often write D_J as D_e, risking confusion with the potential-well depth D_e in Fig. 16.1. Centrifugal distortion is usually not very significant until rotational quantum numbers around $J = 30$ are reached. A more important effect is the dependence of the rotational constant on the vibrational quantum number. This can be parametrized as

$$B_v = B_e - \alpha_e(v + \tfrac{1}{2}), \tag{16.22}$$

where the vibration-rotation interaction constant α_e is typically in the range of 0.1 to 1 cm^{-1}.

16.4 Rotation-vibration spectra

Taking into account corrections to the harmonic oscillator-rigid rotor model, vibrational-rotational energies for a diatomic molecule can be represented by

$$\frac{E_{vJ}}{hc} = \omega_e(v+\tfrac{1}{2}) - \omega_e x_e(v+\tfrac{1}{2})^2 + B_e J(J+1)$$
$$-\alpha_e(v+\tfrac{1}{2})J(J+1) - D_J J^2(J+1)^2. \qquad (16.23)$$

Fig. 16.7 shows the vapor-phase infrared absorption spectrum of HCl, which involves rotational transitions occurring simultaneously with the fundamental vibrational transition $v = 1 \leftarrow v = 0$.

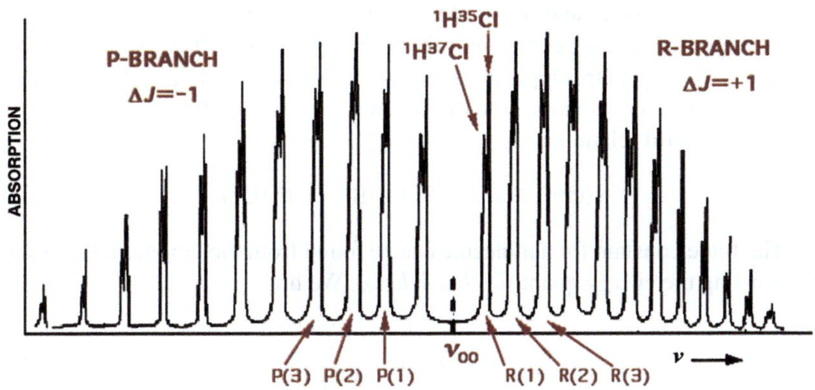

FIGURE 16.7 Infrared rotation-vibration spectrum of gaseous HCl in the region 2500-3100 cm^{-1} with ν_{00} at 2886 cm^{-1}.

Since the absorbed photon carries one unit of angular momentum, a pure vibrational transition is not possible. This accounts for the missing line labeled ν_{00}, which would be part of the *Q-branch* of the spectrum. A Q-branch can occur for non-Σ diatomics and for nonlinear polyatomic molecules. The R-branch in the above spectrum consists of transitions with $\Delta J = +1$ simultaneous with $\Delta v = +1$ vibrational transition. Specifically, the R(1) R(2) ... lines represents the transitions $J = 1 \leftarrow J = 0$, $J = 2 \leftarrow J = 1$ ..., respectively. Analogously the P-branch represents transitions with $\Delta J = -1$: P(1) is $J = 0 \leftarrow J = 1$, etc. The spacing between successive lines in the P-branch are slightly greater than those in the R-branch, as determined by differences between $B_0 J(J+1)$ and $B_1 J(J+1)$ according to Eq. (16.23). The variation in intensity of the absorption lines is caused by the relative populations in rotational levels of the ground vibrational state, consistent with (16.18). The $J = 2$ level evidently has the maximum population. Finally, note that each line has an overlapping "satellite" about one-third as intense. This is caused by the $^1H^{37}Cl$ isotopomer present with about 1/3 the abundance of $^1H^{35}Cl$ (remember the atomic weight of Cl is approximately 35.5). Since the reduced mass of the ^{37}Cl-isotopomer is larger, its vibrational constant ω_e is smaller, so that the satellites appear to the *left* of the main peaks. In a deuterium-enriched sample of HCl we could also observe the $^2H^{35}Cl$ and $^2H^{37}Cl$ spectra.

From accurate measurement of the HCl spectrum, the following spectroscopic constants can be assigned. For $^1H^{35}Cl$, $\omega_e = 2989.5$, $\omega_e x_e = 51.88$, $B_e = 10.5910$, $\alpha_e = 0.3027$, all in cm^{-1}. For $^1H^{37}Cl$, $\omega_e = 2987.3$, $\omega_e x_e = 51.81$, $B_e = 10.5739$, $\alpha_e = 0.3021$. For both, $D_J \approx 5 \times 10^{-4}\ \mathrm{cm}^{-1}$.

16.5 Molecular parameters from spectroscopy

Table 16.1 lists spectroscopic constants for some common diatomic molecules. These are taken from several sources and are of varying accuracy. The National Institute of Standards and Technology (NIST) maintains an extensive online compendium of spectroscopic and other chemical data at: http://webbook.nist.gov/chemistry/. Spectroscopic constants are expressed in MHz and GHz rather than cm^{-1}. To convert units, note that $\tilde{\nu}(\mathrm{cm}^{-1}) = \nu(\mathrm{Hz})/c$ where c is the speed of light. A useful relation is

$$\tilde{\nu}(\mathrm{cm}^{-1}) = 3.33564 \times 10^{-5}\, \nu\ (\mathrm{MHz}). \tag{16.24}$$

The force constant for a molecule can be found from the vibrational constant by equating the energy quantities $\hbar\omega = hc\omega_e$. We find

$$\omega = 2\pi c\, \omega_e = \sqrt{\frac{k}{\mu}}. \tag{16.25}$$

Thus

$$k = (2\pi c\, \omega_e)^2 \mu, \tag{16.26}$$

with

$$\mu = \frac{m_A m_B}{m_A + m_B} = \frac{M_A M_B}{M_A + M_B}\, u, \tag{16.27}$$

where $u = 1.66054 \times 10^{-27}$ kg, the atomic mass unit. M_A and M_B are the conventional atomic weights of atoms A and B (on the scale $^{12}C = 12$). Putting in numerical factors

$$k = 58.9 \times 10^{-6}\, (\omega_e/\mathrm{cm}^{-1})^2\, \frac{M_A M_B}{M_A + M_B}\ \mathrm{N/m}. \tag{16.28}$$

This gives 958.6, 512.4, 408.4 and 311.4 N/m for HF, HCl, HBr and HI, respectively. These values do *not* take account of anharmonicity.

The equilibrium internuclear distance R_e is determined by the rotational constant. By definition,

$$hcB_e = \frac{\hbar^2}{2I}. \tag{16.29}$$

TABLE 16.1 Spectroscopic Constants for Diatomic Molecules.

Molecule	Term	D_0/eV	ω_e/cm^{-1}	$\omega_e x_e/\text{cm}^{-1}$	B_e/cm^{-1}	α_e/cm^{-1}
$^1H_2$	$^1\Sigma_g^+$	4.4773	4395.2	117.9	60.86	2.99
$^1H^2H$	$^1\Sigma^+$	4.5128	3812.3	90.91	45.66	2.00
$^2H_2$	$^1\Sigma_g^+$	4.5553	3118.5	64.10	30.44	1.05
$^{14}N_2$	$^1\Sigma_g^+$	9.7598	2359.6	14.46	2.010	0.019
$^{16}O_2$	$^3\Sigma_g^-$	5.1156	1580.4	11.98	1.446	0.016
$^{19}F_2$	$^1\Sigma_g^+$	1.604	891.8	13.6	0.883	0.015
$^{35}Cl_2$	$^1\Sigma_g^+$	2.475	564.9	4.0	0.244	0.0015
$^{79}Br^{81}Br$	$^1\Sigma^+$	1.9708	323.3	1.08	0.0809	0.00032
$^{127}I_2$	$^1\Sigma_g^+$	1.5437	214.5	0.607	0.0374	0.00012
$^7Li^1H$	$^1\Sigma^+$	2.429	1405.6	23.2	7.513	0.21
$^1H^{19}F$	$^1\Sigma^+$	5.86	4138.5	90.07	20.94	0.77
$^1H^{35}Cl$	$^1\Sigma^+$	4.436	2989.7	52.05	10.59	0.30
$^1H^{79}Br$	$^1\Sigma^+$	3.775	2649.7	45.21	8.473	0.23
$^1H^{127}I$	$^1\Sigma^+$	3.053	2309.5	39.73	6.511	0.18
$^{12}C^{16}O$	$^1\Sigma^+$	11.108	2170.2	13.46	1.931	0.017
$^{14}N^{16}O$	$^2\Pi_{1/2}$	6.50	1904.0	13.97	1.704	0.018

Thus

$$B_e = \frac{\hbar}{4\pi c I}, \tag{16.30}$$

with

$$I = \mu R_e^2 = \frac{m_A m_B}{m_A + m_B} R_e^2 = \frac{M_A M_B}{M_A + M_B} u\, R_e^2 \quad \text{kg m}^2. \tag{16.31}$$

Thus

$$R_e = 410.6 \Big/ \sqrt{\frac{M_A M_B}{M_A + M_B} (B_e/\text{cm}^{-1})} \quad \text{pm}. \tag{16.32}$$

For the hydrogen halides, HF, HCl, HBr, HI, we calculate $R_e = 92.0$, 127.9, 142.0, 161.5 pm, respectively.

16.6 Rotation of polyatomic molecules

A linear polyatomic molecule such as HCN, O=C=O or HC≡CH has a rotational spectrum closely analogous to that of a diatomic molecule, if one takes into account the more complicated form of the moment of inertia. Consider the most general case of a linear triatomic molecule:

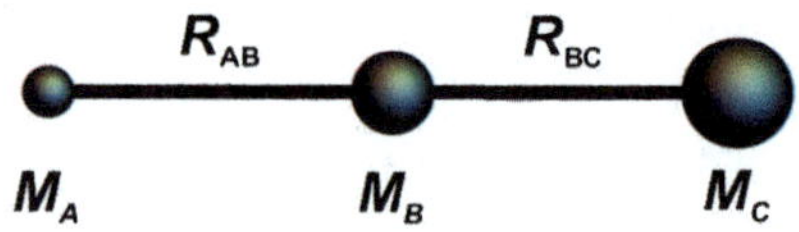

The moment of inertia is given by

$$I = \frac{M_A M_B R_{AB}^2 + M_B M_C R_{BC}^2 + M_A M_C R_{AC}^2}{M_A + M_B + M_C} \tag{16.33}$$

where $R_{AC} = R_{AB} + R_{BC}$. For a symmetric molecule, such as CO_2, with $M_A = M_C = M$ and $R_{AB} = R_{BC} = R$, this reduces to $I = 2MR^2$. For an unsymmetrical linear triatomic molecule such as OCS, the moment of inertia can be calculated from the rotational constant B_e. But this does not provide enough information to determine the individual bond distances R_{CO} and R_{CS}, parameters of greater chemical interest. However, data from two different isotopomers of the molecule—$^{16}O^{12}C^{32}S$ and either $^{16}O^{12}C^{34}S$, $^{16}O^{13}C^{32}S$ or $^{18}O^{12}C^{32}S$—suffice to determine the two bond lengths $R_{CO} = 1.16$ Å and $R_{CS} = 1.56$ Å.

A nonlinear molecule has three moments of inertia about three principal axes, designated I_A, I_B and I_C. The classical rotational energy can be written

$$E = \frac{J_A^2}{2I_A} + \frac{J_B^2}{2I_B} + \frac{J_C^2}{2I_C}, \tag{16.34}$$

where J_A, J_B, J_C are the components of angular momentum about the corresponding principal axes. For example, the moment of inertia about the z-axis of a molecule can be found from

$$I_z = \sum_i m_i (x_i^2 + y_i^2), \tag{16.35}$$

where i runs over all the atoms in the molecule. An atom lying on the z-axis makes no contribution to I_z since $x_i = y_i = 0$.

For a spherical rotor, the three moments of inertia are equal: $I_A = I_B = I_C = I$. Examples are CH_4 and SF_6. The energy simplifies to $J^2/2I$ and the quantum-mechanical Hamiltonian is given by

$$H = \frac{J^2}{2I}. \tag{16.36}$$

The eigenvalues are

$$E_J = \frac{\hbar^2}{2I} J(J+1), \qquad J = 0, 1, 2 \ldots, \tag{16.37}$$

just as for a linear molecule. For a three-dimensional rotor, space-quantization of an angular momentum J gives not only $2J + 1$ possible values of M, the component along the space-fixed z-axis, but also $2J + 1$ possible values of K, which is the component along a molecule-fixed axis. Thus the levels of a spherical rotor have degeneracies of $(2J + 1)^2$ rather than the $(2J + 1)$ for a linear rotor.

A symmetric rotor has two equal moments of inertia, say $I_C = I_B \neq I_A$. The molecules NH_3, CH_3Cl and C_6H_6 are examples. The Hamiltonian takes the form

$$H = \frac{J_A^2}{2I_A} + \frac{J_B^2 + J_C^2}{2I_B} = \frac{J^2}{2I_B} + \left(\frac{1}{2I_A} - \frac{1}{2I_B}\right) J_A^2. \qquad (16.38)$$

Since it its possible to have simultaneous eigenstates of J^2 and its component J_A along the molecule's symmetry axis, the energies of a symmetric rotor have the form

$$E_{JK} = \frac{J(J+1)}{2I_B} + \left(\frac{1}{2I_A} - \frac{1}{2I_B}\right) K^2,$$
$$J = 0, 1, 2 \ldots, \qquad K = 0, \pm 1, \pm 2 \cdots \pm J. \qquad (16.39)$$

There is, in addition, the $(2J + 1)$-fold M degeneracy. A symmetric rotor is classified as *prolate* if $I_A < I_B, I_C$ and *oblate* if $I_A > I_B, I_C$. A linear molecule is a degenerate case of a prolate-symmetric rotor, with $I_A = 0$ and $J_A = 0$. Dipole-allowed transitions of a symmetric rotor have the additional selection rule $\Delta K = 0$ since the electric dipole moment lies in the principal axis and cannot be accelerated to a different orientation within the molecule.

When $I_A \neq I_B \neq I_C$, the molecule is an *asymmetric rotor*. There is then no closed form for the rotational energy eigenvalues but they can be found by interpolation between the energy levels of the prolate and oblate symmetric rotors with $I_A < I_B = I_C$ and $I_A = I_B < I_C$.

16.7 Electronic excitations

The quantum states of molecules are composites of rotational, vibrational and electronic contributions. The energy spacings characteristic of these different degrees of freedom vary over many orders of magnitude, giving rise to very different spectroscopic techniques for studying rotational, vibrational and electronic transitions. Electronic excitations are typically of the order of several electron volts, 1 eV being equivalent to approximately 8000 cm^{-1} or 100 kJ mol^{-1}. As we have seen, typical energy differences are of the order of 1000 cm^{-1} for vibration and 10 cm^{-1} for rotation. Fig. 16.8 gives a general picture of the relative magnitudes of these energy contributions. Each electronic state has a vibrational structure, characterized by vibrational quantum numbers v and each vibrational state has a rotational structure, characterized by rotational quantum numbers J and M.

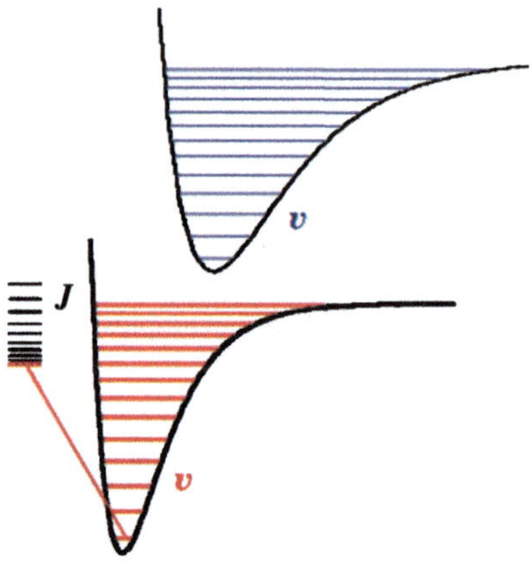

FIGURE 16.8 Schematic representation of the electronic ground state and an excited state of a diatomic molecule. Vibrational levels of the ground state are shown in red, those of the excited state, in blue. The rotational levels for $v = 0$ are also shown.

Every electronic transition in a molecule is accompanied by changes in vibrational and rotational states. Generally, in the liquid state, individual rotational transitions are not resolved, so that electronic spectra consist of broad bands from overlapping rotational transitions. Spectroscopy on the gas phase can often resolve individual rotational, as well as vibrational transitions.

When a molecule undergoes a transition to a different electronic state, the electrons can rearrange themselves much more rapidly than the nuclei. To a very good approximation, the electronic state can be considered to occur instantaneously, while the nuclear configuration remains fixed. This is known as the *Franck-Condon principle*. It has the same physical origin as the Born-Oppenheimer approximation, namely the great disparity in the electron and nuclear masses. In classical terms, the nuclei begin in an equilibrium configuration in the lower electronic state. But after the transition, the nuclei generally find themselves in a nonequilibrium configuration of the new electronic state. Thus they undergo vibrational motion with the original configuration serving as a turning point. Fig. 16.9 shows the energies of the ground and excited states of a diatomic molecule as functions of internuclear distance. Franck-Condon behavior is characterized by *vertical transitions*, in which R remains approximately constant as the molecule jumps from one potential curve to the other.

In a vibrational state $v = 0$ the maximum of probability for the internuclear distance R is near the center of the potential well. For all higher values vibrational states, maxima of probability occur near the two turning points of the potential—where the total energy equals the potential energy. These correspond on the diagram to the end points of the horizontal dashes inside the potential curves. Transitions can actually occur to several excited vibrational levels in the vicinity of v'. The intensity of a transition between the levels v and v' of

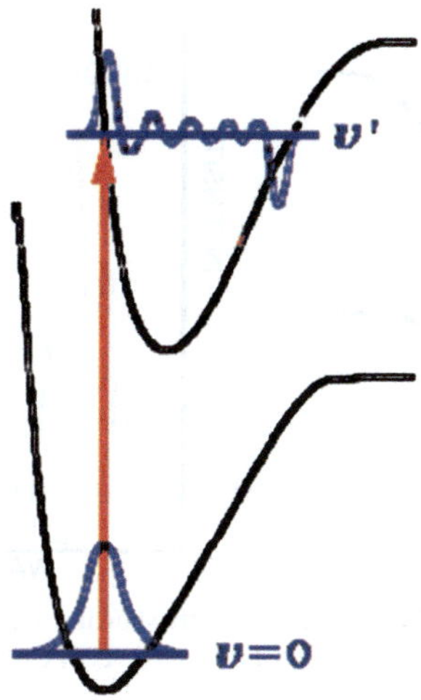

FIGURE 16.9 Most probable transition in absorption from the ground vibrational level according to Franck-Condon principle. The vibrational wavefunctions are shown in blue.

the ground (gnd) and excited (ex) electronic states, respectively, depends on the *Franck-Condon factor*, the overlap of the two vibrational wavefunctions:

$$S_{v'v} = \int \chi_{v'}^{\text{ex}}(R)\, \chi_{v}^{\text{gnd}}(R)\, dR, \tag{16.40}$$

which is a maximum for the vertical transition shown in Fig. 16.9.

A molecule excited to a higher electronic state will, in most instances, spontaneously reemit a photon to return to its ground electronic state. This is called *fluorescence* if it occurs within a few nanoseconds after excitation. Often the molecule will first undergo a sequence of nonradiative decays to the lowest vibrational level of the excited state. The energy involved in this vibrational relaxation is usually converted to thermal energy in surrounding molecules. Transitions from the $v' = 0$ level of the excited state to the several vibrational levels of the ground state are also determined by Franck-Condon factors. The band structure of the fluorescence spectrum is approximately a mirror image of the absorption spectrum, as shown in Fig. 16.10. While the absorption spectrum is characteristic of the vibrational levels of the excited state, the fluorescence spectrum is characteristic of the vibrational levels of the ground state. Fluorescence is familiar in certain dyes which emit brilliant colors between red and green when irradiated with blue or ultraviolet light. Fluorescence spectroscopy is extensively applied in biochemistry and molecular biology. Fluorescent species bound to specific sites in biomolecules can be detected spectroscopically to study complexation and conformational phenomena.

A *Jablonski diagram*, shown in Fig. 16.11, is a simplified representation of some possible absorption and emission processes in molecules. Assuming that

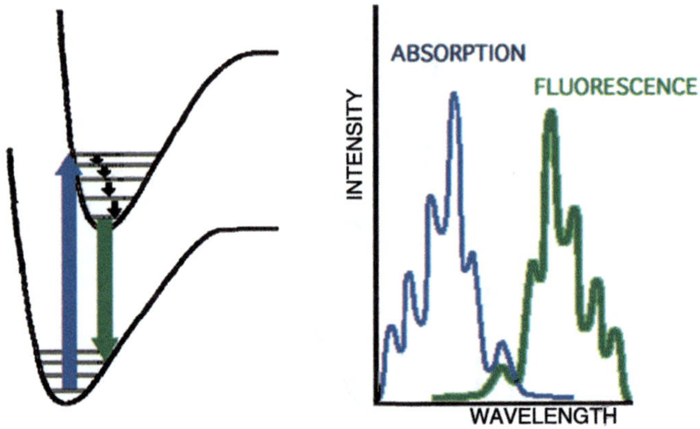

FIGURE 16.10 Absorption and fluorescence spectra showing approximate mirror-image patterns. Black arrows represent relaxation to $\upsilon' = 0$ vibrational level of excited state.

the ground electronic state is a singlet, designated S_0, absorption of radiation can occur to several vibrational levels of the lowest excited singlet state S_1. Several things can then happen to the excited molecule. One possibility, which we described above for a diatomic molecule, is radiationless relaxation to the lowest vibrational level of S_1 followed by emission of a photon, usually within several nanoseconds of the absorption. This is fluorescence, which returns the molecule to one of the vibrational levels of the ground electronic state. Another possible "fate" of the excited state is *intersystem crossing*, a radiationless transition of the molecule to one of the vibrational levels of its lowest electronic triplet state T_1. After relaxation to the lowest vibrational level of T_1, the molecule can again emit a photon to return to the ground state. Since triplet to singlet transitions are forbidden by electric dipole selection rules, the triplet state is *metastable*—it has a relatively long lifetime, typically of the order of 10^{-4} to 100 sec, or even longer. This delayed emission is known as *phosphorescence*. In contrast to fluorescence, phosphorescence persists even after the exciting radiation is discontinued. Transitions between singlet and triplet states, in both intersystem crossing and phosphorescence, are mediated by spin-orbit coupling. By this mechanism, states of different spin and orbital angular momentum can mix, so that the singlet acquires some triplet character and vice versa. Spin-orbit coupling can become significant for organic compounds containing heavier atoms, such as P, S or Cl.

Another possible process is *internal conversion*, radiationless transition to an electronic state of the *same* spin multiplicity, for example $S_1 \leftarrow S_2$, involving excited singlet states. This can be followed by delayed fluorescence. Chemical bonds in electronically excited molecules can also dissociate or rearrange themselves, thereby taking part in *photochemical* reactions.

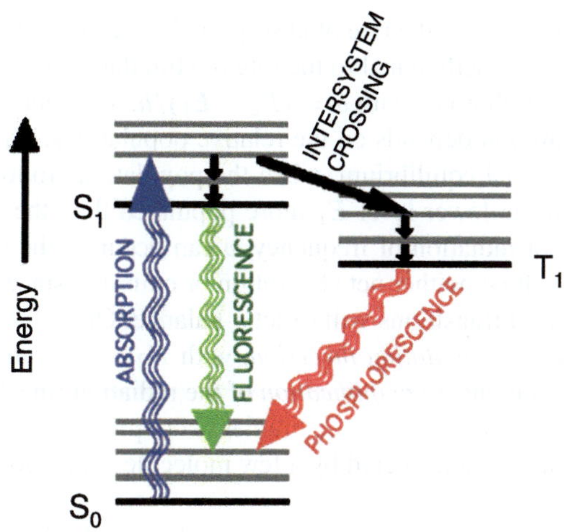

FIGURE 16.11 Jablonski diagram showing radiative processes in color and radiationless processes in black. Short downward arrows represent vibrational relaxation.

16.8 Lasers

The development of lasers has revolutionized several branches of chemistry and spectroscopy, along with its well-known contributions to commerce and industry, not to mention CD players. LASER is an acronym for "Light Amplification by Stimulated Emission of Radiation."

The three fundamental processes involving the interaction of molecules with photons can be represented as generalized chemical reactions. Thus, absorption of a photon can be written

$$X + h\nu \rightarrow X^*,$$

where X^* indicates a molecule excited to a higher quantum state. The inverse process is spontaneous emission

$$X^* \rightarrow X + h\nu.$$

Also possible is stimulated emission

$$X^* + h\nu \rightarrow X + 2h\nu,$$

as first recognized by Einstein in 1917. What is remarkable is that stimulated emission produces a *clone* of the incoming photon. The two emerging photons are identical in frequency, direction and polarization. They are said to be *coherent*. By contrast, photons emitted by different molecules, though they might have the same frequency, will generally have relative phases which are random—they are said to be *incoherent*. The output of a laser is intense coherent radiation produced by stimulated emission.

As we saw in Sect. 6.5, the rate of absorption between two molecular energy levels E_1 and E_2 is exactly equal to the rate of stimulated emission, for a given density of resonant photons, with $\omega = (E_2 - E_1)/\hbar$. Whether there will be *net* absorption or emission depends on the relative populations, N_1 and N_2, of the two levels. At thermal equilibrium, when the populations follow a Boltzmann distribution, with the lower level E_1 more populated than the upper level E_2, a net absorption of radiation of frequency ω can occur. If the two populations are equal, there will be neither net absorption or emission since the rates of upward and downward transitions will exactly balance. Only if we can somehow contrive to achieve a *population inversion*, with $N_2 > N_1$, can we achieve net emission, which amounts to *amplification* of the radiation (the "A" in LASER). Thus to construct a laser, the first requirement is to produce a population inversion. Laser action can be triggered by a few molecules undergoing spontaneous emission.

The principle of laser action was suggested around 1953 by Townes and independently by Basov and Prokhorov. The earliest realizations were *masers*, producing *microwave* amplification by stimulated emission. The first successful optical laser was built by Theodore Maiman in 1960, using synthetic ruby. Crystalline ruby is mostly corundum, Al_2O_3, with approximately one out of every thousand Al^{3+} ions replaced by a Cr^{3+} ion. The relevant energy levels are those of chromium in a trigonal crystal field. The essentials of the ruby laser's operation are represented in Fig. 16.12. A cylindrical rod of ruby is illuminated by a helical xenon flashlamp. Light in the blue and green regions of the spectrum is absorbed by Cr^{3+} ions, pumping a significant fraction of them from their ${}^4A_2$ ground state to several ${}^4F_1$ and ${}^4F_1$ excited states. The excited Cr^{3+} ions subsequently undergo rapid nonradiative transitions to the metastable ${}^2E$ level (lifetime $\approx 4 \times 10^{-3}$ sec) to set up a population inversion with respect to the ground state. A photon produced by spontaneous emission will then initiate stimulated emission of highly monochromatic red light of wavelength 694.3 nm. The ruby rod is made to act as a resonant optical cavity by placing mirrors at both ends so that the radiation bounces back and forth to produce a cascade of stimulated emissions. Since each photon leads to the emission of two photons, intensity of the laser transition will build up rapidly. The mirror on the right in Fig. 16.12 is partially transmitting (about 5%) to allow the laser beam to emerge. The ruby laser must actually operate in a pulsed mode since the population inversion is very difficult to sustain continuously. Note that the laser transition involves a change in spin multiplicity ${}^4A_2 \leftarrow {}^2E$. This becomes an allowed transition in atoms as heavy as Cr because of the significant spin-orbit coupling. The difference in spin multiplicity does however favor the metastability of the ${}^2E$ state.

The ruby laser is the prototype of a *three-level laser*, the most rudimentary arrangement for laser action. A significant improvement is achieved in a *four-level laser*, for example neodymium (Nd^{3+} ions) in YAG (yttrium aluminum garnet), with an energy-level arrangement similar to Fig. 16.13. This has the

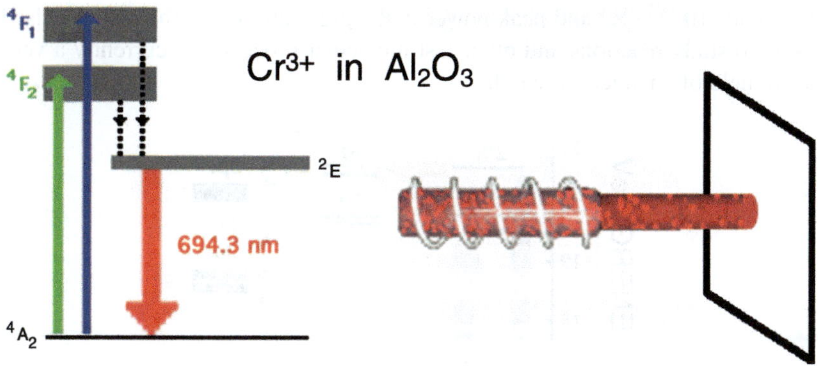

FIGURE 16.12 Essentials of a ruby laser. A cylindrical rod of ruby is illuminated by a helical xenon flashlamp which pumps a significant fraction of the Cr^{3+} ions into the ^4F excited states. Nonradiative relaxation to the ^2E metastable state creates a population inversion which produces the red laser emission. The right-hand mirror on the ruby rod is partly transmitting to allow output of the laser beam.

advantage that the lower metastable level E_2 is initially unpopulated, so that population inversion with the higher metastable level E_3 is much easier to maintain. The Nd-YAG laser can produce several wavelengths in the infrared, the 1064 nm band being the strongest. The intense laser output can be exploited in various nonlinear optical phenomena, including *frequency doubling*, which converts the radiation to green light of wavelength 532 mn.

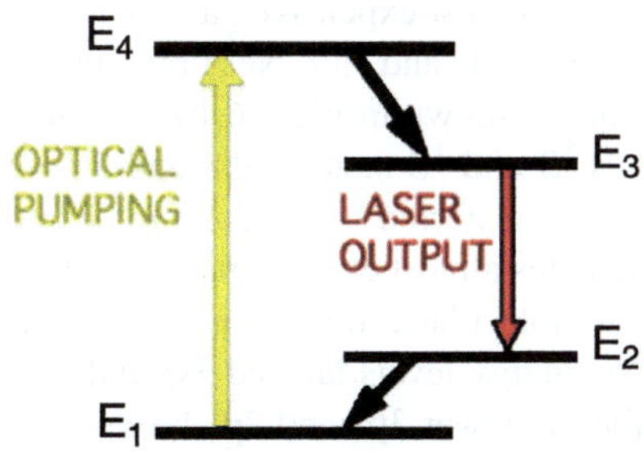

FIGURE 16.13 Schematic energy-level diagram for 4-level laser such as Nd-YAG. Nonradiative processes are shown as black arrows.

The radiation output of a laser is intense, highly monochromatic, sharply collimated and coherent. Laser action has been achieved with thousands of different active media, including solids, liquids and gases, with emitted frequencies spanning the infrared, visible and ultraviolet regions. Most elements in the gas state can be made to lase. Laser action is possible only at wavelengths for which the system has fluorescent emission. The upper levels in laser transitions must be metastable states, with lifetimes typically of the order of 10^{-3} sec. The laser output can be continuous wave (CW) or pulsed, with bursts as short as a picosecond (10^{-12} sec) and peak power in the gigawatt range. The use of pulsed lasers to study reactions and other fast molecular processes is currently a very active field of chemical research.

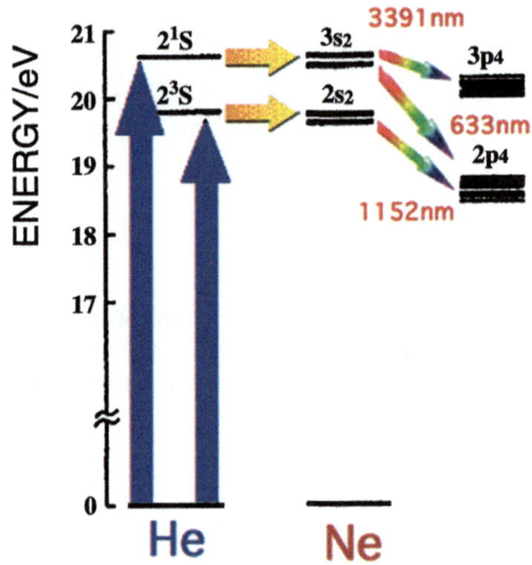

FIGURE 16.14 Helium-neon laser, showing some relevant energy levels. Helium atoms are excited by electric discharge (blue arrows) and transfer energy to neon atoms via collisions (yellow arrows). The most important laser transitions are shown as rainbow arrows.

Of greater interest in chemistry are lasers in which the active medium is a gas. The most common and least expensive gas laser is the *helium-neon laser*, which uses a mixture of 85% He and 15% Ne at pressures of the order of 10 torr. An energy-level diagram is shown in Fig. 16.14. An electric discharge excites He atoms to their $1s\,2s\,{}^3S$ and ${}^1S$ states, 19.8 and 20.6 eV above the ground state. This energy can be transferred by collision to Ne atoms, which very conveniently have excited states almost exactly matching the He levels. Transitions via electric discharge or molecular collisions are *not* restricted by electric-dipole selection rules. The Ne energy levels labeled $3s_2$ and $2s_2$ thus become significantly populated, while the lower $3p_4$ and $2p_4$ levels are initially unpopulated. This inversion allows laser action at several frequencies, the most important being a red line at 632.8 nm and infrared transitions at 1152.3 and 3391.3 nm. The helium-neon laser can also be configured to produce green light at 543.5 nm, yellow at 594.1 nm, orange at 611.9 nm and a multitude of other frequencies. The lower energy levels are depopulated principally by collisions of Ne atoms with the walls of the glass tube, thus maintaining a population inversion which allows the laser to operate in continuous mode.

The *carbon dioxide laser* (see Fig. 16.15) also operates by collisional energy exchange, in this case with nitrogen gas. The N_2 molecules are vibrationally excited by an electric discharge and transfer their energy to excite the ν_3 antisymmetric stretching mode of CO_2. This is enabled by the near coincidence

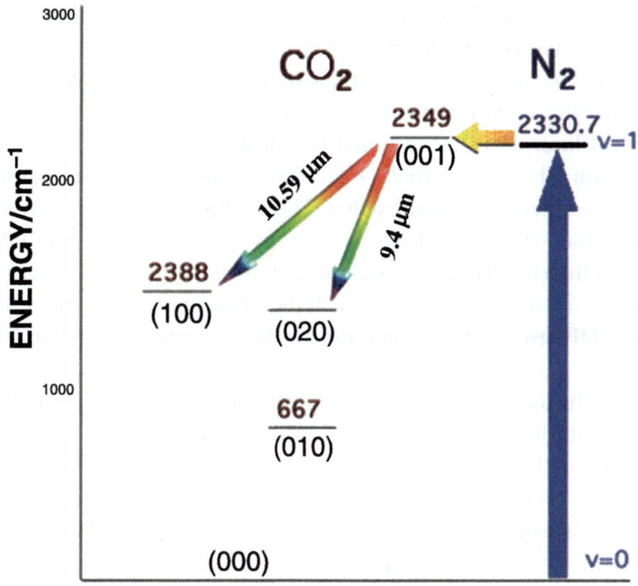

FIGURE 16.15 Carbon dioxide laser, based on population inversion of CO_2 vibrational states $(v_1 v_2 v_3)$.

of the N_2 $v = 1$ level at 2331 cm^{-1} with the CO_2 (001) level ($v_3 = 1$) at 2349 cm^{-1}. The strongest laser emission is the transition from (001) to (100), the first excited level ($v_1 = 1$) of the symmetric stretch ν_1, producing intense infrared radiation at 10.59 μm. A second laser emission is (001) to (020) at 9.4 μm. Transitions between states of different symmetry are made possible by small anharmonicities in the vibrational Hamiltonian. The high-intensity infrared radiation of the CO_2 laser makes it ideal for cutting metals and other materials in manufacturing operations—they essentially melt through a target. The CO_2 laser is also used in skin surgery and other medical applications.

In the *argon-ion laser*, argon gas at about 1 torr is ionized by an electric discharge to produce Ar^+ and Ar^{2+}. These ions can then undergo laser transitions to lower electronic states. The two strongest emissions from Ar^+ produce a blue line at 488 nm and a green line at 514 nm.

Exciplex lasers (also called *excimer lasers*) use reactive halogen atoms to form excited pseudo-molecules with noble gas atoms. Molecules such as XeF^* are stable only in excited electronic states and quickly dissociate after transition to the ground state. This makes possible a large population inversion and produces laser action in the ultraviolet region. A simple prototype for such behavior is the He_2^* excimer, which is an entry in Table 12.1.

Dye lasers are based on organic dye molecules, such as Rhodamine 6G, in liquid solution or suspension. These have the advantage of being tunable over a broad range of wavelengths covering a large number of overlapping $\pi^* \leftarrow \pi$ transitions. Dye lasers are often pumped by other lasers.

Chemical lasers exploit chemical reactions to produce population inversion, thereby converting chemical energy directly into electromagnetic radiation. The first chemical laser was demonstrated by Kasper and Pimentel in 1965, based on the reaction $Cl + H_2 \rightarrow HCl^* + H$. The reaction favors formation of HCl molecules in excited vibrational states. The resulting population inversion can produce laser action in the infrared region. Two devices of current interest are the hydrogen fluoride (HF) laser and the chemical oxygen-iodine laser (COIL), both were considered in connection with the Strategic Defense Initiative ("Star Wars"). The HF laser exploits the most exothermic reaction known, $F + H_2 \rightarrow HF^* + H$, to produce excited HF molecules. Laser emission occurs at 2.7 μm in the infrared. The analogous DF laser (using ^{2}H rather than ^{1}H atoms) shifts the laser wavelength to 3.5 μm, where the atmosphere is more transparent. (Note that the wavelength is proportional to the square root of the reduced mass.) The COIL laser achieves a population inversion in iodine atoms by energy exchange with excited $^1\Delta$ oxygen molecules: $I + O_2^* \rightarrow I^* + O_2$. The laser transition in iodine is $^2P_{3/2} \leftarrow {}^2P_{1/2}$ at 1.315 μm.

Semiconductor lasers (*laser diodes*) now dominate in everyday applications, such as laser printers, CD players, bar-code readers, laser pointers, optical-fiber transmission and numerous other modern devices. These are compact, mobile, low-cost and use very little power. A semiconductor laser is built around a junction of p-type and n-type layers. When electric current is passed through, electrons from the n-region fall into holes in the p-region, producing laser light from the junction. The most commonly used semiconductors are gallium arsenide and other III-V compounds. (Silicon-based semiconductors do not lase since their energy is dissipated as heat.) Aluminum gallium arsenide lasers, used in CD players, produce infrared radiation around 720 nm.

16.9 Raman spectroscopy

The spectroscopic processes we have considered thus far involve absorption, stimulated emission and spontaneous emission of photons, accompanied by transitions among molecular energy states. It is also possible for photons to be *scattered* by molecules. Most scattering events are *elastic*, meaning that the energy of the photon is unchanged, although it can be deflected into a different direction. This process, in which the emitted photon has the same wavelength as the incident photon, is called *Rayleigh scattering*. The intensity of Rayleigh scattering is proportional to λ^{-4} and is thus more effective for shorter wavelengths—the blue end of the visible spectrum. The color of the sky is in fact caused by preferential scattering of the blue wavelengths in sunlight by molecules in the atmosphere. About 1 in every 10^7 scattering events involves *inelastic scattering* in which the incident and emitted photons have different wavelengths. This is known as the *Raman effect*, discovered by the C. V. Raman in 1928. The increase or decrease of photon energy is associated with a simultaneous change in the rotational-vibrational state of the molecule.

Inelastic scattering of photons is exploited in *Raman spectroscopy*, a complementary technique to absorption and emission spectroscopy for studying the energy levels of molecules. Raman spectra use visible light rather than infrared or microwaves to obtain vibrational and rotational parameters. This is especially useful for studying molecules in aqueous solution since the incident frequency can be chosen where there is no absorption by water. The development of lasers has revolutionized Raman spectroscopy. The high intensity and directionality of lasers can compensate for the low probability of Raman scattering. *Stokes lines* are those for which the energy of the scattered radiation is lower than the incident radiation, while *anti-Stokes lines* are those for which the energy of the scattered radiation is higher. The energy shift from the Rayleigh line is equal to some rotational-vibrational energy difference in the molecule. The Raman spectrum of CCl_4 is shown in Fig. 16.16. Note that the Stokes and anti-Stokes lines are equally displaced from the Rayleigh line. Stokes lines are the more intense since anti-Stokes lines can occur only in molecules which are initially in excited vibrational states. Because of the high monochromaticity of the incident laser beam, transitions differing by even a fraction of a wavenumber from the Rayleigh frequency can be detected, which makes pure rotational transitions accessible to laser-Raman spectroscopy.

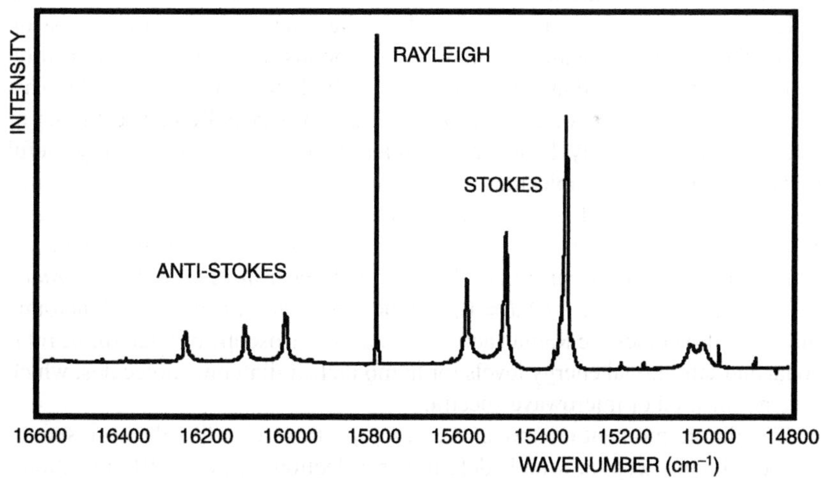

FIGURE 16.16 Vibrational Raman spectrum of CCl_4 using 632.8 nm radiation. Note that wavenumber scale increases right to left.

Infrared and Raman spectroscopy both measure vibrational energies of molecules but according to different selection rules. For a vibration to be infrared (IR) active, the dipole moment of the molecule must change. For example, the asymmetric stretch ν_3 and the bending mode ν_2 in carbon dioxide are IR active while the symmetric stretch ν_1 is not (see Fig. 16.3). For a transition to be Raman active, the *polarizability* of the molecule must change in the vibration. This occurs in the symmetric stretch, as the molecule is expanded and compressed, but not in the other two modes. Thus only ν_1 in CO_2 is Raman active. For molecules like CO_2 possessing a center of inversion, it is observed that no normal mode can be both infrared and Raman-active (it might be inactive to both). This is known as the *mutual exclusion rule*. In molecules lacking a center of symmetry, modes are likely to be active in both infrared and Raman spectroscopy. For example, all the vibrations in H_2O and NH_3 are *both* IR and Raman-active.

For a molecule interacting with a radiation field, the perturbation of the Hamiltonian can be represented by

$$V = -\boldsymbol{\mu}\cdot\mathbf{E} + \tfrac{1}{2}\mathbf{E}\cdot\boldsymbol{\alpha}\cdot\mathbf{E} = -\sum_i \mu_i E_i + \tfrac{1}{2}\sum_{ij} \alpha_{ij} E_i E_j, \tag{16.41}$$

where i, j run over cartesian coordinates x, y, z. This is a generalization of Eq. (6.53), now including contributions quadratic in the electric field of the radiation. The two parts of the perturbation Hamiltonian can be interpreted in terms of the quantum theory or radiation (Sect. 6.5). Terms linear in the electric field $\mathbf{E}$ produce single-photon processes such as absorption and emission, while terms quadratic in $\mathbf{E}$ are the source of two-photon processes, including Rayleigh and Raman scattering. Scattering can be visualized as absorption of a photon as the molecule makes a transition to a "virtual" excited state, followed by reemission of the photon. If the incident radiation corresponds to an electronic transition frequency, the intermediate state is a real excited state of the molecule. This is known as *resonance Raman spectroscopy* and produces Raman scattering of much enhanced intensity. Resonance Raman spectroscopy is particularly useful in selectively studying molecules which are part of complex biological systems.

Since the Raman effect involves *two* spin-one photons, the angular momentum selection rule becomes $\Delta J = 0, \pm 2$. This gives rise three distinct branches in the rotation-vibration spectra of diatomic and linear molecules: the *O-branch* ($\Delta J = -2$), the *Q-branch* ($\Delta J = 0$) and the *S-branch* ($\Delta J = +2$). All diatomic and linear molecules are Raman active. Raman spectroscopy can determine rotational and vibrational energy levels for homonuclear diatomic molecules, which have no infrared or microwave spectra.

The dipole-moment components μ_x, μ_y, μ_z transform like the corresponding coordinates x, y, z, which determines selection rules for IR transitions. Analogously, the components of the polarizability tensor $\alpha_{xx}, \alpha_{xy}\ldots$ transform like the cartesian quadratic forms $x^2, xy\ldots$. Accordingly, a normal mode will be Raman-active if its irreducible representations is the same as one of the cartesian products x^2, xy, etc. This is usually shown in the right-hand column of the group character table. For example, CCl_4 and other tetrahedral molecules have normal modes of A_1, E and T_2 symmetry. The $\mathcal{T}_d$ character table (see Exercise 15.8) shows that these are all Raman-active, while only T_2 is IR-active.

Chapter 16. Exercises

16.1. The Lennard-Jones potential for the interaction of two atoms has the form

$$V(R) = 4\epsilon\left[\left(\frac{\sigma}{R}\right)^{12} - \left(\frac{\sigma}{R}\right)^{6}\right].$$

Determine the value of R_e and D_e. What is the significance the internuclear distance $R = \sigma$?

16.2. From the spectroscopic constants in Table 16.1, calculate the force constant k and the internuclear distance R_e for each of the hydrogen halides. The answers are given in the text.

16.3. Refer to the data on NO in Table 16.1. (i) Calculate the wavenumber of the $J = 0$ to $J = 1$ transition. (ii) Calculate the equilibrium internuclear distance in pm. (iii) Calculate the force constant in N/m. (iv) Taking account of the anharmonic correction, calculate the wavenumber of the $v = 0$ to $v = 1$ transition.

16.4. Estimate the number of bound vibrational states in each of the following molecules: H_2, HD and D_2.

16.5. From the spectroscopic results for $^1H^{35}Cl$, predict ω_e and B_e for $^1H^{37}Cl$, $^2H^{35}Cl$ and $^2H^{37}Cl$.

16.6. For N_2 molecules in their ground electronic and vibrational states at 300 K, determine the most highly populated rotational level J_{max}.

16.7. Derive a formula for the most highly populated level J_{max} of a spherical rotor. Determine its value for CH_4 at 300 K given that $B = 5.24\ \text{cm}^{-1}$.

16.8. Derive Eqs. (16.20) and (16.21) for centrifugal distortion in a diatomic molecule. Assume a molecule with reduced mass μ and equilibrium extension R obeying Hooke's law with force constant k. Let the molecule be stretched from R to $R + \Delta R$ by a centrifugal force

$$m_1\omega^2 r_1 + m_2\omega^2 r_2 = 2\mu\,\omega^2 R,$$

where ω is the angular velocity. Introduce the quantum-mechanical angular momentum using $|J| = I\omega = \sqrt{J(J+1)}\,\hbar$. Determine the rotational energy of the centrifugally-distorted molecule and identify the constant D_J. You may assume that $\Delta R \ll R$ so that

$$\frac{1}{2\mu(R+\Delta R)^2} \approx \frac{1}{2\mu R^2} - \frac{\Delta R}{\mu R^3}.$$

16.9. Referring to the character tables of $\mathcal{C}_{2v}$, $\mathcal{C}_{3v}$ and $\mathcal{D}_{\infty h}$, identify the symmetry species of infrared and Raman-active normal modes of H_2O, NH_3 and CO_2.

16.10. Referring to the $\mathcal{D}_{6h}$ character table, determine the symmetry species of normal modes in benzene C_6H_6 which can be infrared and Raman-active.

Chapter 17

Statistical thermodynamics

The molecular parameters determined by spectroscopy, described in the last Chapter, enable calculation of thermodynamic properties of those molecules, using the methods of statistical thermodynamics. This Chapter will be mainly concerned with ideal gases composed of monatomic or small polyatomic molecules.

Statistical thermodynamics (a branch of statistical mechanics) relates the macroscopic with the microscopic properties of matter. Thermodynamics is based on four inductive generalizations on the behavior of macroscopic matter—the zeroth, first, second and third laws—drawn from extensive observation of physical and chemical phenomena. Applying the laws of thermodynamics, an immense body of experimental facts can be organized and interrelated. An adjunct to the laws of thermodynamics are equations of state and heat capacities, which are macroscopic properties of individual substances. Thermodynamics is concerned principally with the description of physical and chemical phenomena—with the "hows" of the behavior of matter and energy. The underlying mechanisms—the "whys"—are beyond its scope. By contrast, statistical mechanics probes more deeply; it seeks to *explain* as well as describe—to deduce the behavior of macroscopic matter from the known facts of atomic and molecular structure. To handle the immense number of atoms in a macroscopic sample, atomic principles must be supplemented by principles of probability and the laws of large numbers, thus accounting for the "statistical" aspect of the subject.

17.1 Quantum mechanics

It is possible to base statistical thermodynamics on a purely *classical* theory of molecules, treating them as particles which are governed by Newton's laws of motion. This is usually sufficient to account for much of high-temperature behavior, such as the equations of state of gases. But more accurate results depend on quantum mechanics. Actually, a very few results from the quantum theory of molecules suffice to provide all the necessary background for statistical thermodynamics. For easy reference, we review the required quantum mechanical formulas.

The translational motion of gas molecules can be modeled after the quantum-mechanical particle-in-a box. For a particle of mass m in a box of dimensions $a \times b \times c$, the energy levels are given by

Introduction to Quantum Mechanics. https://doi.org/10.1016/B978-0-12-822310-9.00025-2

$$\epsilon_{n_1,n_2,n_3} = \frac{h^2}{8m}\left(\frac{n_1^2}{a^2} + \frac{n_2^2}{b^2} + \frac{n_3^2}{c^2}\right), \quad n_1, n_2, n_3 = 1, 2, 3, \ldots. \tag{17.1}$$

Since the motion is three-dimensional, the energy depends on three quantum numbers.

The rotational energies of a diatomic molecule, in the rigid-rotor approximation, are given by

$$\epsilon_{J,M} = \frac{\hbar^2}{2I}J(J+1), \quad J = 0, 1, 2, \ldots, \quad M = 0, \pm 1, \pm 2, \ldots, \pm J, \tag{17.2}$$

where I is the moment of inertia of the molecule. For a diatomic molecule with atomic masses m_1, m_2, separated by a distance R,

$$I = \mu R^2, \quad \mu = \frac{m_1 m_2}{m_1 + m_2}, \tag{17.3}$$

where μ is the reduced mass. Rotation in a plane involves two degrees of freedom; thus the energy involves two quantum numbers, J and M. However the value of the energy depends only on J, so the different values of M represent degenerate quantum states with the same energy. The energy level ϵ_J is thus $(2J+1)$-fold degenerate. For example, the $J=0$ level is nondegenerate, the $J=1$ level is 3-fold degenerate, with $M = 0, \pm 1$, the $J=2$ level is 5-fold degenerate, with $M + 0, \pm 1, \pm 2$, and so forth.

The vibrational energy levels of a diatomic, in the harmonic-oscillator approximation, are given by

$$\epsilon_v = \left(v + \frac{1}{2}\right) h\nu_0, \quad v = 0, 1, 2, \ldots, \tag{17.4}$$

where ν_0 is the fundamental vibrational frequency.

17.2 Thermodynamic functions

A brief review of mathematical thermodynamics will be useful in developing the results of statistical thermodynamics. The first and second laws of thermodynamics can be encapsulated in the *fundamental equation of thermodynamics*:

$$dU = TdS - pdV. \tag{17.5}$$

This implies that the energy U can be regarded as a natural function of entropy S and volume V. For other combinations of independent variables, it is convenient to define the *thermodynamic potentials* enthalpy $H(S, p) = U + pV$, Helmholtz free energy $A(T, V) = U - TS$ and Gibbs free energy $G(T, p) = U + pV - TS$. These lead to the relations:

$$dH = TdS + Vdp,$$

$$dA = -SdT - pdV,$$
$$dG = -SdT + Vdp. \tag{17.6}$$

Our development of statistical thermodynamics is based on specified values of volume V, temperature T and number of molecules N. Thus the most appropriate thermodynamic potential is the Helmholtz free energy function $A(T, V, N)$. For present purposes, the number of molecules N can be assumed to be fixed. From the second equation of (17.6) it follows that

$$S = -\left(\frac{\partial A}{\partial T}\right)_{V,N} \tag{17.7}$$

and

$$p = -\left(\frac{\partial A}{\partial V}\right)_{T,N}. \tag{17.8}$$

Writing the energy as

$$U = A + TS = A - T\left(\frac{\partial A}{\partial T}\right)_{V,N}, \tag{17.9}$$

we arrive at one of the Gibbs-Helmholtz equations

$$U = -T^2\left(\frac{\partial A/T}{\partial T}\right)_{V,N} = \left(\frac{\partial A/T}{\partial 1/T}\right)_{V,N}. \tag{17.10}$$

We will later compute the heat capacities at constant volume and at constant pressure, given by

$$C_{\mathrm{V}} = \left(\frac{\partial U}{\partial T}\right)_{V,N} \tag{17.11}$$

and

$$C_{\mathrm{p}} = \left(\frac{\partial H}{\partial T}\right)_{p,N}. \tag{17.12}$$

17.3 The Boltzmann distribution

Now we come to the fundamental question in statistical mechanics: given an assembly of N weakly interacting systems (usually molecules), in a fixed volume V at absolute temperature T, each of which has the allowed energy levels $\epsilon_0, \epsilon_1, \epsilon_2, \ldots$, how will the molecules distribute themselves among the energy levels? Denote by $n_0, n_1, n_2, \ldots$ the number of molecules with energies

$\epsilon_0, \epsilon_1, \epsilon_2, \ldots$, respectively. These are known as *occupation numbers*. For a system with N molecules, we have

$$\sum_k n_k = N. \tag{17.13}$$

The total energy of the system is then given by

$$\sum_k n_k \epsilon_k = E. \tag{17.14}$$

If you just want to jump ahead to the final result, the Boltzmann distribution is given by

$$p_k = \frac{e^{-\epsilon_k/kT}}{q}. \tag{17.15}$$

We will go through the details of the derivation in this and the following Sections.

Keep in mind that, for a macroscopic system, N is of the order of 10^{23}. A (hypothetical) detailed enumeration of the energies of each of the N molecules describes what is known as a *microstate*. We wish to calculate the number of possible microstates $W(n_0, n_1, n_2, \ldots)$, $W(n_k)$ for short, which correspond to a *distribution*, described by a set of occupation numbers n_k. This can be derived as follows. First note that the total number of permutations of N objects is given by $N!$. This is far greater than the number of microstates, since it does not take into account the possible permutations among each of the occupation numbers: $n_0!, n_1!, n_2!, \ldots$. However, we can deduce that

$$N! = W(n_k) n_0! n_1! n_2!, \ldots, \tag{17.16}$$

and therefore

$$W(n_k) = \frac{N!}{n_0! n_1! n_2!, \ldots}. \tag{17.17}$$

It is easier to work with the logarithm of this function, which varies monotonically with the function. Now

$$\ln W = \ln(N!) - \ln(n_0!) - \ln(n_1!) - \ln(n_2!) - \cdots = \ln(N!) - \sum_k \ln(n_k!). \tag{17.18}$$

Factorials of large numbers can be represented using Stirling's approximation. A simpler version can be based on the manipulation

$$\ln(n!) = \ln(n) + \ln(n-1) + \ln(n-2) + \cdots = \sum_{k=1}^{n} \ln k, \tag{17.19}$$

but this very long summation can be well approximated by an integral

$$\sum_{k=1}^{n} \ln k \approx \int_0^n \ln x \, dx = n \ln n - n. \tag{17.20}$$

The number of microstates is therefore approximated by

$$\ln W \approx N \ln N - N - \sum_k (n_k \ln n_k - n_k). \tag{17.21}$$

We seek now to maximize of this quantity, subject to the constraints (17.13) and (17.14). A constrained maximum can be found using the method of Lagrange multipliers, whereby we now compute the unconditional maximum of the function

$$F(n_k) = \ln W(n_k) - \alpha \sum_k n_k - \beta \sum_k n_k \epsilon_k, \tag{17.22}$$

with respect to variation in all the n_k. We find

$$\frac{\partial F(n_k)}{\partial n_j} = -\frac{d(n_j \ln n_j - n_j)}{dn_j} - \alpha - \beta\epsilon_j = 0, \quad j = 0, 1, 2, \ldots. \tag{17.23}$$

The result is

$$\ln n_j + \alpha + \beta\epsilon_j = 0, \tag{17.24}$$

or

$$n_j = e^{-\alpha} e^{-\beta\epsilon_j}. \tag{17.25}$$

The Lagrange multiplier α can be determined by the condition $\sum_k n_k = N$, giving

$$e^{-\alpha} = \frac{N}{q}, \quad \text{where} \quad q = \sum_j e^{-\beta\epsilon_k}. \tag{17.26}$$

The probability that a molecule will be in the energy state ϵ_k is given by

$$p_k = \frac{n_k}{N}. \tag{17.27}$$

Thus the Boltzmann distribution takes the form

$$p_k = \frac{e^{-\beta\epsilon_k}}{q}, \tag{17.28}$$

where $\sum_k p_k = 1$ and q is known as the *molecular partition function*. Next we will deduce the explicit forms of β and q and show the connection with thermodynamics.

Ludwig Boltzmann

J. Willard Gibbs

17.4 Molar partition function

The Boltzmann distribution also applies to macroscopic (or molar) systems, in the form

$$P_k = \frac{e^{-\beta E_k}}{Q}, \tag{17.29}$$

where Q is known as the *molar partition function*. For a system of N noninteracting (or weakly interacting) molecules, the total energy is the sum of the individual molecular energies:

$$E_k = \epsilon_{k_1}^{(1)} + \epsilon_{k_2}^{(2)} + \cdots + \epsilon_{k_N}^{(N)}, \tag{17.30}$$

where k_1 represents the state of molecule number (1), k_2, the state of molecule number (2), and so forth. The label k stands for the composite set $\{k_1, k_2, \ldots, k_N\}$. The molar partition function is then given by

$$Q = \sum_k e^{-\beta E_k} = \sum_{k_1} e^{-\beta\epsilon_{k_1}} \sum_{k_2} e^{-\beta\epsilon_{k_2}} \ldots \sum_{k_N} e^{-\beta\epsilon_{k_N}}. \tag{17.31}$$

But since the molecules are identical, the right-hand side is the product of N equal factors, each representing the molecular partition function q, and evidently

$$Q = q^N. \tag{17.32}$$

But wait! We have implicitly assumed that the N individual molecules are distinguishable. But a fundamental principle of quantum mechanics implies that identical particles are indistinguishable. The associations between particle numbers and quantum numbers are merely different aspects of the same quantum state. For N molecules, the overcount would involve a factor of $N!$, the number of ways of permuting N objects. Thus the last equation can be repaired by dividing by $N!$ and we write instead

$$Q = \frac{q^N}{N!}. \tag{17.33}$$

This pertains to molecules in the gaseous or liquid states. However, (17.32) does remain valid for distinguishable molecules, for example, in a perfect crystal by virtue of the fixed positions of the molecules in a crystal lattice.

The average energy of a molar system can be calculated from (17.29) as follows:

$$\overline{E} = \sum_k E_k P_k = \frac{1}{Q}\sum_k E_k e^{-\beta E_k} = -\frac{1}{Q}\frac{\partial Q}{\partial \beta} = -\frac{\partial \ln Q}{\partial \beta}. \tag{17.34}$$

The work of Ludwig Boltzmann and J. Willard Gibbs on statistical thermodynamics concluded that the statistical average values of macroscopic quantities were so overwhelmingly probable that all other values could be neglected. Thus the averages could be identified with macroscopic thermodynamic variables. For example, it could be concluded that

$$\overline{E} = U, \tag{17.35}$$

the thermodynamic energy. We can thus write

$$U = -\frac{\partial \ln Q}{\partial \beta}, \tag{17.36}$$

which is tantalizingly suggestive of the Gibbs-Helmholtz relation (17.10). We will, in fact, conjecture that

$$\beta = 1/kT, \quad \text{and} \quad \ln Q = -A/kT, \tag{17.37}$$

where k is a fundamental constant, first introduced by Boltzmann, with dimensions of energy/temperature. From comparison with experimental results, it is found that Boltzmann's constant is given by

$$k = 1.3806 \times 10^{-23}\ \mathrm{JK}^{-1}. \tag{17.38}$$

Several of the previous formulas can now be written in more explicit form. The Boltzmann distribution for a molecule is given by

$$p_k = \frac{e^{-\epsilon_k/kT}}{q}, \tag{17.39}$$

with the molecular partition function

$$q = \sum_k e^{-\epsilon_k/kT}. \tag{17.40}$$

The Boltzmann distribution has a pyramidal structure with lower energy levels being more heavily populated than higher energy levels. As T is increased, the higher energies increase their fraction of the population. These tendencies are shown schematically in Fig. 17.1.

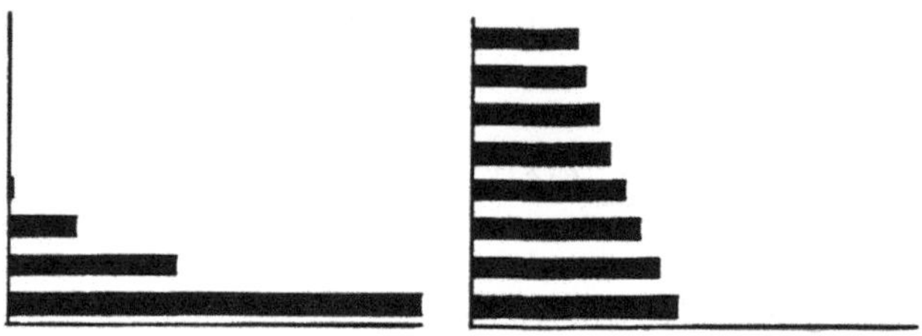

FIGURE 17.1 General forms of the Boltzmann distribution for low temperature (left) and high temperature (right).

The thermodynamic functions described in Sect. 17.2 can be now expressed in terms of the molar partition function $Q(T, V, N)$. The variables T and N are explicitly contained in the formula for Q. The possible dependence of the energies on volume, particularly for gases, accounts for the inclusion of V as an independent variable. As was mentioned earlier, the Helmholtz free energy A is a natural function of T, V and N. The fundamental link between molecular properties and thermodynamics, as suggested by Eq. (17.37), is given by

$$A = -kT \ln Q. \tag{17.41}$$

The other thermodynamic variables then follow. The pressure is determined by

$$p = -\left(\frac{\partial A}{\partial V}\right)_{T,N} = kT\left(\frac{\partial \ln Q}{\partial V}\right)_{T,N}, \tag{17.42}$$

which leads to the equation of state of the system. The energy is given by

$$U = -T^2\left(\frac{\partial A/T}{\partial T}\right)_{V,N} = kT^2\left(\frac{\partial \ln Q}{\partial T}\right)_{V,N}. \tag{17.43}$$

The definition $A = U - TS$ leads to a formula for the entropy:

$$S = \frac{U}{T} + k \ln Q. \tag{17.44}$$

17.5 Ideal monatomic gas

Ideal (or perfect) monatomic gases possess only kinetic energies of translation. Vibration and rotation occur only in polyatomic gases, while intermolecular interactions imply a nonideal gas. We use the particle-in-a-box energies (17.1) to evaluate the molecular partition function. The state of a molecule is determined by the three quantum numbers n_1, n_2, n_3, each running independently from 1 to ∞, thus

$$q = \sum_{n_1}^{\infty}\sum_{n_2}^{\infty}\sum_{n_3}^{\infty} e^{\epsilon_{n_1 n_2 n_3}/kT} = \sum_{n_1}^{\infty} \exp\left(-\frac{h^2}{8ma^2kT}n_1^2\right) \times \sum_{n_2}^{\infty} \exp\left(-\frac{h^2}{8mb^2kT}n_2^2\right) \sum_{n_3}^{\infty} \exp\left(-\frac{h^2}{8mc^2kT}n_3^2\right). \quad (17.45)$$

The three sums are equivalent, differing only in the box dimension a, b, c. Each sum runs over an immense number of terms and can therefore be well approximated by an integral of the form

$$\int_0^{\infty} e^{-\alpha x^2}\, dx = \frac{1}{2}\sqrt{\frac{\pi}{\alpha}}. \quad (17.46)$$

Applying this to each of the three factors above, we obtain

$$q = (2\pi mkT/h^2)^{3/2} a\, b\, c = (2\pi mkT)^{3/2} V/h^3, \quad (17.47)$$

where $V = abc$ is the volume of the box. The molar partition function is then given by $Q = q^N/N!$. Since the thermodynamic functions depend on Q through its logarithm, we evaluate

$$\ln Q = N \ln q - \ln(N!) = N \ln(qN/e), \quad (17.48)$$

using Stirling's approximation: $\ln(N!) = N \ln N - N = N \ln(N/e)$.

The equation of state can be found from (17.42):

$$p = kT\left(\frac{\partial \ln Q}{\partial V}\right)_{T,N}. \quad (17.49)$$

Noting that

$$\ln Q = V^N + \text{terms independent of V}, \quad (17.50)$$

we find

$$p = NkT/V. \quad (17.51)$$

This is just the familiar ideal gas law $pV = nRT$, with $n = N/N_A$; N_A being Avogadro's number 6.022×10^{23}. The Boltzmann constant can thereby be related to the gas constant R by

$$k = R/N_A. \tag{17.52}$$

The energy of an ideal monatomic gas follows from (17.43). Noting that

$$\ln Q = \ln(T^{3N/2}) + \text{terms independent of T}, \tag{17.53}$$

we obtain the well-known result

$$U = \frac{3}{2}NkT = \frac{3}{2}nRT. \tag{17.54}$$

The heat capacities at constant volume and constant pressure follow from

$$C_{\mathrm{V}} = \frac{\partial U}{\partial T} = \frac{3}{2}Nk \tag{17.55}$$

and

$$C_{\mathrm{p}} = \frac{\partial H}{\partial T} = \frac{5}{2}Nk, \tag{17.56}$$

since $H = U + pV = \frac{3}{2}NkT + NkT$.

As noted earlier, the equation of state and energy can be obtained from classical statistical mechanics. The classical analog of the partition function q is the phase integral ω. The phase integral for a single particle in three dimensions is given by

$$\omega = \int\int e^{-H(p,q)/kt} d^3\mathbf{p} d^3\mathbf{q}, \tag{17.57}$$

where $H(p,q)$ is the classical Hamiltonian function. For a monatomic gas molecule

$$H = \frac{p^2}{2m} = \frac{1}{2m}(p_x^2 + p_y^2 + p_z^2). \tag{17.58}$$

Using the integral

$$\int_{-\infty}^{\infty} e^{-p^2/2mkT}\, dp = \sqrt{2\pi mkT} \tag{17.59}$$

and noting that the integral over $d^3\mathbf{q}$ equals the volume V, we obtain

$$\omega = (2\pi mkT)^{3/2} V. \tag{17.60}$$

Thus

$$q = \omega/h^3 = (2\pi mkT)^{3/2} V/h^3, \tag{17.61}$$

in agreement with (17.47). Classical treatments are generally valid at higher temperatures, when the relevant degrees of freedom are said to be *fully excited*.

17.6 The Sakur-Tetrode equation

The entropy of an ideal gas involves the partition function directly, thus the quantum result must be used in order to incorporate the factors h and $N!$. From Eq. (17.44), we find

$$S = \frac{3}{2}Nk + k\ln\left(\frac{q^N}{N!}\right). \tag{17.62}$$

Using Stirling's approximation $N! \approx (N/e)^N$, this can be reduced to

$$S = Nk\ln\left[\frac{(2\pi mkT)^{3/2}Ve^{5/2}}{h^3N}\right], \tag{17.63}$$

a result known as the *Sakur-Tetrode equation.*

The standard molar entropy of an ideal gas can be calculated by setting $N = N_A$, $N_Ak = R$, $V_\mathrm{m}^\ominus = RT/p^\ominus$, with $p^\ominus = 1$ bar. Evaluating the numerical factors, we obtain (per mole)

$$S_\mathrm{m}^\ominus/R = 1.5\ln M + 2.5\ln T - 1.1517, \tag{17.64}$$

where M is the conventional molecular weight in $\mathrm{g\,mol^{-1}}$. For example, the standard molar entropy of argon at 298.15 K, with $M = 39.95$, gives a calculated theoretical value of $S_m^\ominus = 18.6237R = 154.8472\,\mathrm{JK^{-1}\,mol^{-1}}$. The experimental value is reported as $154.84\,\mathrm{JK^{-1}\,mol^{-1}}$. This remarkable agreement represents a dual triumph for both theory and experiment. The experimental determination of absolute entropy involves measurements of heat capacities of the gaseous, liquid and solid states down to a temperature approaching absolute zero, in addition to thermal measurement of phase transitions. Specifically:

$$S_\mathrm{m}^\ominus = \int_0^T \frac{C_\mathrm{p}^\ominus(T)}{T}\,dT + \sum_i \frac{\Delta H_i^\ominus}{T_i}. \tag{17.65}$$

17.7 The Born-Oppenheimer approximation

The energy of a molecule can be rigorously separated into translational energy of the center of mass plus internal energy involving nuclear rotation and vibration, and possibly electronic excitation. In applications to the statistical thermodynamic properties of polyatomic molecules, these degrees of freedom can be well approximated as additive contributions to the energy:

$$\epsilon \approx \epsilon_\mathrm{trans} + \epsilon_\mathrm{rot} + \epsilon_\mathrm{vib} + \epsilon_\mathrm{elec}. \tag{17.66}$$

The molecular partition function correspondingly breaks up into factors for each of these individual degrees of freedom:

$$q = \sum_n e^{-\epsilon_n/kT} \approx \sum_t e^{-\epsilon_t/kT}\sum_r e^{-\epsilon_r/kT}\sum_v e^{-\epsilon_v/kT}\sum_e e^{-\epsilon_e/kT}, \tag{17.67}$$

where t, r, v, e stand for the quantum numbers appropriate to the corresponding degrees of freedom. Within the Born-Oppenheimer approximation, each sum above can be carried out independently:

$$q \approx q_{\text{trans}}\, q_{\text{rot}}\, q_{\text{vib}}\, q_{\text{elec}}, \tag{17.68}$$

where

$$q_{\text{trans}} = \sum_t e^{-\epsilon_t/kT}, \text{ etc.} \tag{17.69}$$

The molar partition function is then given by

$$Q = \frac{q_{\text{trans}}^N}{N!}\, q_{\text{rot}}^N\, q_{\text{vib}}^N\, q_{\text{elec}}^N. \tag{17.70}$$

We associate the $N!$ explicitly with the factor q_{trans} so that all the formulas for the monotonic ideal gas derived above can be carried over unchanged to give the translational contributions for a polyatomic ideal gas. The factorizability of the molecular partition function implies the additivity of contributions to the logarithm:

$$\ln Q = N \ln\left(\frac{q_{\text{trans}} e}{N}\right) + N \ln q_{\text{rot}} + N \ln q_{\text{vib}} + N \ln q_{\text{elec}} \tag{17.71}$$

and, since all the thermodynamic functions depend on $\ln Q$,

$$X = X_{\text{trans}} + X_{\text{rot}} + X_{\text{vib}} + X_{\text{elec}}, \tag{17.72}$$

where X stands for any of the thermodynamic functions A, p, U, S, etc. The translational contributions are given by the monatomic formulas above. It remains to derive the rotational, vibrational and electronic contributions to ideal-gas thermodynamic functions.

The equation of state (17.51) is determined by the dependence of Q on V. However, only the translational contribution contains V, through the particle-in-a-box energy levels. The internal degrees of freedom are independent of V, thus the equation of state $p = NkT/V$ apparently pertains to polyatomic as well as monatomic ideal gases.

17.8 Rotation of diatomic molecules

The energy levels of a diatomic molecule (17.2), expressed in wavenumber units, can be written

$$\frac{\epsilon_{J,M}}{hc} = BJ(J+1), \tag{17.73}$$

where B is the rotational constant in cm^{-1}, determined from spectroscopy. The rotational factor in the molecular partition function is found by summing over all values of J and M:

$$q_{\text{rot}} = \sum_{J=0}^{\infty} \sum_{M=-J}^{J} e^{-hcBJ(J+1)/kT} = \sum_{J=0}^{\infty} (2J+1) e^{-hcBJ(J+1)/kT}. \tag{17.74}$$

Since the energy is independent of M, the sum over M introduces the degeneracy $(2J + 1)$. The above sum cannot be evaluated in closed form. However, since it is generally true that $hcB/kT << 1$, the sum can be well approximated by an integral, as was the case for the translational partition function. The only situation in which this inequality does not hold is for the H_2 molecule and its isotopic variants at low temperatures, ca 20 K. This will be considered separately. In all other cases we can write

$$q_{\text{rot}} = \int_0^{\infty} (2J+1) e^{-hcBJ(J+1)/kT} \, dJ. \tag{17.75}$$

The integral is easily evaluated after a change of variables $u = J(J+1)$, $du = (2J+1)dJ$, leading to

$$q_{\text{rot}} = \frac{kT}{hcB}. \tag{17.76}$$

This result is valid for a heteronuclear molecule, such as HCl or CO. However, for a homonuclear molecule, such as H_2 or Cl_2, interchanging the two atoms, by a rotation of 180°, gives an orientation indistinguishable from the original. Again, quantum mechanics says that these cannot be counted separately, just as there is no way to distinguish identical particles. Analogously to the case of the $N!$ factor for translation, the error can be redressed by dividing the rotational partition function by 2.

The *symmetry number* σ can be defined as the number of indistinguishable orientations which can be obtained by rotations of the molecule. For diatomic molecules, we have either $\sigma = 1$ for the heteronuclear case or $\sigma = 2$ for the homonuclear case. The rotational partition function can thus be written more generally as

$$q_{\text{rot}} = \frac{kT}{\sigma hcB}. \tag{17.77}$$

Linear polyatomic molecules can also be assigned a symmetry number. For example, CO_2 (O=C=O) has $\sigma = 2$ while HCN has $\sigma = 1$.

A symmetry number of 2 actually means that the partition function runs only over even or only over odd values of J. We must consider the exchange of nuclear spins, which can give a symmetric or antisymmetric function. Even rotational quantum numbers $J = 0, 2, \ldots$ are symmetric wrt interchange of nuclei while odd rotational quantum numbers $J = 0, 2, \ldots$ are antisymmetric. The overall symmetry wrt to exchange of nuclei must be antisymmetric for nuclei which are fermions, with nuclear spin $I = \frac{1}{2}, \frac{3}{2}, \dots$ (e.g., protons), and symmetric for nuclei which are bosons, with nuclear spin $I = 0, 1, \dots$ (e.g., deuterons, ^{12}C, ^{16}O).

Rotational contributions to the thermodynamic functions can now be derived. The key results, per mole, are

$$U_m^{\text{rot}} = RT, \quad C_{V,m}^{\text{rot}} = R \quad \text{and} \quad S_m^{\text{rot}} = R \ln\left(\frac{ekT}{\sigma hcB}\right). \tag{17.78}$$

Recall that the translational energy of a gas equals $\frac{3}{2}RT$ per mole, which works out to an energy of $\frac{1}{2}kT$ per translational degree of freedom (3 per molecule). This is a known as the *equipartition principle* of classical statistical mechanics. The rotational contribution equals $\frac{2}{2}RT$, since rotation of a diatomic molecule involves two degrees of freedom. The equipartition result works for translation and rotation but generally fails when applied to molecular vibration (except at high temperature).

It is sometimes useful to express these results in terms of a *rotational characteristic temperature*

$$\Theta_{\text{rot}} = \frac{hcB}{k}. \tag{17.79}$$

The rotational partition function can then be written

$$q_{\text{rot}} = \frac{T}{\sigma \Theta_{\text{rot}}}. \tag{17.80}$$

17.9 Rotation of polyatomic molecules

The rotational Hamiltonian for a nonlinear polyatomic molecule is given by

$$H_{\text{rot}} = \frac{J_1^2}{I_1} + \frac{J_2^2}{I_2} = \frac{J_3^2}{I_3}, \tag{17.81}$$

where I_1, I_2, I_3 represent the three principal moments of inertia and J_1, J_2, J_3, the corresponding angular momentum components. The rotational degrees of freedom are almost always fully excited so that classical statistical mechanics can be used. The phase integral can be written

$$\omega_{\text{rot}} = \int\int e^{-J_1^2/2I_1kT} e^{-J_2^2/2I_2kT} e^{-J_3^2/2I_3kT} \, dJ_1 \, dJ_2 \, dJ_3 \, d(\text{angles}). \tag{17.82}$$

Each integral over an angular-momentum component gives a factor

$$\int_{-\infty}^{\infty} e^{-J^2/2IkT} \, dJ = (2\pi IkT)^{1/2}, \tag{17.83}$$

while the integrals over angles θ, ϕ, ψ gives $4\pi \times 2\pi = 8\pi^2$. The result is

$$\omega_{\text{rot}} = 8\pi^2(2\pi I_1 kT)^{1/2}(2\pi I_2 kT)^{1/2}(2\pi I_3 kT)^{1/2}. \tag{17.84}$$

Taking account of the three degrees of freedom and the appropriate symmetry number σ, the rotational partition function can be written

$$q_{\text{rot}} = \frac{\omega_{\text{rot}}}{\sigma h^3} = \frac{8\pi^2(I_1 I_2 I_3)^{1/2}(kT)^{3/2}}{\sigma h^3}. \tag{17.85}$$

In terms of three spectroscopic constants A, B, C in cm^{-1}, this can be written

$$q_{\text{rot}} = \frac{1}{\sigma}\left[\frac{\pi}{ABC}\left(\frac{kT}{hc}\right)^3\right]^{1/2}. \tag{17.86}$$

When the rotational constants are all different, $A \neq B \neq C$, the molecule is called an *asymmetric rotor*. For a *symmetric rotor*, $C = B \neq A$, while for a *spherical rotor*, $A = B = C$. CH_4 is the classic example of a spherical rotor. Nonlinear polyatomic molecules can have symmetry numbers greater than 2. For example $\sigma = 3$ for NH_3, $\sigma = 12$ for CH_4 and C_6H_6, $\sigma = 24$ for SF_6.

17.10 Molecular vibration

The harmonic-oscillator approximation for vibration of a diatomic molecule gives the energy levels (17.4). Expressed in wavenumber units, this can be written

$$\frac{\epsilon_v}{hc} = \left(v + \frac{1}{2}\right)\tilde{\nu}, \quad v = 0, 1, 2, \ldots, \tag{17.87}$$

where $\tilde{\nu}$ (alternatively written ω_e) is the vibrational constant in cm^{-1} determined by spectroscopy. It is convenient to choose the energy origin such that the lowest vibrational energy level is $\epsilon_0 = 0$. The zero-point energy is now included in the electronic energy of the molecule. The vibrational partition function is then given by

$$q_{\text{vib}} = \sum_{v=0}^{\infty} e^{-hc\tilde{\nu}v/kT}. \tag{17.88}$$

This has the form of a geometric progression:

$$1 + x + x^2 + \cdots = \sum_{v=0}^{\infty} x^v = \frac{1}{1-x}, \tag{17.89}$$

where $x = e^{-u}$, with $u = hc\tilde{\nu}/kT$. This can also be written $u = \Theta_{\text{vib}}/T$, in terms of the *vibrational characteristic temperature* $\Theta_{\text{vib}} = hc\tilde{\nu}/k$. Thus

$$q_{\text{vib}} = (1 - e^{-u})^{-1} = (1 - e^{-hc\tilde{\nu}/kT})^{-1}. \tag{17.90}$$

A useful parameter for computations of both rotations and vibrations is the *second radiation constant*:

$$\frac{hc}{k} = 1.438775 \text{ Kcm (degree per cm}^{-1}). \tag{17.91}$$

(This constant appears in Planck's theory of blackbody radiation.)

The vibrational contributions to the thermodynamic functions follow from

$$\ln Q_{\text{vib}} = N \ln q_{\text{vib}} = -N \ln(1 - e^{-u}). \tag{17.92}$$

Derivatives wrt T can be transformed using

$$\frac{d}{dT} = -\frac{hc\omega_e}{ku^2}\frac{d}{du} = -\frac{T}{u}\frac{d}{du}. \tag{17.93}$$

This leads to the following formulas for the thermodynamic functions per mole:

$$U_{\text{m}}^{\text{vib}} = RT\frac{u}{e^u - 1}, \tag{17.94}$$

$$C_{\text{V,m}}^{\text{vib}} = R\frac{u^2 e^u}{(e^u - 1)^2}, \tag{17.95}$$

$$S_{\text{m}}^{\text{vib}} = R\left[\frac{u}{e^u - 1} - \ln(1 - e^{-u})\right]. \tag{17.96}$$

These are called *Einstein functions*, since they were first obtained by Einstein in connection with the low-temperature thermodynamic behavior of crystals.

Fig. 17.2 shows the temperature dependence of the vibrational contribution to heat capacity. Let us deduce the high- and low-temperature limits. As $T \to \infty$, $u \to 0$ and $e^u \approx 1 + u$. Thus $C_{\text{V,m}}^{\text{vib}} \approx R$ per mole at high temperature. This actually accords with the classical equipartition principle, since the molecular Hamiltonian contains two quadratic contributions: giving $\frac{1}{2}kT$ for the kinetic energy and another $\frac{1}{2}kT$ for the harmonic-oscillator potential energy. In the context of crystals made up of monatomic atoms or ions, this is known as the *law of Dulong and Petit.*

For the low-temperature limit, as $T \to 0$, $u \to \infty$ and $C_{\text{V}}^{\text{vib}} \to 0$, in disagreement with the law of Dulong and Petit and a puzzling result at the beginning of the Twentieth Century. Einstein's explanation of the limiting low-temperature behavior of crystals was an early success of the quantum theory. The limiting dependence as $T \to 0$, $C_{\text{V}}^{\text{vib}} \approx \alpha T^3$, was later derived by Debye.

For polyatomic molecules, vibrations can be represented as the superposition of a set of dynamically-independent concerted atomic motions known as the *normal modes of vibration*. For a molecule with n atoms, there are $3n-6$ normal modes ($3n-5$ for linear molecules). Each normal mode contributes a vibrational frequency $\tilde{\nu}_i$, most often determined from infrared or Raman spectroscopy. The

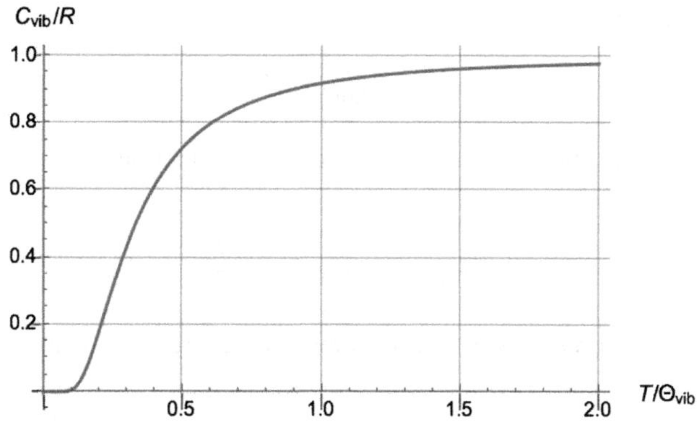

FIGURE 17.2 Vibrational heat capacity over a range of temperatures.

vibrational partition function can be well approximated as a product of normal-mode contributions:

$$q_{vib} = q_1 q_2 \cdots q_{3n-6 \text{ or } 5}, \tag{17.97}$$

and the thermodynamic functions are a corresponding sum of contributions.

17.11 Electronic contributions

Energy differences between electronic states of atoms and molecules, particularly between the ground and first excited states, are typically much greater than kT at ordinary temperatures:

$$\Delta\epsilon_{elec} >> kT. \tag{17.98}$$

Thus, in most cases, Boltzmann factors for excited electronic states are entirely negligible. Notable exceptions are a few molecules with multiplet ground states, including NO, O_2, NO_2 and ClO_2, which have low-lying excited electronic states. The electronic partition function

$$q_{elec} = g_0 e^{-\epsilon_0/kT} + g_1 e^{-\epsilon_1/kT} + \ldots, \tag{17.99}$$

where g_0, g_1, etc. are the degeneracies of the electronic levels. Unlike q_{rot} and q_{vib}, this partition function cannot be explicitly evaluated in any general way. However, in view of the fact that the second (and following) terms are negligible, one can usually include just the first term. For atoms, it is simplest to choose the energy origin such that $\epsilon_0 = 0$. For diatomic molecules, recall that the $v = 0$ vibrational energy level lies at an energy $D_e - D_0$, approximately $\frac{1}{2}h\nu$, above the minimum of the potential-energy curve. It is convenient to define the origin as the energy of the separated atoms at rest, which shifts the ground state to $\epsilon_0 = -D_0$. In view of the preceding discussion, we can usually write, for a diatomic molecule,

$$q_{\text{elec}} = g_0 e^{D_0/kT}. \tag{17.100}$$

Electronic contributions to the thermodynamic functions can then be found from

$$\ln Q_{\text{elec}} = N \ln g_0 + \frac{N D_0}{kT}, \tag{17.101}$$

where D_0 is interpreted as zero for atoms. The energy and heat capacity are given by

$$U_{\text{elec}} = -N D_0, \quad C_V^{\text{elec}} = 0, \tag{17.102}$$

excluding the exceptional cases of low-lying excited states. The electronic contribution to entropy is given by

$$S_{\text{elec}} = Nk \ln g_0. \tag{17.103}$$

For example, the entropy of O_2 is $S^{\ominus}_{\text{elec}} = R \ln 3$ per mole.

17.12 Summary

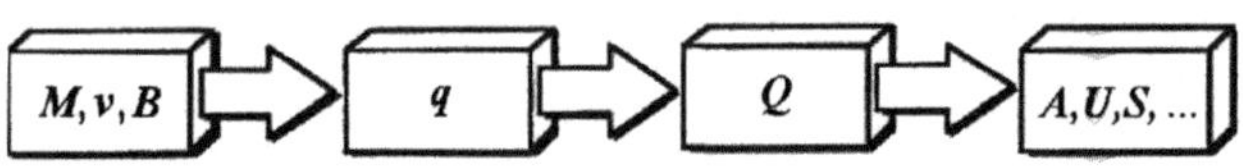

FIGURE 17.3 Thermodynamic functions from molecular parameters.

Fig. 17.3 shows a schematic flowchart for computation of thermodynamic functions from molecular parameters. Table 17.1 summarizes our key results for the statistical thermodynamics of monatomic and polyatomic ideal gases.

TABLE 17.1 Statistical thermodynamics formulas

	Translation	Rotation (linear)	Rotation (nonlinear)	Vibration	Electronic
q	$(2\pi mkT)^{3/2}V/h^3$	$\frac{kT}{\sigma hcB}$	$\frac{1}{\sigma}\left[\frac{\pi}{ABC}\left(\frac{kT}{hc}\right)^3\right]^{1/2}$	$(1-e^{-u})^{-1}$	$g_0 e^{D_0/kT}$
U_m	$\frac{1}{2}RT$	RT	$\frac{3}{2}RT$	$RT\frac{u}{e^u-1}$	$-N_A D_0$
$C_{V,m}$	$\frac{1}{2}R$	R	$\frac{3}{2}R$	$R\frac{u^2e^u}{(e^u-1)^2}$	0
S_m	$R\ln\left[\frac{(2\pi mkT)^{3/2}V_m e^{5/2}}{h^3 N_A}\right]$	$R\ln\left(\frac{ekT}{\sigma hcB}\right)$	$R\ln\left[\frac{1}{\sigma}\left(\frac{\pi}{ABC}\right)^{1/2}\left(\frac{ekT}{hc}\right)^{3/2}\right]$	$R\left[\frac{u}{e^u-1}-\ln(1-e^{-u})\right]$	$R\ln g_0$

Vibrational contributions per normal mode i: $u = hc\tilde{\nu}_i/kT$.

Supplement 17A. Low-temperature heat capacity of hydrogen molecules

H_2 is the lowest boiling point molecular species, remaining in the gas phase down to 20 K. The rotational and vibrational characteristic temperatures are $\Theta_{rot} = 87.5$ K and $\Theta_{vib} = 5986$ K, respectively. At and above room temperature, $T \approx 300$ K, vibration is unexcited while rotation is almost fully excited. The rotational contribution to heat capacity accordingly approaches its equipartition value, R per mole. Owing to its exceptionally small moment of inertia, rotation becomes inactive at temperatures below about 50 K. Below room temperature, the heat capacity as a function of temperature behaves anomalously, as was first explained by Dennison in 1927. Since H_2 is a homonuclear molecule, only half of its rotational states are accessible. In the singlet nuclear-spin state, known as parahydrogen (p-H_2), only even-J rotational states are accessible, while in the triplet nuclear-spin state, known as orthohydrogen (o-H_2), only odd-J rotational states are accessible. The molecular partition functions representing the rotational and nuclear spin degrees of freedom are thereby given by

$$q_{rot}^{para} = \sum_{J \text{ even}} (2J+1)e^{-J(J+1)\Theta_{rot}/T} \tag{17.104}$$

or

$$q_{rot}^{ortho} = 3\sum_{J \text{ odd}} (2J+1)e^{-J(J+1)\Theta_{rot}/T}, \tag{17.105}$$

where the factors 1 and 3 represent the degeneracies of the para and ortho nuclear-spin states, respectively.

The rotational contribution to heat capacity per mole can be calculated using

$$C_{rot}(T) = \frac{\partial}{\partial T}\left(RT^2 \frac{\partial \ln q_{rot}}{\partial T}\right). \tag{17.106}$$

Fig. 17.4 shows plots for o-H_2 and p-H_2. At room temperature, hydrogen gas consists of a 3:1 mixture. The two forms do not interconvert unless a catalyst, such as activated charcoal or platinum is present, so that ratio will persist as the temperature is lowered. The results for the isotopomer HD, a heteronuclear diatomic molecule with $\sigma = 1$ and $\Theta_{rot} = 65.7$ K, is also shown in the figure.

Chapter 17. Exercises

17.1. Boltzmann's famous formula for the entropy of a system at equilibrium is

$$S = k \ln \Omega, \tag{17.107}$$

where Ω represents the number of microstates consistent with a macroscopic state. The molar partition function can be written

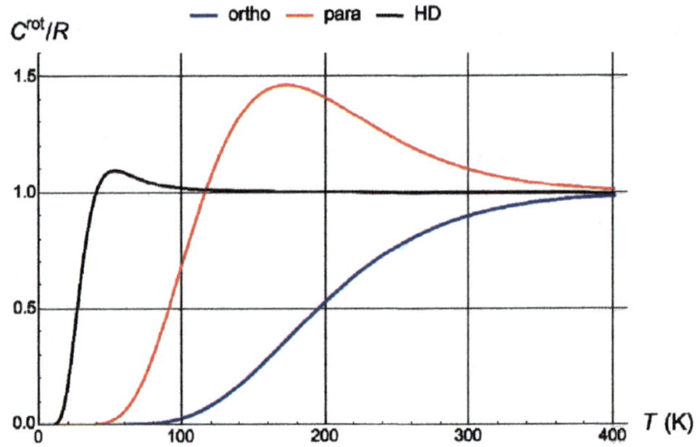

FIGURE 17.4 Heat capacities of ortho and para H_2 and HD.

$$Q = \sum_n g_n e^{-E_n/kT}, \tag{17.108}$$

where the summation is over energy levels, rather than individual states, and g_n represents the degeneracy of the energy level E_n. Assume that the most probable energy is so overwhelmingly dominant that all other states can be neglected. Setting $g_n = \Omega$ and $E_n = U$, show that the statistical mechanical formula for entropy $S = \frac{U}{T} + k \ln Q$ (Eq. (17.44)) leads to Boltzmann's formula.

17.2. The molecules in the atmosphere, mostly N_2 and O_2 experience a gravitational potential energy

$$E(h) = mgh, \tag{17.109}$$

where m is the molecular mass (assume, for simplicity, an value consistent with the statistical average molecular weight of N_2 and O_2, $M \approx .029$), $g \approx 9.8$ m/sec^2 is the acceleration of gravity and h is the altitude above sea level, in meters. Assuming that the atmosphere is at a constant temperature, with a pressure $p(0)$ at sea level, and neglecting kinetic energy and other contributions to the gas molecules, use the Boltzmann distribution to derive the barometer formula for the pressure at an elevation of h meters:

$$p(h) = p(0)e^{-Mgh/RT}. \tag{17.110}$$

Estimate the average atmospheric pressure in Denver, with an elevation $h \approx$ 1600 m (1 mile).

17.3. The classical kinetic energy of a molecule moving in the x-direction with a speed v_x is given by

$$\epsilon(v_x) = \frac{1}{2} m v_x^2. \tag{17.111}$$

Derive an expression for the distribution of molecular speeds, taking $-\infty < v_x < \infty$. Normalizing the distribution to a total probability of 1, show that the probability density is given by

$$\rho(v_x) = \left(\frac{m}{2\pi kT}\right)^{1/2} e^{-mv_x^2/2kT}. \tag{17.112}$$

Next, derive the distribution of molecular speeds $v = \sqrt{v_x^2 + v_y^2 + v_z^2}$. Write the composite Boltzmann distribution for v_x, v_y, v_z and normalize, using polar coordinates in velocity space. The result is the famous Maxwell-Boltzmann distribution:

$$\rho(v) = 4\pi v^2 \left(\frac{m}{2\pi kT}\right)^{3/2} e^{-mv^2/2kT}. \tag{17.113}$$

The distribution is sketched at three different temperatures:

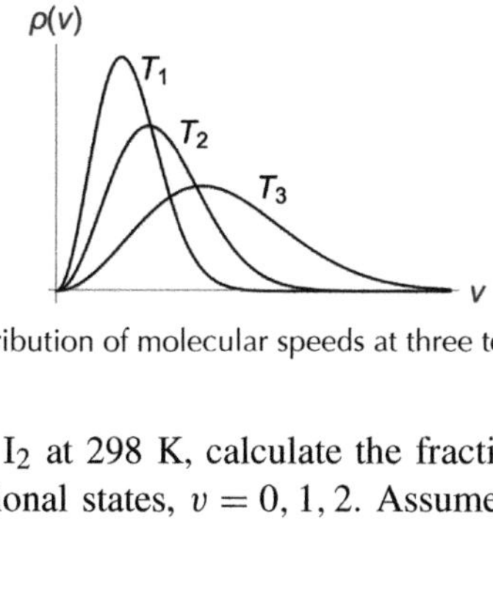

Maxwell-Boltzmann distribution of molecular speeds at three temperatures: $T_3 > T_2 > T_1$.

17.4. For iodine vapor I_2 at 298 K, calculate the fraction of the molecules in the lowest three vibrational states, $v = 0, 1, 2$. Assume a vibrational constant $\tilde{\nu} = 214.5\ \text{cm}^{-1}$.

17.5 The Jth rotational level of a diatomic molecule has a degeneracy of $2J + 1$, taking account of the possible values of M. At a given temperature T, which level is the most populated?

17.6 Apply the preceding result to the rotational levels of N_2 at 298 K. The rotational constant is $B = 1.9987\ \text{cm}^{-1}$.

17.7 Iodine vapor $^{127}I_2$ has the spectroscopic constants $\tilde{\nu}$=214.5 cm^{-1} and B=0.03737 cm^{-1}. Assuming the harmonic vibration and rigid rotor models, calculate the molar value of $C^{\ominus}_{\text{p,m}}$ at 298 K. The experimental value is 36.90 JK^{-1} mol^{-1}.

17.8 Calculate the molar entropy $S^{\ominus}_{\text{m}}$ of iodine vapor at 298 K. The experimental value is $S^{\ominus}_{\text{m}} = 260.69$ JK^{-1} mol^{-1}.

17.9 Calculate the molar heat capacity $C^{\ominus}_{p,m}$ of ammonia NH_3 at 298 K. The fundamental vibrational frequencies are $\nu_1 = 3337$, $\nu_2 = 950$, $\nu_3 = 3414$ (two-fold), $\nu_4 = 1627.5$ (two-fold) cm^{-1}. The experimental value of $C^{\ominus}_{p,m}$ is $35.06\ JK^{-1}\ mol^{-1}$.

Chapter 18

Nuclear magnetic resonance

Nuclear magnetic resonance (NMR) is a versatile and highly-sophisticated spectroscopic technique which has been applied to a growing number of diverse applications in science, technology and medicine. We will consider, for the most part, magnetic resonance involving ^{1}H and ^{13}C nuclei.

18.1 Magnetic properties of nuclei

In all our previous work, it has been sufficient to treat nuclei as structureless point particles characterized fully by their mass and electric charge. On a more fundamental level, as was discussed in Chap. 1, nuclei are actually composite particles made of nucleons (protons and neutrons), which are themselves made of quarks. The additional properties of nuclei which will now become relevant are their spin angular momenta and magnetic moments. Recall that electrons possess an intrinsic or spin angular momentum **s** which can have just two possible projections along an arbitrary direction in space, namely $\pm\frac{1}{2}\hbar$. Since $\hbar$ is the fundamental quantum unit of angular momentum, the electron is classified as a particle of *spin one-half*. The electron's spin state is described by the quantum numbers $s = \frac{1}{2}$ and $m_s = \pm\frac{1}{2}$. A circulating electric charge produces a magnetic moment $\boldsymbol{\mu}$ proportional to the angular momentum **J**. This is written

$$\boldsymbol{\mu} = \gamma\, \mathbf{J}, \tag{18.1}$$

where the constant of proportionality γ is known as the *magnetogyric ratio*. The z-component of μ then has the possible values

$$\mu_z = \gamma \hbar m_J \quad \text{where} \quad m_J = -J, -J+1, \ldots, +J, \tag{18.2}$$

determined by space quantization of the angular momentum **J**. The energy of a magnetic dipole in a magnetic field **B** is given by

$$E = -\boldsymbol{\mu}\cdot\mathbf{B} = -\mu_z B, \tag{18.3}$$

where magnetic field defines the z-axis. The SI unit of magnetic field (more correctly, *magnetic induction*) is the *tesla*, designated T. Electromagnets used in NMR produce fields in excess of 10 T. Small iron magnets have fields around .01 T, while some magnets containing rare-earth elements such as NIB (niobium-iron-boron) reach 1 T. The Earth's magnetic field is approximately 5×10^{-5} T

Introduction to Quantum Mechanics. https://doi.org/10.1016/B978-0-12-822310-9.00026-4

(0.5 gauss in alternative units), depending on geographic location. At the other extreme, a neutron star, which is really a giant nucleus, has a field predicted to be of the order of 10^8 T. The energy relation (18.3) determines the most conveniently units for magnetic moment, namely joules per tesla, $\mathrm{J\,T^{-1}}$.

For orbital motion of an electron, where the angular momentum is ℓ, the magnetic moment is given by

$$\mu_z = -\frac{e\hbar}{2m} m_\ell = -\mu_B \, m_\ell, \tag{18.4}$$

where the minus sign reflects the *negative* electric charge. The *Bohr magneton* is defined by

$$\mu_B = \frac{e\hbar}{2m} = 9.274 \times 10^{-24} \ \mathrm{J\,T^{-1}}. \tag{18.5}$$

The magnetic moment produced by electron spin is written

$$\mu_z = -g \, \mu_B \, m_s, \tag{18.6}$$

with introduction of the *g-factor*. Eq. (18.4) implies $g=1$ for orbital motion. For electron spin, however, $g=2$ (more exactly, 2.0023). The factor 2 compensates for $m_s = \frac{1}{2}$ such that spin and $\ell = 1$ orbital magnetic moments are both equal to one Bohr magneton.

Many nuclei possess spin angular momentum, analogous to that of the electron. The nuclear spin, designated I, has an integral or half-integral value: $0, \frac{1}{2}, 1, \frac{3}{2}$, and so on. Table 18.1 lists some nuclei of importance in chemical applications of NMR. The proton and the neutron both are spin $\frac{1}{2}$ particles, like the electron. Complex nuclei have angular momenta which are resultants of the spins of their component nucleons. The deuteron ^{2}H, with $I = 1$, evidently has parallel proton and neutron spins. The ^{4}He nucleus has $I = 0$, as do ^{12}C, ^{16}O, ^{20}Ne, ^{28}Si and ^{32}S. These nuclei contain filled shells of protons and neutrons with the vector sum of the component angular momenta equal to zero, analogous to closed shells of electrons in atoms and molecules. In fact, all even-even nuclei have spins of zero. Nuclear magnetic moments are of the order of a *nuclear magneton*

$$\mu_N = \frac{e\hbar}{2M} = 5.051 \times 10^{-27} \mathrm{J\,T^{-1}}, \tag{18.7}$$

where M is the mass of the proton. The nuclear magneton is smaller than the Bohr magneton by a factor $m/M \approx 1836$.

In analogy with Eqs. (18.2) and (18.6), nuclear moments are represented by

$$\mu_z = g_I \, \mu_N \, m_I = \hbar \gamma_I \, m_I, \tag{18.8}$$

where g_I is the nuclear g-factor and γ_I, the magnetogyric ratio. Most nuclei have positive g-factors, as would be expected for a rotating positive electric

TABLE 18.1 Some common nuclei in NMR spectroscopy.

nuclide	I	g_I	μ/μ_N	$\gamma_I/10^7 T^{-1} s^{-1}$	abundance %
^1_0n	$\frac{1}{2}$	−3.8260	−1.9130	−18.324	
$^1_1$H	$\frac{1}{2}$	5.5857	2.7928	26.752	99.98
$^2_1$H	1	0.8574	0.8574	4.1067	0.0156
$^{11}_5$B	$\frac{3}{2}$	1.7923	2.6886	8.5841	80.4
$^{13}_6$C	$\frac{1}{2}$	1.4046	0.7023	6.7272	1.1
$^{14}_7$N	1	0.4038	0.4038	1.9338	99.634
$^{15}_7$N	$\frac{1}{2}$	−0.5664	−0.2832	−2.7126	0.366
$^{17}_8$O	$\frac{5}{2}$	−0.7572	−1.894	−3.627	0.037
$^{19}_9$F	$\frac{1}{2}$	5.2567	2.628	25.177	100
$^{31}_{15}$P	$\frac{1}{2}$	2.2634	1.2317	10.840	100

charge. It was long puzzling that the neutron, although lacking electric charge, has a magnetic moment. It is now understood that the neutron is a composite of three charged quarks, udd. The negatively-charged d-quarks are predominantly in the outermost regions of the neutron, thereby producing a negative magnetic moment, like that of the electron. The g-factor for ^{15}N, ^{17}O, and other nuclei dominated by unpaired neutron spins, is consequently also negative.

18.2 Nuclear magnetic resonance

The energy of a nuclear moment in a magnetic field, according to Eq. (18.3), is given by

$$E_{m_I} = -\hbar \gamma_I \, m_I \, B. \tag{18.9}$$

For a nucleus of spin I, the energy of a nucleus in a magnetic field is split into $2I + 1$ *Zeeman* levels. A proton, and other nuclei with spin $\frac{1}{2}$, have just two possible levels:

$$E_{\pm 1/2} = \mp \frac{1}{2} \hbar \gamma \, B, \tag{18.10}$$

with the α-spin state ($m_I = -1/2$) lower in energy than the β-spin state ($m_I = +1/2$) by

$$\Delta E = \hbar \gamma \, B. \tag{18.11}$$

Fig. 18.1 shows the energy of a proton as a function of magnetic field. In zero field ($\mathbf{B} = 0$), the two spin states are degenerate. In a field B, the energy splitting corresponds to a photon of energy $\Delta E = \hbar\omega = h\nu$ where

$$\omega_L = \gamma B \quad \text{or} \quad \nu_L = \frac{\gamma B}{2\pi}, \tag{18.12}$$

known as the *Larmor frequency* of the nucleus. For the proton in a field of 1 T, $\nu_L = 42.576$ MHz, as the proton spin orientation flips from $+\frac{1}{2}$ to $-\frac{1}{2}$. This transition is in the radiofrequency region of the electromagnetic spectrum. NMR spectroscopy consequently exploits the technology of radiowave engineering.

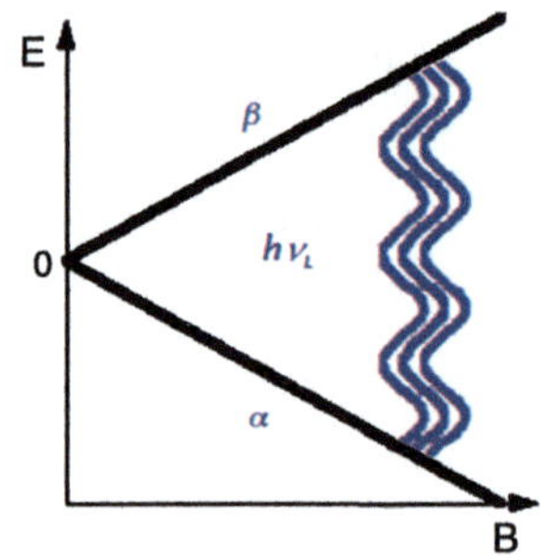

FIGURE 18.1 Energies of spin-$\frac{1}{2}$ nucleus in magnetic field showing NMR transition at Larmor frequency ν_L.

A transition cannot occur unless the values of the radiofrequency and the magnetic field accurately satisfy Eq. (18.12). This is why the technique is categorized as a *resonance* phenomenon. No radiation can be absorbed or emitted by the nuclear spins unless some resonance condition is satisfied. In the earlier techniques of NMR spectroscopy, it was found more convenient keep the radiofrequency fixed and sweep over values of the magnetic field B to detect resonances. These have been largely supplanted by Fourier-transform techniques, to be described later.

The transition probability for the upward transition (absorption) is equal to that for the downward transition (stimulated emission). The contribution of spontaneous emission is negligible at radiofrequencies. Thus if there were equal populations of nuclei in the α and β spin states, there would be zero *net* absorption by a macroscopic sample. The possibility of observable NMR absorption depends on the lower state having at least a slight excess in population. At thermal equilibrium, the ratio of populations follows a Boltzmann distribution

$$\frac{N_\beta}{N_\alpha} = \frac{e^{-E_\beta/kT}}{e^{-E_\alpha/kT}} = e^{-\hbar\gamma B/kT}. \tag{18.13}$$

Thus the relative population difference is approximated by

$$\frac{\Delta N}{N} = \frac{N_\alpha - N_\beta}{N_\alpha + N_\beta} \approx \frac{\hbar\gamma B}{2kT}. \tag{18.14}$$

Since nuclear Zeeman energies are so small, the populations of the α and β spin states differ very slightly. For protons in a 1 T field, $\Delta N/N \approx 3 \times 10^{-6}$. Although the population excess in the lower level is only of the order of parts per million, NMR spectroscopy is capable of detecting these weak signals. Higher magnetic fields and lower temperatures enhanced NMR sensitivity.

18.3 The chemical shift

NMR has become such an invaluable technique for studying the structure of atoms and molecules because nuclei represent ideal noninvasive probes of their electronic environment. If all nuclei of a given species responded at their characteristic Larmor frequencies, NMR might then be useful for chemical analysis, but little else. The real value of NMR to chemistry comes from minute differences in resonance frequencies dependent on details of the electronic structure around a nucleus. The magnetic field induces orbital angular momentum in the electron cloud around a nucleus, thus, in effect, partially shielding the nucleus from the external field B. The actual or *local* value of the magnetic field at the position of a nucleus is expressed

$$B_{\text{loc}} = (1 - \sigma)B, \tag{18.15}$$

where the fractional reduction of the field is denoted by σ, the *shielding constant*, typically of the order of parts per million. The actual resonance frequency of the nucleus in its local environment is then equal to

$$\nu = (1 - \sigma)\frac{\gamma B}{2\pi}. \tag{18.16}$$

A classic example of this effect is the proton NMR spectrum of ethanol CH_3CH_2OH, shown in Fig. 18.2. The three peaks, with intensity ratios 3:2:1 can be identified with the three chemically-distinct environments in which the protons find themselves: three methyl protons (CH_3), two methylene protons (CH_2) and one hydroxyl proton (OH).

The variation in resonance frequency due to the electronic environment of a nucleus is called the *chemical shift*. Chemical shifts on the *delta scale* are defined by

$$\delta = \frac{\nu - \nu^{\text{o}}}{\nu^{\text{o}}} \times 10^6, \tag{18.17}$$

where ν^{o} represents the resonance frequency of a reference compound, usually tetramethylsilane $Si(CH_3)_4$, which is rich in highly-shielded chemically-equivalent protons, as well as being unreactive and soluble in many liquids. By definition $\delta = 0$ for TMS and almost everything else is "downfield" with positive values of δ. Most compounds have delta values in the range of 0 to 12 (hydrogen halides have negative values, up to $\delta \approx -13$ for HI). The hydrogen atom has $\delta \approx 13$ while the bare proton would have $\delta \approx 31$. Conventionally, the

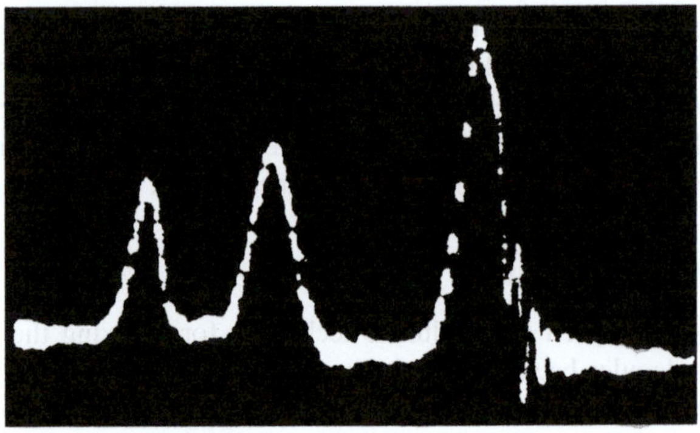

FIGURE 18.2 Oscilloscope trace showing the first NMR spectrum of ethanol, taken at Stanford University in 1951. Courtesy Varian Associates, Inc.

δ-scale is plotted as increasing from right to left, in the opposite sense to the magnitude of the magnetic field. Nuclei with larger values of δ are said to be more *deshielded*, with the bare proton being the ultimate limit. Fig. 18.3 shows some representative values for proton and ^{13}C chemical shifts in organic compounds.

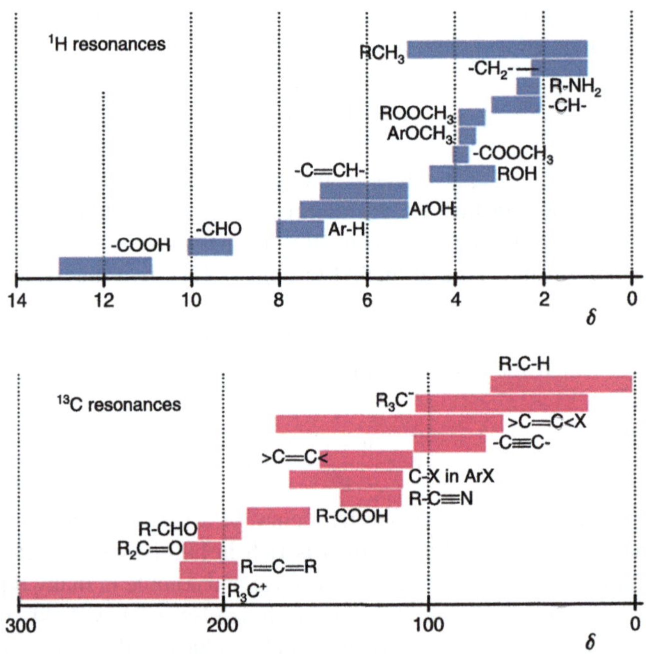

FIGURE 18.3 Ranges of ^{1}H and ^{13}C chemical shifts for common organic functional groups. From P. Atkins, *Physical Chemistry*, (Freeman, New York, 2002).

The δ scale for ^{13}C is based on the ^{13}C resonance in isotopically-enriched TMS. The natural abundance of ^{13}C is about 1.1%. Since carbon nuclei are surrounded by a larger number of electrons, ^{13}C chemical shifts are about an order of magnitude larger than those for protons.

Fig. 18.2 shows a high-resolution NMR spectrum of ethanol, including a δ-scale. The "fine structure" splittings of the three chemically-shifted components will be explained in the next Section. The chemical shift of a nucleus is very difficult to treat theoretically. However, certain empirical regularities, for example those represented in Fig. 18.2 provide clues about the chemical environment of the nucleus. We will not consider these in any detail except to remark that often increased deshielding of a nucleus (larger δ) can often be attributed to a more *electronegative* neighboring atom. For example the proton in the ethanol spectrum (Fig. 18.4) with $\delta \approx 5$ can be identified as the hydroxyl proton, since the oxygen atom can draw significant electron density from around the proton.

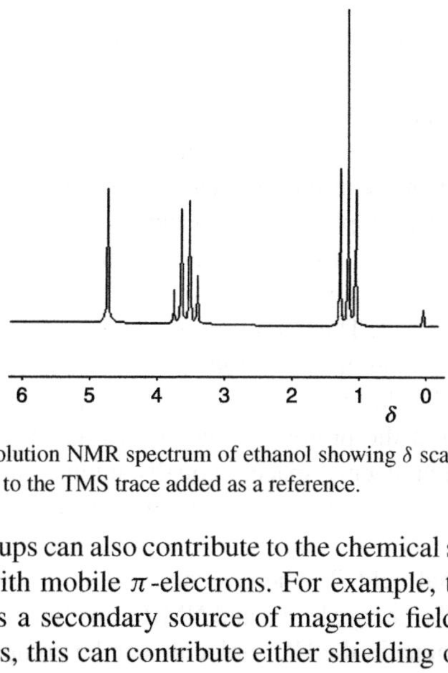

FIGURE 18.4 High-resolution NMR spectrum of ethanol showing δ scale of chemical shifts. The line at $\delta = 0$ corresponds to the TMS trace added as a reference.

Neighboring groups can also contribute to the chemical shift of a given atom, particularly those with mobile π-electrons. For example, the *ring current* in a benzene ring acts as a secondary source of magnetic field. Depending on the location of a nucleus, this can contribute either shielding or deshielding of the external magnetic field, as shown in Fig. 18.5. Where the arrows are parallel to the external field **B**, including protons directly attached to the ring, the effect is *deshielding*. However, any nuclei located within the return loops will experience a *shielding* effect. The interaction of neighboring groups can be exploited to obtain structural information by using *lanthanide shift reagents*. Lanthanides (elements 58 through 71) contain $4f$-electrons, which are not usually involved in chemical bonding and can give large paramagnetic contributions. Lanthanide complexes which bind to organic molecules can thereby spread out proton resonances to simplify their analysis. A popular chelating complex is Eu(dpm)$_3$, tris(dipivaloylmethanato)europium, where dpm is the group $(CH_3)_3C–CO{=}CH–CO–C(CH_3)_3$.

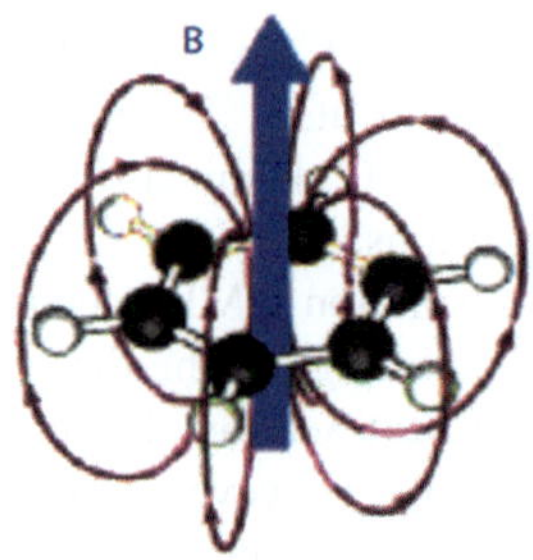

FIGURE 18.5 Magnetic fields, shown as red loops, produced by ring current in benzene.

18.4 Spin-spin coupling

Two of the resonances in the ethanol spectrum shown in Fig. 18.4 are split into closely-spaced multiplets—one triplet and one quartet. These are the result of spin-spin coupling between magnetic nuclei which are relatively close to one another, usually separated by no more than three covalent bonds. Identical nuclei in identical chemical environments are said to be *equivalent*. They have equal chemical shifts and do *not* exhibit spin-spin splitting. Nonequivalent magnetic nuclei, on the other hand, can interact and thereby affect one another's NMR frequencies. A simple example is the HD molecule, in which the spin-$\frac{1}{2}$ proton can interact with the spin-1 deuteron, even though the atoms are chemically equivalent. The proton's energy is split into two levels by the external magnetic field, as shown in Fig. 18.1. The neighboring deuteron, itself a magnet, will also contribute to the local field at the proton. The deuteron's *three* possible orientations in the external field, with $M_I = -1, 0, +1$, with different contributions to the magnetic field at the proton, as shown in Fig. 18.6. The proton's resonance is split into three evenly spaced, equally intense lines (a triplet), with a separation of 42.9 Hz. Correspondingly the deuteron's resonance is split into a 42.9 Hz doublet by its interaction with the proton. These splittings are *independent* of the external field B, whereas chemical shifts are *proportional* to B. Fig. 18.6 represents the energy levels and NMR transitions for the proton in HD. Nuclear-spin phenomena in the HD molecule can be compactly represented by a *spin Hamiltonian*

$$\hat{H} = -\hbar\gamma_H M_H(1-\sigma_H)B - \hbar\gamma_D M_D(1-\sigma_D)B + hJ_{HD}\mathbf{I}_H \cdot \mathbf{I}_D. \quad (18.18)$$

The shielding constants σ_H and σ_D are, in this case, equal since the two nuclei are chemically identical. For sufficiently large magnetic fields B, the last term is effectively equal to $hJ_{HD}M_H M_D$. The *spin-coupling constant* J can be directly equated to the splitting expressed in Hz.

We consider next the case of two equivalent protons, for example, the CH_2 group of ethanol. Each proton can have two possible spin states with $M_I = \pm\frac{1}{2}$, giving a total of four composite spin states. Just as in the case of electron

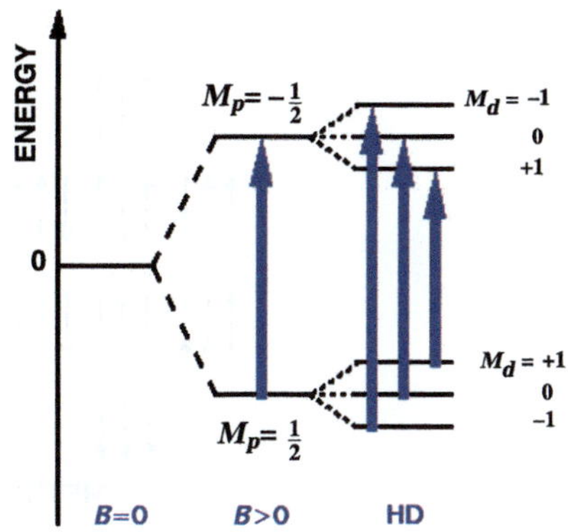

FIGURE 18.6 Nuclear energy levels for proton in HD molecule. The two Zeeman levels of the proton when $B > 0$ are further split by interaction with the three possible spin orientations of the deuteron $M_d = -1, 0, +1$. The proton NMR transition, represented by blue arrows, is split into a triplet with separation 42.9 Hz.

spins, these combine to give *singlet* and *triplet* nuclear-spin states with $M = 0$ and 1, respectively. Also, just as for electron spins, transitions between singlet and triplet states are forbidden. The triplet state allows NMR transitions with $\Delta M = \pm 1$ to give a single resonance frequency, while the singlet state is inactive. As a consequence, spin-spin splittings do *not* occur among identical nuclei. For example, the H_2 molecule shows just a single NMR frequency. And the CH_2 protons in ethanol do not show spin-spin interactions with one another. They *can* however cause a splitting of the neighboring CH_3 protons. Fig. 18.7 (left side) shows the four possible spin states of two equivalent protons, such as those in the methylene group CH_2, and the triplet with intensity ratios 1:2:1 which these produce in nearby protons. Also shown (right side) are the eight possible spin states for three equivalent protons, say those in a methyl group CH_3, and the quartet with intensity ratios 1:3:3:1 which these produce. In general, n equivalent protons will give a splitting pattern of $n + 1$ lines in the ratio of binomial coefficients $1 : n : n(n-1)/2 \ldots$ (These are also rows in Pascal's triangle.) The tertiary hydrogen in isobutane $(CH_3)_3CH^*$, marked with an asterisk, should be split into 10 lines by the 9 equivalent methyl protons.

The NMR spectrum of ethanol CH_3CH_2OH (Fig. 18.4) can now be interpreted. The CH_3 protons are split into a 1:2:1 triplet by spin-spin interaction with the neighboring CH_2. Conversely, the CH_2 protons are split into a 1:3:3:1 quartet by interaction with the CH_3. The OH (hydroxyl) proton evidently does not either cause or undergo spin-spin splitting. The explanation for this is hydrogen bonding, which involves rapid exchange of hydroxyl protons among neighboring molecules. If this rate of exchange is greater than or comparable to the NMR radiofrequency, then the splittings will be "washed out." Only one line with a motion-averaged value of the chemical shift will be observed. NMR has consequently become a useful tool to study intramolecular motions.

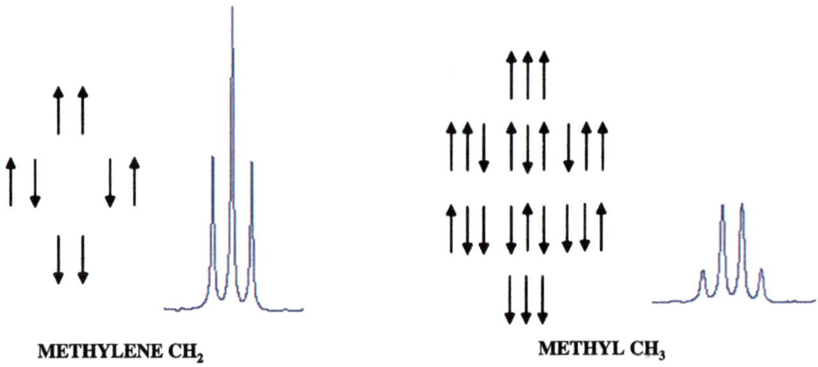

FIGURE 18.7 Splitting patterns from methylene and methyl protons.

18.5 Mechanism for spin-spin interactions

Two magnetic nuclei separated by a distance $\mathbf{r}$ have a mutual potential energy given by

$$V = \frac{\boldsymbol{\mu}_1 \cdot \boldsymbol{\mu}_2}{r^3} - \frac{3(\boldsymbol{\mu}_1 \cdot \mathbf{r})\,(\boldsymbol{\mu}_2 \cdot \mathbf{r})}{r^5}. \tag{18.19}$$

These interactions are responsible for the broad NMR spectra observed in solids. In liquids and gases, these contributions are averaged to zero by rapid tumbling of molecules through all possible orientations. Nuclei in liquids and gases can however still interact through covalent bonds by a second-order mechanism involving *electron* magnetic moments. The dominant contribution is the *Fermi contact interaction*, which comes from the instantaneous juxtaposition of electron and nuclear spins. The interaction energy is given by

$$\Delta E = -\frac{8\pi}{3}|\psi(0)|^2\, \boldsymbol{\mu}_S \cdot \boldsymbol{\mu}_I, \tag{18.20}$$

where $\psi(0)$ represents the electron wavefunction at the position of the nucleus. The energy is lowest when the electron and nuclear spins are *antiparallel*. For an isolated hydrogen atom, the transition between the parallel and antiparallel spin configurations corresponds to 1420 MHz or 21 cm. This is the source of the faint microwave radiation from hydrogen atoms in interstellar space observed in radio astronomy.

Molecular orbitals must have some s character to order to contribute to the Fermi contact interaction since only s-atomic orbitals have nonzero density at the nucleus. When a nuclear spin interacts with a spin-singlet electron pair in a doubly-occupied molecular orbital, it induces a very small *spin polarization* in the electron distribution. In the immediate vicinity of the magnetic nucleus, the density of electron spin *antiparallel* to the nuclear spin is slightly greater, as represented in Fig. 18.8. This is compensated by a larger density of parallel spin elsewhere, such that the net electron-spin angular momentum remains zero.

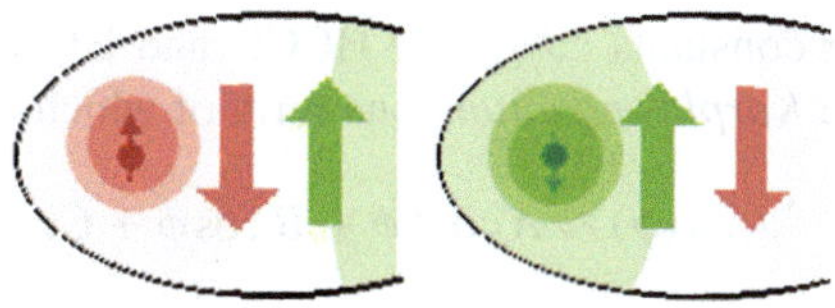

FIGURE 18.8 Spin polarization induced by Fermi contact interaction. Regions shown in green [red] have slight excess of α [β] electron spin density.

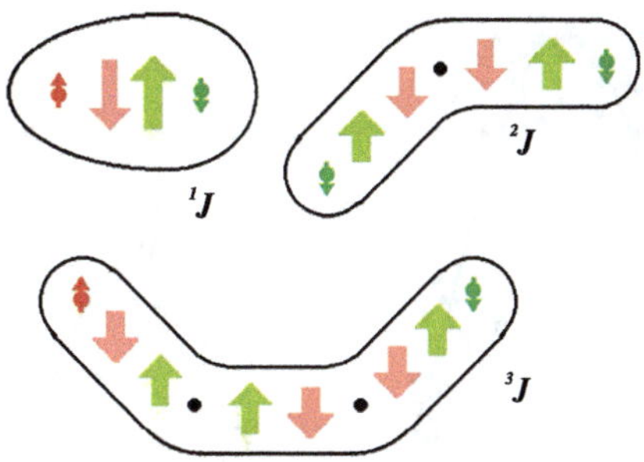

FIGURE 18.9 Mechanisms for electron-coupled nuclear spin-spin interactions for nuclei separated by one, two and three bonds. Only the lowest-energy nuclear-spin states are shown.

Spin polarization produced by Fermi contact interactions makes possible *electron-coupled nuclear spin-spin interactions*. For spin-$\frac{1}{2}$ nuclei in atoms directly bonded to one another, the *antiparallel* nuclear spin state evidently has the lower energy, as shown in Fig. 18.9. The state with parallel nuclear spins (not shown) has the higher energy of the spin-spin doublet. The coupling constant for adjacent nuclei is designated ^1J. For nuclei separated by two or three bonds, the constants are designated ^2J for geminal coupling and ^3J for vicinal coupling. When the coupling goes through more than one bonding orbital, the adjacent electron spin densities around an intermediate atom tend to be *parallel*, thus maximizing contributions to exchange energy. This is the same mechanism which leads to Hund's rule for maximum electron-spin multiplicity. It is clear from Fig. 18.9 that the parallel nuclear spin configuration is more stable in ^2J coupling but the antiparallel configuration is again favored in ^3J coupling. Remember that alternative nuclear-spin states are also allowed, but with higher energy. Since the spin-spin couplings correspond to terms $hJ\mathbf{I}_X \cdot \mathbf{I}_Y$ in the spin Hamiltonian (18.18), the ^1J and ^3J coupling constants tend to be negative while the ^2J coupling constant tends to be positive. These generalizations apply for most couplings involving protons and ^{13}C nuclei in organic compounds. Coupling between nuclei separated by more than three bonds is generally weak but can sometimes be detected. For nuclei heavier than hydrogen, the situation can become much more complex, with additional dipolar interactions between nuclear spins and electron spin and orbital angular momenta.

Vicinal coupling constants $^3J_{HH}$ for HCCH and HCNH group have been approximated by the *Karplus equation*, one form of which is

$$^3J_{HH}(\phi) \approx A\cos^2\phi + B\cos\phi + C, \tag{18.21}$$

where ϕ is the dihedral angle between the two protons. Fig. 18.10 shows a series of $HC_\alpha NH$ $^3J_{HH}$ couplings determined in the study of a protein structure.

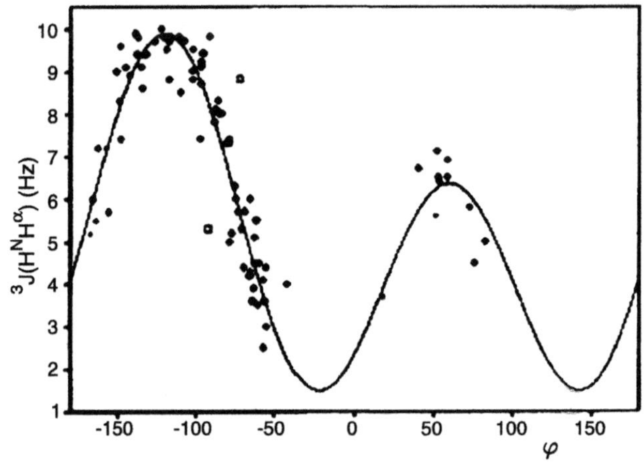

FIGURE 18.10 $^3J_{HH}$ couplings in the enzyme staphylococcal nuclease (SNase) showing least-squares fit to a Karplus equation.

18.6 Magnetization and relaxation processes

A magnetic field $\mathbf{B}_0$ acting on a nuclear moment $\boldsymbol{\mu}$ produces a torque $\boldsymbol{\tau}$, where

$$\boldsymbol{\tau} = \boldsymbol{\mu} \times \mathbf{B}_0. \tag{18.22}$$

From Newton's second law, it follows that the torque equals the rate of change of angular momentum $\mathbf{I}$, so that

$$\boldsymbol{\tau} = \frac{d\mathbf{I}}{dt} = \boldsymbol{\omega} \times \mathbf{I}, \tag{18.23}$$

representing a precessional motion sweeping out a cone around the axis of the field with a frequency ω. Since $\boldsymbol{\mu} = \gamma_I \mathbf{I}$, the precession is seen to occur at the Larmor frequency (18.12)

$$\omega = \omega_L = \gamma_I B_0. \tag{18.24}$$

The individual nuclear magnetic moments in a macroscopic sample add vectorially to give a *net magnetization* $\mathbf{M}$. This will be proportional to the population difference $N_\alpha - N_\beta$, according to Eq. (18.14), and thus be larger at lower temperatures. A magnetic field $\mathbf{B}_0$ in the Z-direction will produce a net *longitudinal* magnetization M_Z. The *transverse* components M_X and M_Y will average to zero when the individual nuclei precess with random phases. It is possible to change the net magnetization by exposing the nuclear spin system to radiation at the resonance frequency. If enough energy is put in, the spin system can become *saturated*, with the α and β populations becoming equal, so that $M_Z = 0$. The magnetization subsequently returns to its equilibrium value M_0 with a time constant T_1 called the *spin-lattice relaxation time*, according to

$$M_Z = M_0(1 - e^{-t/T_1}). \tag{18.25}$$

Spin-lattice relaxation is induced by time-varying magnetic fields in the environment of the spin system, principally from molecular rotations.

If the net magnetization were suddenly rotated into the XY plane (we will learn how shortly), $\mathbf{M}$ would precess about the Z-axis at the Larmor frequency ω_L, with all the spins initially rotating in phase with one another. In time, however, the spins will begin to dephase, as well as recovering their longitudinal magnetization. The time constant T_2 which describes the return to equilibrium of the transverse magnetization, M_{XY}, is called the spin-spin relaxation time, so that

$$M_{XY} = M_0 e^{-t/T_2}. \tag{18.26}$$

The net magnetization in the XY plane then goes to back to zero while the longitudinal magnetization along Z returns to M_0. Both processes occur simultaneously with T_1 always greater than or equal to T_2.

Thus far, we have considered the motion of spins with respect to the *laboratory* frame of reference. It is convenient to define a *rotating* frame of reference which follows the precession about the Z-axis at the Larmor frequency. The rotating coordinates are designated as X' and Y'. A magnetization vector rotating at the Larmor frequency in the laboratory frame appears to be stationary in the rotating frame, while the relaxation of M_Z magnetization to its equilibrium value looks the same in both frames. A transverse magnetization vector rotating faster or slower than the Larmor frequency appears to be rotating clockwise or counterclockwise, respectively, in the $X'Y'$-frame. Fig. 18.11 shows how the transverse magnetization simultaneously dephases and relaxes toward its equilibrium longitudinal magnetization. When viewed in the rotating frame, the transverse component fans out over a sector of a circle as the phases of the individual spins lose coherence. Concurrently, the magnetization vector becomes concentrated in narrower and narrower cones of precession. The time constants T_2 and T_1 govern the respective processes.

18.7 Pulse techniques and Fourier transforms

Radiation which induces NMR transitions is usually produced by radiofrequency (RF) coils wound around the sample. These are energized by alternating

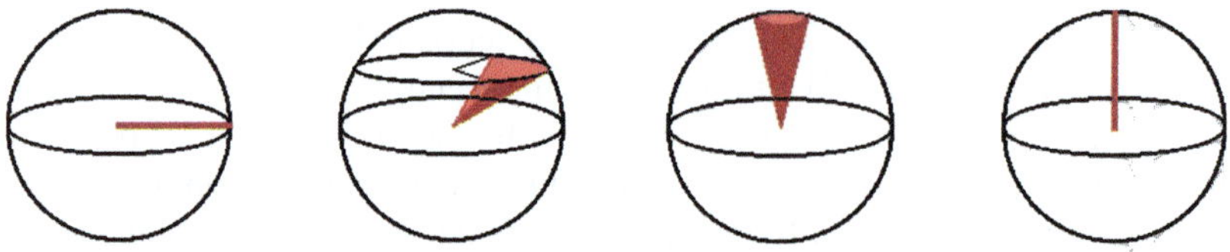

FIGURE 18.11 Relaxation of transverse magnetization as viewed in the rotating frame.

currents approximately matching the Larmor frequency. The coils are arranged so that the magnetic field of the radiation $\mathbf{B}_1$ is perpendicular to the static field $\mathbf{B}_0$. In contrast to other types of spectroscopy, the electric field of the radiation plays no role here. The simplest case is a circularly polarized RF field, such that, in the laboratory frame

$$B_{1X} = B_1 \cos \omega t, \qquad B_{1Y} = B_1 \sin \omega t. \tag{18.27}$$

Transformed to a frame rotating at frequency ω, the radiation field simplifies to

$$B_{1X'} = B_1 = \text{const}, \qquad B_{1Y'} = 0, \tag{18.28}$$

equivalent a constant magnetic field in the rotating frame. A magnetization initially in the Z-direction can then precesses around the Y' axis with an angular frequency

$$\omega_1 = \gamma_I B_1, \tag{18.29}$$

known as the *Rabi frequency*. If the radiation consists of a finite pulse of duration Δt, Rabi precession turns the magnetization through an angle $\theta = \omega_1 \Delta t$. Rotation of $\mathbf{M}$ into the XY-plane, imagined in the preceding Section, can be thus accomplished with a 90° pulse, for which $\theta = \pi/2$. A 180° pulse, with $\theta = \pi$, causes a reversal of the magnetization vector from $\mathbf{M}$ to $-\mathbf{M}$.

The earliest NMR spectra were obtained by varying the radiofrequency in a constant magnetic field or by varying the field with a fixed frequency and recording the individual resonances. The slow and laborious continuous-wave (CW) approach has been largely supplanted by Fourier-transform techniques (FT-NMR) in which all the nuclei are excited at the same time by a radiofrequency pulse. This can produce the entire spectrum in a single step, often in less than a second. Andrew Derome suggested [in *Modern NMR Techniques for Chemistry Research* (Pergamon, New York, 1987)] the following analogy for FT-NMR, comparing it to tuning a bell. In principle, you could measure each resonance frequency by exciting the bell with a series of tuning forks. But this would be extremely tedious. A better way is to hit the bell with a hammer, exciting all of its resonances at once, and Fourier-analyzing the BOINNNGGG to identify the spectrum of frequencies.

How does pulsed Fourier-transform NMR spectroscopy work? An RF coil generally produces radiation at a single frequency ω_0. However, if the radiation is emitted as a very short pulse, of the order of microseconds, this is equivalent to a superposition of many frequencies in a broad band around ω_0 and it can excite the resonances of all the spins in a sample at the same time. The frequency spread and pulse length are related by $\Delta\omega\Delta t \approx 1$, which has the same form as the energy-time uncertainty principle. After a high-intensity 90° pulse, as described above, the nuclear magnetizations evolve much like that shown in Fig. 18.11. Back in the laboratory frame, we would see an *oscillating* transverse magnetization. This causes *reemission* of radiofrequency radiation, which can be picked up by detecting coils positioned around the sample. The resulting signal, during the time the magnetization is returning to its equilibrium state, is called *free-induction decay* (FID). It is a superposition of *all* the resonance frequencies of the sample. Fourier transformation, carried out by a computer built into the NMR spectrometer, translates the signal from the time domain t to the frequency domain ω. The resulting FT-NMR spectrum is similar in appearance to earlier CW spectra but with resolution and sensitivity improved by several orders of magnitude. Fig. 18.12 shows free-induction decays and Fourier transforms for model systems with just one and two resonance frequencies. Fig. 18.13 shows a more realistic example, the FID signal for ethanol. A frequency ω_0 gives an exponentially-decaying sinusoidal function

$$f(t) = f(0)\cos\omega_0 t\, e^{-t/T_2}, \tag{18.30}$$

where T_2 is the transverse relaxation time. The complete FID signal is a superposition of all contributing frequency components. The spectral distribution is obtained from the real Fourier transform of the FID signal

$$S(\omega) = \int_0^\infty f(t)\cos\omega t\, dt. \tag{18.31}$$

18.8 Two-dimensional NMR

For a complex molecule such as a protein, a high-resolution NMR spectrum might contain hundreds of lines and its analysis would be essentially hopelessly. In the 1970s Richard Ernst and coworkers developed sophisticated techniques involving *sequences* of RF pulses, followed by Fourier analysis in more than one dimension (Ernst received the Nobel Prize in 1991). This has made it possible to determine the structure of proteins and other biomolecules with a facility rivaling X-ray crystallography. We will consider only the simplest such techniques, a method for two-dimensional NMR known as correlation spectroscopy (COSY). Conventional (one-dimensional) NMR spectra are plots of intensity vs. frequency. In two-dimensional spectroscopy, intensity is plotted as a function of *two* frequencies, in what resembles a topographical map. In one-dimensional FT-NMR, the signal is recorded as a function of time and then Fourier transformed to the frequency domain. In two-dimensional NMR, the

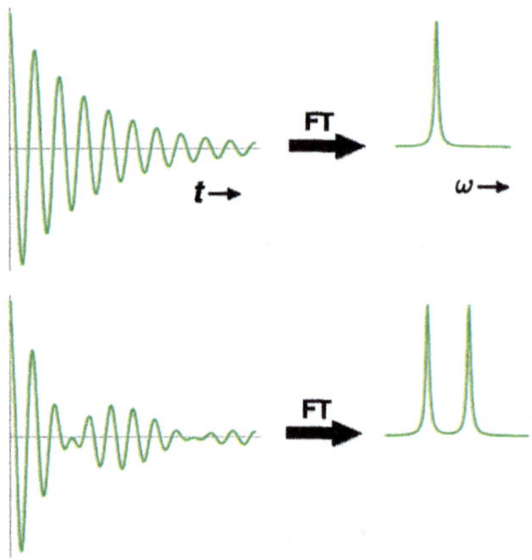

FIGURE 18.12 Free-induction decay signals and their Fourier transforms for simple systems with one and two resonance frequencies.

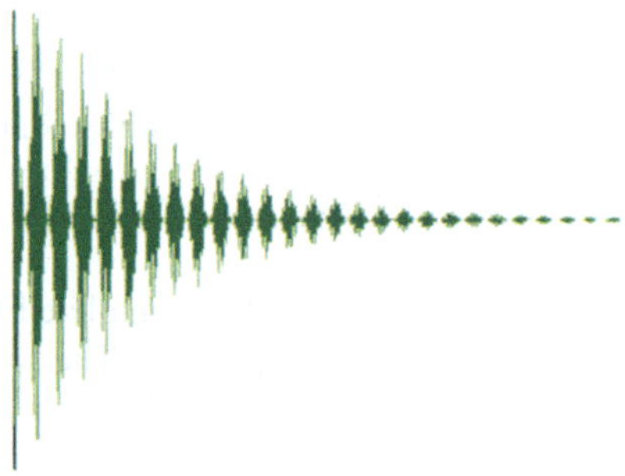

FIGURE 18.13 Free-induction decay (FID) signal for the proton resonances in ethanol. Fourier transformation gives the NMR spectrum shown in Fig. 18.4.

signal is recorded as a function of two time variables, t_1 and t_2, and the result Fourier transformed twice to yield a function of two frequency variables. The general scheme for two-dimensional spectroscopy is a Preparation-Evolution-Mixing-Detection (PEMD) sequence of operations, as shown in Fig. 18.14. In the preparation phase, the sample allowed to come to equilibrium and is then excited by a 90° pulse. The resulting magnetization is allowed to evolve for time t_1. During the mixing phase, the system is subjected to another 90° pulse. The signal is then recorded as a function of the second time variable t_2. The cycle is repeated for incrementally increased time intervals t_1, beginning with $t_1 = 0$, and the resulting function of t_1 and t_2 is Fourier transformed.

The two-dimensional COSY spectrum for ethanol is shown in Fig. 18.15, with the two frequency coordinates expressed in terms of chemical shifts δ_1 and δ_2. In Sects. 18.3–18.4, we had considered in detail the NMR spectrum for this molecule. The one-dimensional spectrum, drawn at the top, consists of three chemical-shifted components for the CH_3, OH and CH_2 protons, with two of

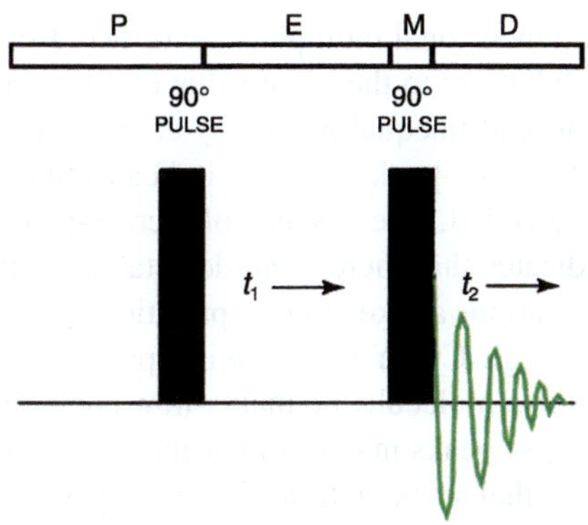

FIGURE 18.14 Sequence of operations in correlation spectroscopy (COSY): Preparation, Evolution, Mixing, Detection. The signal is analyzed by a two-dimensional Fourier transform.

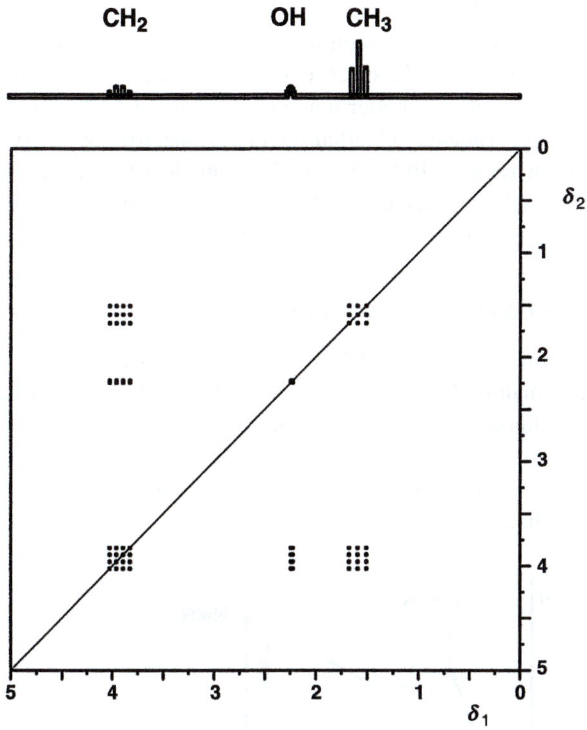

FIGURE 18.15 COSY spectrum for ethanol. Frequencies are expressed in terms of chemical shifts δ. The corresponding 1D-NMR spectrum is shown at the top.

them further split into multiplets. The COSY spectrum is symmetrical about the diagonal, with the diagonal peaks iterating the one-dimensional spectrum. In the two-dimensional spectrum, multiplets appear as square or rectangular arrays of peaks. These two-dimensional multiplets are of two distinct types: diagonal-peak multiplets which belong to the same value of δ and cross-peak multiplets which are centered around unequal δ_1 and δ_2 coordinates. The appearance in a COSY spectrum of a cross-peak multiplet indicates that the protons at shifts δ_1 and δ_2 must be J-coupled. The absence of a cross-peak connecting the OH and CH_3 protons indicates that there is no detectable coupling between them. These simple generalizations are key to interpretation of two-dimensional NMR spectra. Thus, from a single COSY spectrum it is possible to trace out the whole network of couplings in a molecule, no matter how complex.

What causes the cross-peaks in two-dimensional spectra? If a peak occurs at some δ_1, δ_2, this means that an excitation of frequency δ_1 was present during the evolution time t_1. And during the mixing time some magnetization was transferred to another excitation of frequency δ_2, which was detected during time t_2. Only nuclei which are connected by spin-spin coupling can experience such transfers of magnetization.

Many other pulse-sequence techniques besides COSY can be used to produce multidimensional NMR spectra. It will suffice here to simply list the acronyms of some of the better known methods: EXSY (exchange spectroscopy), NOESY (nuclear Overhauser effect spectroscopy), TOCSY (total correlation spectroscopy), ROESY (rotational nuclear Overhauser effect spectroscopy). The nuclear Overhauser effect (NOE) refers to a change in intensity of one NMR peak when another peak is irradiated.

18.9 Magnetic resonance imaging

Magnetic resonance imaging (MRI) is a noninvasive technique for viewing the inside of the human body. The level of detail it provides is extraordinary compared with the alternative imaging methods for medical diagnosis and anatomical studies. In contrast to X-rays, MRI can image the body's soft tissue. Images of the brain (Fig. 18.17) can enable early detection of cerebrovascular or neurodegenerative disease.

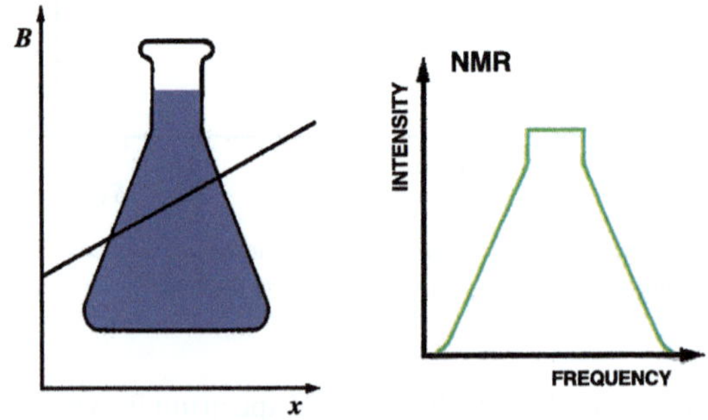

FIGURE 18.16 Principle of MRI. A flask of water in a magnetic field with a linear gradient gives the NMR spectrum shown at right.

The basic idea of MRI is that a system of protons in an *inhomogeneous* magnetic field will exhibit a variation in resonance frequency proportional to the strength of the field. Thus the geometry of an object can be translated into a frequency map. The principle is illustrated in Fig. 18.16, applied to a flask of water. With field gradients applied successively along the x, y and z axes, computer analysis can reconstruct a three-dimensional image. Image contrast in a biological system can be enhanced by using pulse sequences which exploit differences in T_1 and T_2 for protons in various types of tissue. With a time dimension added, *functional MRI* can detect such things as changes in different parts of the brain during perceptive or cognitive activity.

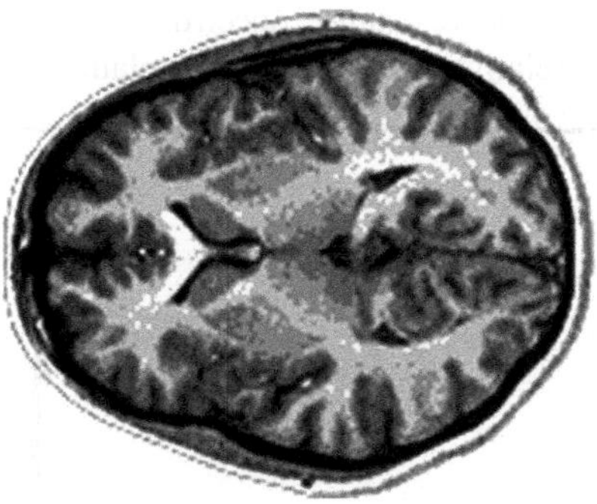

FIGURE 18.17 MRI of axial cross-section of human brain. Extensive cortical folding is correlated with high intelligence.

Chapter 18. Exercises

18.1. Analyze the proton NMR spectrum of diethylketone, shown below.

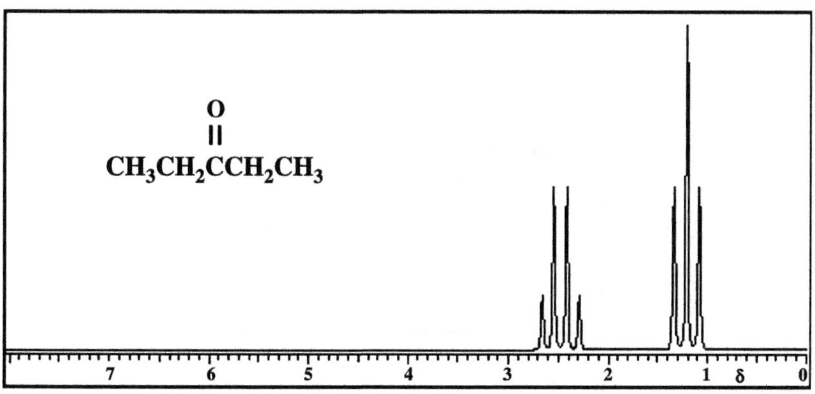

18.2. The NMR spectrum of methane CH_4 shows just a single peak. Explain why. Now explain the proton NMR spectrum of the isotopically-substituted dideuteromethane, shown below.

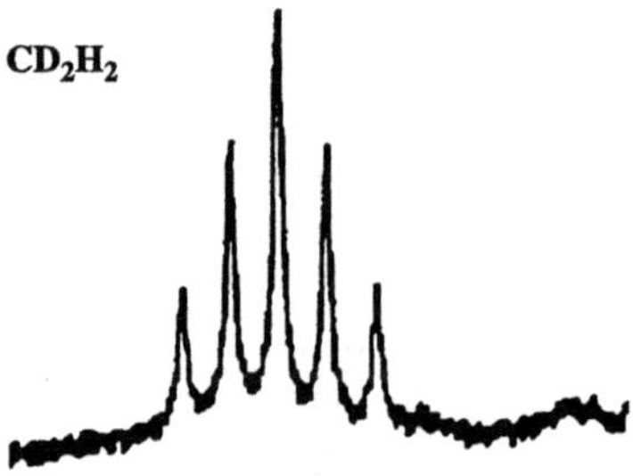

18.3. The proton magnetic resonance spectrum of toluene (methylbenzene) shows two peaks with relative intensities 5:3. Explain this spectrum.

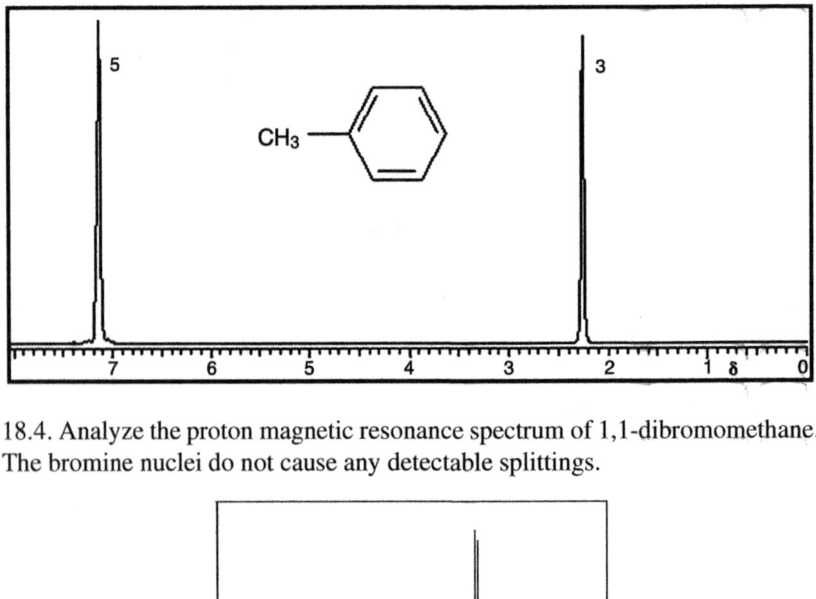

18.4. Analyze the proton magnetic resonance spectrum of 1,1-dibromomethane. The bromine nuclei do not cause any detectable splittings.

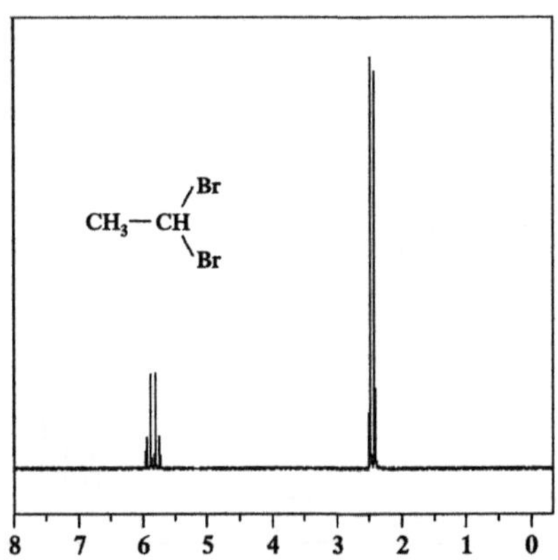

18.5. Assume that the magnetic moments μ_1 and μ_2 in Eq. (18.19) are both directed along the z-axis. Show that the angular dependence of the dipole-interaction potential energy then contains the factor $(3\cos^2\theta - 1)$. Show that this quantity vanishes when averaged over all orientations.

18.6. After a 90° pulse, the spins in a sample will begin to dephase by spin-spin relaxation. Show that this process can be reversed for a time by application of a 180° pulse. This is exploited in a number of *spin-echo* techniques.

Chapter 19

Wonders of the quantum world

After going through Chapters 1–18, it is hoped that the reader is convinced of the essential validity of the quantum theory and of its success in providing a conceptual framework for the fundamental phenomena of physics, chemistry and biology. It is by any measure a highly successful physical theory, capable of making correct predictions for innumerable physical phenomena. This includes statistical predictions involving large numbers of observations. But honestly, you still don't really "understand" quantum mechanics (at least I don't). But we're in good company. Richard Feynman wrote "I think it is safe to say that no one understands quantum mechanics." Niels Bohr said that "Anyone who can contemplate quantum mechanics without getting dizzy hasn't properly understood it." According to Roger Penrose, "while the theory agrees incredibly well with experiment and while it is of profound mathematical beauty, it makes absolutely no sense." Feynman very succinctly summarized the situation this way: "We cannot make the mystery go away ... we will just tell you how it works." The instinctive disbelief often experienced when first confronted with the novelties of quantum mechanics is aptly described by the biologist Peter Medawar: "The human mind treats a new idea the way the body treats a strange protein—it rejects it." No less an eminence than Albert Einstein could never buy into the worldview of quantum mechanics, notwithstanding his own role as one of its creators. Maxwell believed that "...the effectual studies of the sciences must be ones of simplification and reduction of the results of previous investigations to a form in which the mind can grasp them." Thus "Clockwork Universe" is a metaphor which succinctly captures the essence of classical physics. But quantum mechanics lacks a simple metaphor accessible to everyday experience and common sense.

In thinking about quantum mechanics, it is well to keep in mind the schema relating appearance and reality central to the metaphysics of Immanuel Kant. Appearance, what he called *phenomena*, represents our observations and experiences, both external and internal. The reality beyond phenomena, which he called *noumena*, represents ultimate causation, which is forever hidden from our perception. Theories are models we create in attempts to make connections between appearance and reality.

19.1 The Copenhagen interpretation

Fig. 19.1 is a portrait of the participants in the Fifth Solvay International Conference in 1927, including all of the founding fathers of quantum mechanics.

Introduction to Quantum Mechanics. https://doi.org/10.1016/B978-0-12-822310-9.00027-6

Immanuel Kant

They came together to contemplate the foundations of the newly formulated theory. Here the long-running dialog between Niels Bohr and Albert Einstein first began.

FIGURE 19.1 The 1927 Solvay Congress on the quantum theory. Colorized version from the American Physical Society historical collection.

The *Copenhagen interpretation* of quantum mechanics was the working philosophy developed largely by Neils Bohr, Werner Heisenberg and Max Born in the late 1920s. According to the Copenhagen interpretation, a quantum system exists in some sort of nebulous existential fog until a *measurement* is carried out. It is pragmatically asserted that only after a measurement can a value be assigned to a dynamical variable. It is considered meaningless to assume some value *before* it is measured. This is of course much at odds with notions of *objective reality*, which contends that there exist attributes which exist independently of whether or when they are observed. This outlook is central to the classical picture of Nature ("The Moon is there whether or not we look at it"). For Einstein, among many other thinkers, "... belief in an external world independent of the perceiving subject is the basis of all natural science."

Another contentious issue associated with the Copenhagen interpretation is the so called *measurement problem.* Everyone agrees that a quantum system will evolve deterministically in accordance with the time-dependent Schrödinger equation *until* an observation is made. According to the orthodox Copenhagen interpretation, the observation forces a discontinuous change in which the wavefunction settles into one of the eigenstates of the measured dynamical variable. This somewhat mystical transition is known as *collapse of the wavefunction.* It is *not* described by any conventional Schrödinger equation, although attempts have been made to treat the system plus the measuring apparatus as parts of an entangled quantum system.

Phenomena in the submicroscopic quantum world inevitably create apparent paradoxes from the viewpoint of classical macroscopic experience. We will focus in this Chapter on two of the most counterintuitive aspects of quantum theory: superposition (Schrödinger's Cat) and entanglement (EPR and Bell's theorem).

19.2 Superposition

The inescapable implications of the double-slit and similar experiments is the capability of a quantum system to exist in a state which simultaneously partakes of alternative realities. Such a state can moreover exhibit the effects of *interference* between its component realities. Chemists generally agree that a benzene molecule can be represented as a *resonance hybrid* of at least its two Kekulé structures and that a hydrogen bond is an intermediate between two ordinary covalent bonds. The basic idea of superposition is that, if a quantum system is capable of existing in the individual states Ψ_1 and Ψ_2, then it can also exist in a linear combination of these two states, which can be written

$$|\Psi\rangle = c_1|\Psi_1\rangle + c_2|\Psi_2\rangle. \tag{19.1}$$

This can be generalized for any number of contributing states. According to the Copenhagen interpretation, the wavefunction for state Ψ can under certain circumstances *collapse* to one of the component states Ψ_1 or Ψ_2, with a probability $|c_1|^2$ or $|c_2|^2$, respectively. When this happens, all other components of the superposition essentially disappear.

It is relatively easy to produce simple superposition phenomena with photons. This will also serve to introduce some devices which we will encounter later in some key experiments. Fig. 19.2A shows the operation of a *beamsplitter.* This is a glass plate, partially-silvered on one surface, so as to reflect half of an incident light beam and transmit the other half. A wave reflected directly off the silvered side has its phase shifted by 180° with respect to the transmitted wave. However, a wave which passes through the glass layer before being reflected, has its phase unchanged. After passing through the beamsplitter, an

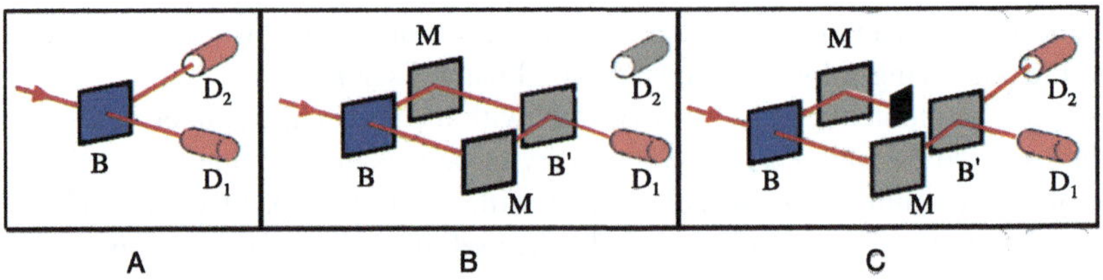

FIGURE 19.2 Mach-Zehnder interferometer demonstrating photon superposition. M denotes a mirror. B and B' are beamsplitters, with reflecting surface on the gray side. D_1 and D_2 are detectors.

incident photon has a 50–50 chance of entering detectors D_1 and D_2. The setup in Fig. 19.2B, with two oppositely-oriented beamsplitters and two mirrors, is known as a *Mach-Zehnder interferometer*. If the path lengths are precisely the same, all photons will go to detector D1 and none will go to detector D2. To explain this result, a photon must be treated as a wave which travels along *both* paths at each beamsplitter. The photon's wavefunction is thus represented by a superposition similar to (19.1). Taking into account the phase shifts in the successive reflections, the two waves arriving at detector D1 are in phase and will reinforce, while the two waves arriving at detector D2 are 180° out of phase and will cancel.

Now suppose that we block the light beam after the mirror in the upper path, as shown in Fig. 19.2C. This will ensure that all of the light arriving at the second beamsplitter had traversed only the lower path. In this case, there is no interference, and the second beamsplitter sends equal components of the incident wave into the two detectors. This result is counterintuitive, in that blocking out some of the light seems to *increase* its detection. But the explanation based on photon superposition is very clear.

19.3 Schrödinger's Cat

In 1935 Erwin Schrödinger published an essay questioning whether strict adherence to the Copenhagen interpretation can cause the "weirdness" of the quantum world to creep into everyday reality. He speculated on how the principle of superposition, which is so fundamental for the quantum-mechanical behavior of microscopic systems, might possibly affect the behavior of a large scale object.

Schrödinger proposed a rather diabolical *Gedankenexperiment* (thought experiment) known as *Schrödinger's Cat*. A more humane version of the experiment makes use of the apparatus sketched in Fig. 19.3. A cat is confined to a opaque box while a weak radioactive source is monitored by a Geiger counter. Detection of a decaying atom during a specified time period triggers a spray of catnip (hydrogen cyanide in Schrödinger's original atrocity) into the box occupied by our cat. If the spray is activated, the cat will relax into a blissful quantum state. Otherwise the cat, annoyed to be cooped up in a dark box, will become excited into a perturbed state. Assume a 50% probability for each outcome.

Then—if you believe the Copenhagen interpretation—until the box is opened, the cat's quantum state must be described by a superposition

$$|\Psi\rangle = \frac{1}{\sqrt{2}}\left(|☺\rangle + |☹\rangle\right). \tag{19.2}$$

Only after the box is opened and the cat *observed* will the wavefunction collapse to a recognizable state of bliss or annoyance. In an extension of this experiment, an observer known as *Wigner's friend* is invited into a closed room with the apparatus. And until the door is opened, Wigner's friend himself will also become part of a quantum superposition, which is starting to become ridiculous! We might be able to accept the concept of an atomic system being described as a superposition of quantum states. But a cat?

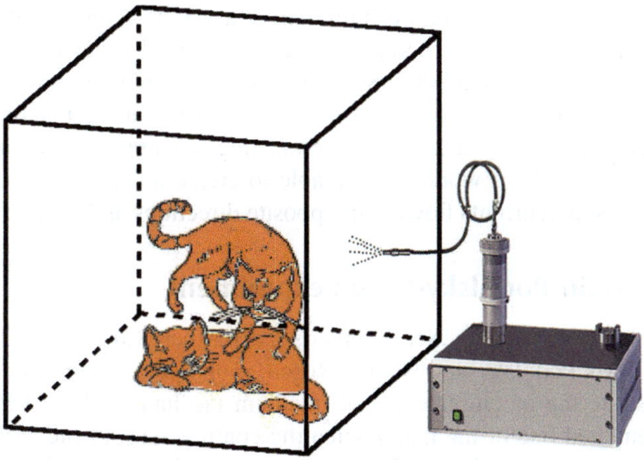

FIGURE 19.3 Schrödinger Cat experiment. A spray of catnip is released when a decay from a small radioactive sample is detected. Version suggested by Sarah Blinder and Amy Blinder, *Am. J. Phys.* **69**, 633 (2001).

It might perfectly acceptable for an atom to be in a superposition of radioactively decayed and undecayed states. The difficulty arises when you start to wonder what is happening inside the box after the radioactive process has run its course. Can superposition on a microscopic level can be amplified to apply to a macroscopic object? Can our cat be temporarily suspended in a superposed state of contentment and annoyance? Gell-Mann and others have suggested that the coherence of superposed quantum states can be compromised by the irreducible interaction of a system with the surroundings. Our cat is not a small element of the microscopic quantum world but a large complex system made up of trillions of atoms. These can occupy an immense number of possible quantum states which are indistinguishable macroscopically. Moreover, the very strong interactions with the environment soon washes out the cat's quantum behavior. The modern consensus resolves the Schrödingers Cat paradox by invoking *decoherence*. Decoherence can be viewed as a continual process of "self-measurement" brought about by interactions within a quantum system and with the surroundings. Thus anything as complicated as a cat will certainly be well described as a classical object. This implies that the quantum superposition collapses the instant a nucleus decays, and everything thereafter follows classical determinism.

Decoherence in quantum systems is somewhat akin to transverse relaxation in NMR (T_2 processes), in which the nuclear spins lose their phase coherence. The proposal that *consciousness* is somehow connected with collapse of the wavefunction, although intriguing, does not appear to be relevant.

Quantum superposition does remain alive and well on the atomic scale. Christopher Monroe and coworkers in 1996 were able to prepare a single beryllium ion as a superposition of wavepackets representing two different electronic states spatially separated by as much as 80 nm. By an appropriate sequence of laser pulses, they were able to detect interference between the two wavepackets. Inevitably, this experiment has been referred to as "Schrödinger's cation."

A superconducting "Schrödinger's Cat" has been demonstrated by David Wineland and coworkers in 2000. Supercurrents, containing billions of electron pairs all residing in a single quantum state, can move around a macroscopically sized superconductor, such as a superconducting quantum interference device (SQUID) circuit. These workers were able to create a superposition of states consisting of supercurrents flowing in opposite directions at the same time.

19.4 Einstein-Podolsky-Rosen experiment

In 1935 (the same year as Schrödinger's Cat) Albert Einstein, in collaboration with Boris Podolsky and Nathan Rosen, proposed a *Gedankenexperiment* to demonstrate the incompleteness of quantum mechanics. The "EPR experiment" stimulated one of the major scientific controversies of the 20th Century and continues to be a subject of intense contemplation and analysis. EPR focuses on the quantum-mechanical pronouncement (the Heisenberg uncertainty principle) that the position and momentum of a subatomic particle cannot be exactly known simultaneously. The particle can be in a state of definite momentum, but then we cannot know where it is located. Conversely, we can put the particle at a definite position, but then its momentum is completely indeterminate. It is also possible to create states in which we have limited knowledge of both observables, consistent with $\Delta x \Delta p \geq \hbar/2$. This state of affairs is not due to any inadequacy of measurement techniques but is rather an inescapable feature of the quantum-mechanical description of Nature. In the original form of the EPR experiment, shown schematically in Fig. 19.4, two particles A and B from a common source fly apart in opposite directions, perhaps as a result of radioactive disintegration. At some subsequent time, the position x of particle A is measured, which can in principle be done exactly. At the same instant, the momentum p_x of particle B is measured, also exactly. By conservation of momentum, which is also valid in quantum mechanics, the momentum of A ought to be the negative of B's. Thus the position and momentum of particle A are apparently determined simultaneously!

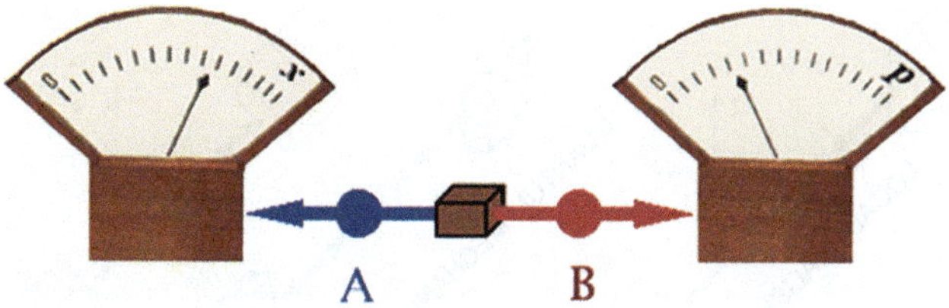

FIGURE 19.4 Einstein-Podolski-Rosen (EPR) Gedankenexperiment.

The Copenhagen response to EPR was that the two measurements *cannot* be regarded as independent since particles A and B remain correlated as parts of a single indivisible quantum system. Thus measuring the position of A will perturb the momentum of B just as surely as it would perturb the momentum of A. The states of the two particles are said to be *entangled.* Entanglement persists no matter how great the separation of the two particles. Einstein argued that this was tantamount to faster-than-light communication between A and B ("spooky action at a distance"), which is contrary to *locality* (or local realism) implied by the theory of relativity. He thought that subatomic particles must possess yet undiscovered *hidden variables* which gives their quantum states *objective reality* even after they separate. This conflicts of course with the Copenhagen viewpoint that the value of an observable does not even *exist* until a measurement is made. Einstein, among others, couldn't swallow such violations of objective reality and locality implied by quantum mechanics. Einstein agreed that quantum mechanics was a correct theory *phenomenologically*. But he objected that it gave an incomplete account of physical reality, which is well summarized in the title of the EPR paper: "Can quantum-mechanical description of physical reality be considered complete?"

The concept of hidden variables can be illustrated by a simple example. The result of a coin toss—heads or tails—can, on the simplest level, be regarded as a random occurrence. Yet, if the coin's complicated trajectory were analyzed in detail, the result would become completely determinate. The "hidden variables" are the coordinates and momenta describing the motion of the penny. This is certainly a highly challenging problem in mechanics. But it is, in principle, solvable so that the result of the coin toss is, despite appearances, far from random. Moreover, the apparently statistical nature of coin tosses—approximately 50% heads, 50% tails—is actually a consequence of a distribution among the large number of possible initial configurations when the coin is released.

David Bohm in 1951 proposed a modified version of the EPR experiment which is conceptually equivalent but easier to analyze mathematically. This makes use of *spin* correlations rather than momentum correlations between particles. Bohm pictured a pair of spin-1/2 particles in a singlet state blown apart in opposite directions. The spin components are measured by two Stern-Gerlach detectors, as shown in Fig. 19.5. The spins of the two particles are antiparallel and remain so even after they are separated. Thus if particle A is spin-up, particle

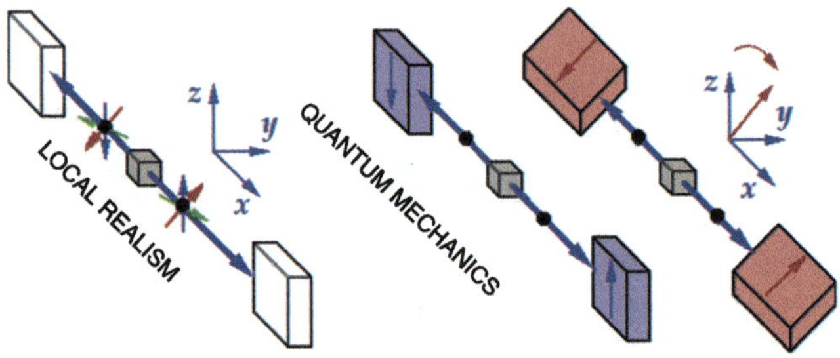

FIGURE 19.5 Bohm's modification of the EPR experiment based on a pair of correlated spin-$1/2$ particles. Left: interpretation according to local realistic models. Right: quantum-mechanical interpretation for two different orientations of Stern-Gerlach detectors.

B must be spin-down, and vice versa—the two spins are thus *correlated*. This is true if the two detectors are initially oriented in the z-direction. Remarkably, the perfect correlation persists even when the detectors are *both* rotated by an arbitrary angle from the z-axis. The result registered by each individual detector is completely random. But in every case the other detector will give the opposite reading. The states of the two particles are *random* but *correlated*. Measuring the spin state of one particle will, in a sense, instantaneously *force* the other particle into the other spin state. Such entanglement takes place no matter how far apart the particles have moved. Rejection of the possibility for such instantaneous "communication" or "telepathy" is the core of the EPR argument that quantum mechanics is incomplete.

It is a common misconception that the EPR experiment was intended to disprove quantum mechanics by means of a paradox. In fact, nothing in the EPR argument suggests doubt about the correctness of quantum mechanics. Their intent was to show that the complete description of a microscopic state is richer than that provided by quantum mechanics. They hoped to stimulate discovery of a more fundamental theory able to predict more than just statistical behavior in atomic phenomena. EPR were perceptive enough to have zeroed in on the one feature of quantum mechanics that is most counterintuitive for the classical worldview—entanglement.

According to the viewpoint of local realism, the recurring correlations in the Bohm experiment can be attributed to the existence of *hidden variables* which determine the spin state in every possible direction. It is as if each particle carried a little code book containing all this detailed information, a situation something like the left-hand side in Fig. 19.5. It must be concluded—so far—that *both* local realism *and* the quantum mechanical picture of the world are separately capable of giving consistent accounts of the EPR and Bohm experiments. In what follows, we will refer to the two competing worldviews as local realism (LR) and quantum mechanics (QM). By QM we will understand the conventional formulation of the theory, complete as it stands, *without* hidden variables or other auxiliary constructs.

19.5 Bell's theorem

In 1964, J. S. Bell derived a remarkable relation capable of *experimentally* deciding between local realism and quantum mechanics. His idea involves measurements for *different* orientations of the Stern-Gerlach detectors. Thus far we have considered only the case of two detectors oriented in opposite directions. Let $P(\theta)$ represent the probability of recording *antiparallel* spins when the angle between the detectors is θ. Only for $\theta = 180°$ (oppositely aligned detectors) would we have $P(\pi) = 1$, meaning certainty (unit probability). For $\theta = 0$ (aligned detectors) the probability drops to zero, so that $P(0) = 0$. For detectors oriented 90° apart, there would be equal probabilities for parallel or antiparallel spins, so that $P(\pi/2) = 1/2$. To find a general relation for $P(\theta)$, we make use of the quantum-mechanical operators S_x, S_y and S_z for spin-1/2. It is convenient to express these in terms of dimensionless Pauli spin operators σ_x, σ_y and σ_z such that

$$\mathbf{S} = \frac{\hbar}{2}\boldsymbol{\sigma}. \tag{19.3}$$

Eqs. (7.60) imply the operator relations

$$\begin{aligned} &\sigma_x|\alpha\rangle = |\beta\rangle, \qquad \sigma_y|\alpha\rangle = i|\beta\rangle, \qquad \sigma_z|\alpha\rangle = |\alpha\rangle, \\ &\sigma_x|\beta\rangle = |\alpha\rangle, \qquad \sigma_y|\beta\rangle = -i|\alpha\rangle, \qquad \sigma_z|\beta\rangle = -|\beta\rangle. \end{aligned} \tag{19.4}$$

The Pauli spin operator for the component at an angle θ with respect to the z-axis is given by

$$\sigma_\theta = \cos\theta\,\sigma_z - \sin\theta\,\sigma_y. \tag{19.5}$$

Suppose one detector is rotated by θ, while the other detector remains vertical at $\theta = 0$. The vertical detector will select one of the particles to be in the α-spin state (spin-up) with respect to the z-axis. Therefore the other particle must be in the β state (spin-down). The expectation value for σ_θ for a β-spin is given by

$$\langle\beta|\sigma_\theta|\beta\rangle = \cos\theta\langle\beta|\sigma_z|\beta\rangle - \sin\theta\langle\beta|\sigma_y|\beta\rangle = -\cos\theta. \tag{19.6}$$

This expectation value can be related to the spin-up and spin-down probabilities $P{\uparrow}(\theta)$ and $P{\downarrow}(\theta)$ by

$$\langle\sigma_\theta\rangle = P{\uparrow} - P{\downarrow} = 2P{\uparrow} - 1, \tag{19.7}$$

noting that $P{\uparrow} + P{\downarrow} = 1$. From (19.6) and (19.7), the probability for detecting spin-up in the θ direction is given by

$$P{\uparrow}(\theta) = \frac{1-\cos\theta}{2} = \sin^2(\theta/2). \tag{19.8}$$

Let us, following Bell, provisionally assume the point of view of local realism. Referring to Fig. 19.6, we suppose our particles are initially observed with both detectors parallel to the z-axis. We would then find zero probability for correlated spins since $P(0) = 0$. Now, if one of the detectors were rotated clockwise by an angle θ_1, the detection probability would increase from zero to $P(\theta_1) = \sin^2(\theta_1/2)$. Likewise, the other detector rotated counterclockwise by θ_2, would register an average value $P(\theta_2) = \sin^2(\theta_2/2)$. Suppose now that *both* detectors are rotated. From the perspective of local realism, we can reason as follows. The probability $P(\theta_1)$ is increased from zero because a number of spins becomes parallel. Rotation of the other detector gives an analogous increase to $P(\theta_2)$. Assuming that these two results are independent of one another, a first guess might be that the coincidence probability after rotating *both* detectors to a relative angle $\theta = \theta_1 + \theta_2$ is given by $P(\theta_1) + P(\theta_2)$. But this would be an overcount, since those instances in which *both* detectors register spins antiparallel to their paired detectors, they must be *parallel* to one another, and thus *not* be appropriately correlated. This must reduce $P(\theta_1 + \theta_2)$ to a value *less* than the sum, which can be expressed as an inequality

$$P(\theta_1 + \theta_2) \leq P(\theta_1) + P(\theta_2). \tag{19.9}$$

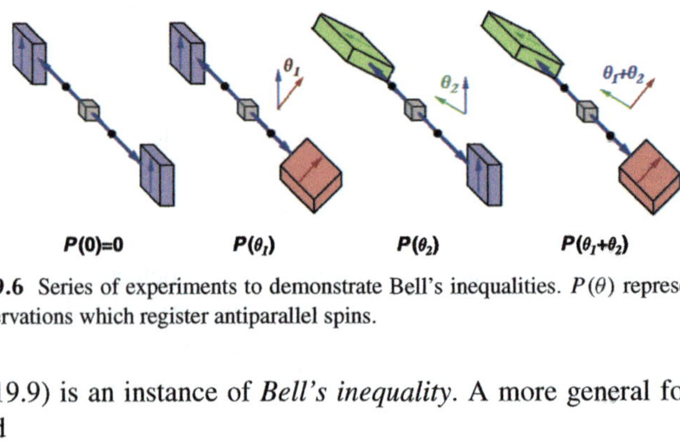

FIGURE 19.6 Series of experiments to demonstrate Bell's inequalities. $P(\theta)$ represents the fraction of observations which register antiparallel spins.

Eq. (19.9) is an instance of *Bell's inequality*. A more general form can be expressed

$$P(a, b) \leq P(b, c) + P(c, a), \tag{19.10}$$

representing correlations involving three different axes in space, designated a, b and c. Bell's inequality can be demonstrated by assuming that definite spin components along all three directions exist for every pair of particles, in accordance with objective reality. This can be described by a set of $2^3 = 8$ triply-composite probability functions $P(a\uparrow, b\uparrow, c\uparrow)$, $P(a\uparrow, b\uparrow, c\downarrow)$, etc. Clearly, each correlation probability in (19.10) is equal to a sum of four triple composites, for example,

$$P(a, b) = P(a\uparrow, b\downarrow, c\uparrow) + P(a\downarrow, b\uparrow, c\downarrow) + P(a\downarrow, b\uparrow, c\uparrow) + P(a\downarrow, b\uparrow, c\downarrow), \tag{19.11}$$

and analogously for the other two. Fig. 19.7 provides a graphical proof of Bell's inequality, which is actually a very general result in probability theory, not limited to spin variables.

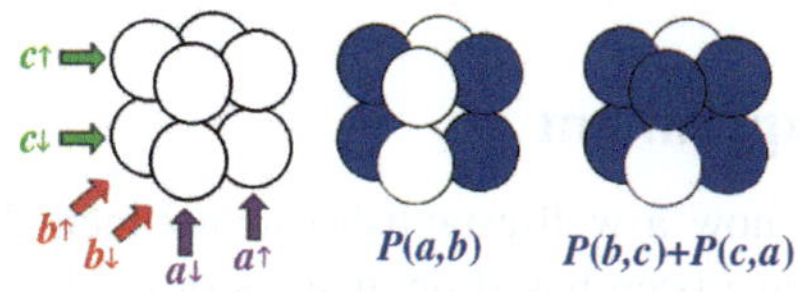

FIGURE 19.7 Graphical demonstration of Bell's inequality (19.10) using a three-dimensional Venn diagram. Each of 8 regions represents one combination of the variables a, b and c. It is clear that $P(a, b)$ is a subset of $P(b, c) + P(c, a)$.

It is readily apparent that Bell's inequality in the form (19.7) is in *disagreement* with quantum mechanics. With the coincidence probabilities (19.6), the sense of the inequality is, in fact, *reversed* since

$$\sin^2[(\theta_1 + \theta_2)/2] \geq \sin^2(\theta_1/2) + \sin^2(\theta_2/2) \quad \text{for} \quad 0 \leq \theta_1, \theta_2 \leq \pi. \tag{19.12}$$

For example, when $\theta_1 = 20°, \theta_2 = 30°$, Eq. (19.12) gives $0.1786 > 0.0971$. This means that quantum mechanics predicts greater than expected correlations between events that are out of range by classical causality. On a practical note, even if a small fraction of coincidences fails to resister because of imperfections in the apparatus, the validity of inequality (19.12) is not likely to be compromised.

The essential inconsistency of quantum mechanics with any local realistic picture is also evident from the form of the two-spin singlet state wavefunction

$$|\Psi\rangle = \frac{1}{\sqrt{2}}\Big(|\alpha\rangle_1|\beta\rangle_2 - |\beta\rangle_1|\alpha\rangle_2\Big). \tag{19.13}$$

There is no way in which such a function can be written as a simple product of the form $|\Psi\rangle = |\psi\rangle_1|\phi\rangle_2$, representing two noninteracting particles. Thus there is no way in which two particles acting independently of one another can simulate the entanglement of the singlet state.

Bell's inequality provides a clear-cut test of local reality *vs.* quantum mechanics. The unambiguous answer, from a variety of experiments which we describe in the next Section, is that quantum mechanics wins! Thus we can conclude that we live in Universe which does not respect local reality. Quantum *entanglement*—the term used by Schrödinger—really happens! In drawing this conclusion we are actually glossing over a number of still-unresolved hair-splitting metaphysical arguments. This remarkable result is often summarized as *Bell's theorem*:

The local realistic model is violated by quantum mechanics.

Henry Stapp regards Bell's theorem as "the most revolutionary scientific discovery of the Twentieth Century." The inherent nonlocality of quantum mechanics means that two particles once having been together might continue to assert instantaneous influence on one another—even if they are in different galaxies. It is even possible that all the matter in the Universe, having originated in the Big Bang, is in some way mutually entangled.

19.6 Aspect's experiment

Bell's theorem is by now a well-established experimental fact. The most accurate experiments have been based on analogs of the EPR-Bohm experiment measuring *photon* polarizations rather than spins of massive particles. Instead of spin-up and spin-down states, photons can have right and left circular polarizations. In certain processes, two photons with correlated polarizations—one left, one right—can be emitted in opposite directions. Wheeler had proposed in 1946 that the pair of photons emitted in the annihilation of positronium were entangled with opposite polarizations. This was experimentally confirmed by Wu and Shaknov in 1949.

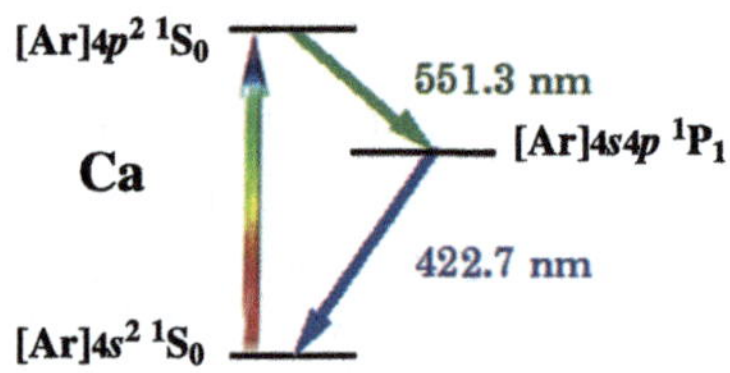

FIGURE 19.8 Emission of two polarization-correlated photons by calcium atom after laser excitation.

Circular polarization is equivalent to a superposition of linear polarizations in two perpendicular directions, say "horizontal" (H) and "vertical" (V). Whereas two quantized states of spin-1/2 particles differ in orientation by 180°, for photons, the two orthogonal states of polarization differ by 90°. Formulas derived for spin-1/2 particles can generally be applied to photons with $\theta/2$ replaced by θ. It is useful to define the expectation value for correlation between two photons $E(a,b)$ such that $E = 1$ (perfect correlation) if one is H and the other V, $E = -1$ (perfect anticorrelation) if they are both H or both V. Expressed as a function of θ, the angle between the polarization vectors of photons a and b, we would have $E(90°) = 1$ and $E(0°) = -1$. For $\theta = 45°$, equal probabilities of correlation and anticorrelation would give the statistical result $E(45°) = 0$. For arbitrary θ, the quantum-mechanical result is

$$E(\theta) = \sin^2\theta - \cos^2\theta. \tag{19.14}$$

For angles $\theta = 0°$, 45° and 90° the predictions of local realism and quantum mechanics agree. For the general case, a modification of Bell's inequality appropriate for photon polarizations was derived by Clauser, Horne, Shimony and Holt. The CHSH relation predicts $|S| \leq 2$, where

$$S \equiv E(a,b) - E(a,b') + E(a',b) + E(a',b'). \tag{19.15}$$

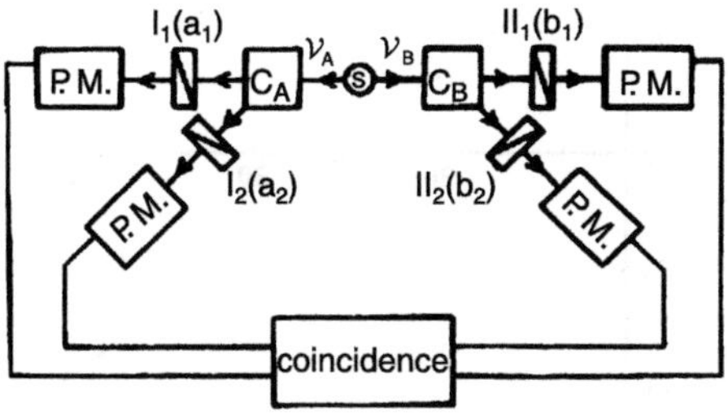

FIGURE 19.9 Schematic diagram of Aspect's experiment: *Phys. Rev. D* **14**, 1944 (1976). C_A and C_B are two-channel polarizers which direct linearly polarized photons to the photomultipliers PM.

Violations of Bell's inequality were successfully demonstrated by Clauser, Horne, Shimony and coworkers in the 1970s. The definitive experiments are those of Alain Aspect and coworkers in 1982. Correlated photons were produced by emission from doubly-excited calcium atoms, as shown in Fig. 19.8. An atomic beam irradiated by two lasers excites calcium atoms to their $[\mathrm{Ar}]4p^2\ {}^1\mathrm{S}_0$ level by two-photon absorption. Spontaneous emission of a 551.3 nm (green) photon accompanies a transition to $[\mathrm{Ar}]4s4p\ {}^1\mathrm{P}_1$. This is followed very rapidly by emission of a second 422.7 nm (blue) photon, as the calcium atom returns to its $[\mathrm{Ar}]4s^2\ {}^1\mathrm{S}_0$ ground state. Each photon carries one unit of angular momentum, with their vector sum equal to zero. The two photons are accordingly entangled with opposite circular polarizations. In Aspect's experiment, schematically represented in Fig. 19.9, polarization detectors for orientations a_1, a_2 and b_1, b_2 measure the linear polarizations of the two photons and coincidences are electronically monitored. Aspect's experimental results, plotted in Fig. 19.10, show correlation statistics as a function of angle between polarimeters. These agree perfectly with the quantum-mechanical result (19.12). The maximum deviation from local realism occurs for a series of measurements with detectors separated by 22.5° increments. Aspect's result gives $S = 2.697 \pm 0.015$, in contrast to the CHSH condition $S \leq 2$. Thus violation of Bell's inequality is verified with a confidence level greater than 40 standard deviations.

A further modification addresses a remote loophole for local realism in which the polarization detectors might somehow be able to signal one another at subluminal speed. Aspect designed an experiment in which mirrors switched at high speed could direct the two photons to different detectors, *after they were already in flight*. It was verified in this "delayed choice" experiment that Bell's inequality is still violated, in what is currently considered the most conclusive test of non-locality.

19.7 Multiple photon entanglement

The general principles of entanglement are valid for all quantum-mechanical observables, including momentum, spin and so on. In practice however, the most

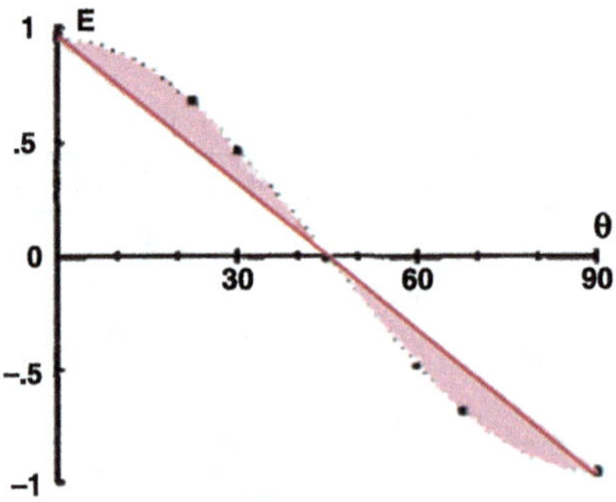

FIGURE 19.10 Results of Aspect experiment. Dotted curve shows prediction of quantum mechanics (multiplied by 0.955 to correct for detection efficiency). Shaded regions show where Bell's inequality is violated. From A. Aspect, P. Grangier, and G. Roger, *Phys. Rev. Lett.* **49**, 91 (1982).

accessible entanglement phenomena involve photon polarizations. This is true because of the availability of highly efficient instrumentation for controlling and detecting optical photons. A technique for producing polarization-entangled photons pairs—more efficient than atomic cascade sources—has been used since the 1980s. Known as *parametric down conversion*, this is based on the nonlinear optical properties of certain crystals, most notably β-barium borate (BBO). As shown in Fig. 19.11, photons of frequency ν_0 from a high-intensity argon-ion UV laser are incident on a specially-prepared BBO crystal, a small fraction of the photons are down converted to a pair of photons of frequency $\nu_0/2$, with linear polarizations H and V. The H and V photons exiting the crystal in two particular directions can form entangled pairs, which show up at the circle intersections in Fig. 19.11. By appropriate manipulation by optical devices, four possible *EPR-Bell* states can be obtained, namely

$$|\Psi^{\pm}\rangle = \frac{1}{\sqrt{2}}\Big(|H\rangle_1|V\rangle_2 \pm |V\rangle_1|H\rangle_2\Big),$$
$$|\Phi^{\pm}\rangle = \frac{1}{\sqrt{2}}\Big(|H\rangle_1|H\rangle_2 \pm |V\rangle_1|V\rangle_2\Big). \tag{19.16}$$

For the case of two correlated particles, the decision between local realism and quantum mechanics hinged on statistical violations of (Bell's) inequalities. It was subsequently suggested (around 1989) by Greenberger, Horne and Zeilinger (GHZ) that for three or more correlated particles inequalities are no longer necessary. They showed that certain measurements give unambiguous results which can decide between LR and QM. Zeilinger and coworkers were able to produce GHZ states of three entangled photons in 1999 and four entangled photons in 2001. We will describe the four-photon experiments.

As shown in Fig. 19.12, two pairs of entangled photons are produced by sources A and B, which are actually realized by passing a UV pulse twice through the same BBO crystal. Source A emits photons number 1 and 2, while

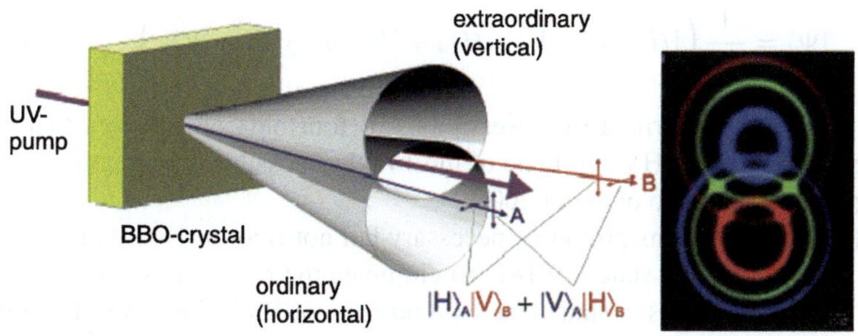

FIGURE 19.11 Experimental setup for producing entangled photons by type-II parametric down conversion. Conical streams of H and V polarized photons are emitted from BBO crystal. A colorized image of the output infrared radiation is shown at right. Entangled beams of photons are created at the cone intersections. Based on P. G. Kwiat, *et al.*, *Phys. Rev. Lett.* **75** 4337 (1995).

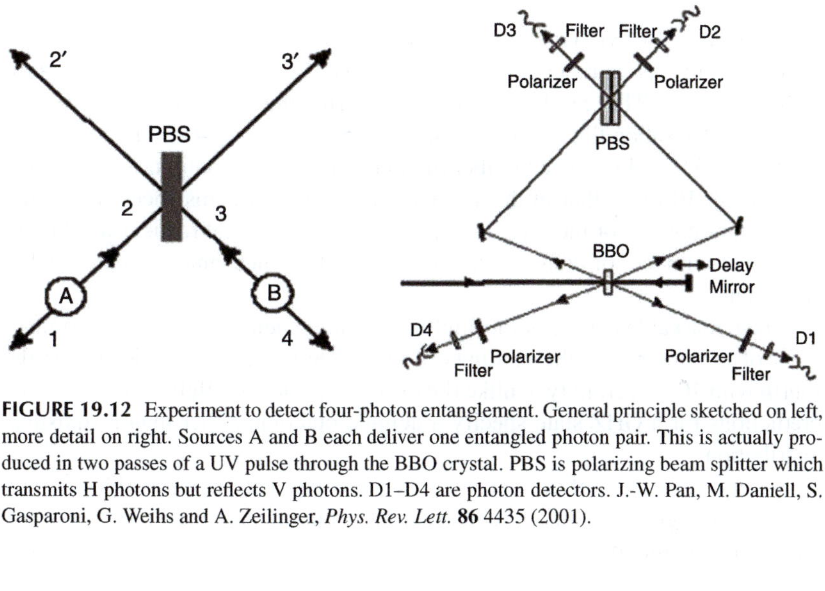

FIGURE 19.12 Experiment to detect four-photon entanglement. General principle sketched on left, more detail on right. Sources A and B each deliver one entangled photon pair. This is actually produced in two passes of a UV pulse through the BBO crystal. PBS is polarizing beam splitter which transmits H photons but reflects V photons. D1–D4 are photon detectors. J.-W. Pan, M. Daniell, S. Gasparoni, G. Weihs and A. Zeilinger, *Phys. Rev. Lett.* **86** 4435 (2001).

source B emits photons number 3 and 4. The state of the four-photon system at this point can be represented as

$$|\Psi\rangle = \frac{1}{\sqrt{2}}\Big(|H\rangle_1|V\rangle_2 - |V\rangle_1|H\rangle_2\Big) \otimes \frac{1}{\sqrt{2}}\Big(|H\rangle_3|V\rangle_4 - |V\rangle_3|H\rangle_4\Big). \quad (19.17)$$

One photon from each pair (numbers 2 and 3) are directed to the inputs of a polarizing beam splitter (PBS) which always transmits H but reflects V polarization. Detectors D2 and D3 then record coincidences in which one and only one photon enters each. (Events in which both photons enter the same detector are rejected.) Consequently, photons 2′ and 3′ must be both H or both V. A GHZ state of four particles results with

$$|\Psi\rangle = \frac{1}{\sqrt{2}}\Big(|H\rangle_1|V\rangle_{2'}|V\rangle_{3'}|H\rangle_4 + |V\rangle_1|H\rangle_{2'}|H\rangle_{3'}|V\rangle_4\Big). \tag{19.18}$$

This state was confirmed by observations of fourfold coincidences that were either HVVH or VHHV, but none of the 14 other possible combinations—with a signal-to-noise ratio of about 200:1.

These observations provide a necessary but not rigorously sufficient condition for the coherent state (19.18). To eliminate the remote possibility that the four photons were just a statistical superposition of HVVH and VHHV, further measurements were performed with the polarizers rotated by ±45°. The basis functions for ±45° diagonal linear polarization are given by

$$|\pm 45°\rangle = \frac{1}{\sqrt{2}}\big(|H\rangle \pm |L\rangle\big). \tag{19.19}$$

Transforming (19.18) to the diagonal basis results in a superposition of 8 terms containing an *even* number of $|+45°\rangle$ components, but none with an odd number of $|+45°\rangle$s. This can be tested by comparing the coincidence counts with the four polarization filters set to $(+45°/+45°/+45°/+45°)$ and to $(+45°/+45°/+45°/-45°)$. The number of counts in the first configuration is found to be about 10 times that in the second. Under ideal circumstances, we would expect zero counts for the second configuration. Since local realism would predict equal numbers of counts, support for quantum mechanics is again highly convincing.

The remarkable feature of this multiple entanglement is that, for any arrangement of the polarizers, the readings of any three detectors will determine the fourth with 100% certainty. Unlike the experiments testing Bell's inequality, the predictions for a GHZ state specify a definite outcome rather than a statistical distribution.

A very active field of current research is the exploration of possible applications of entangled photons to teleportation, communication, cryptography and quantum computation.

19.8 Philosophical problems of quantum theory

The contents of this book, in common with most texts on quantum mechanics, has been focused on recipes for computing the properties of systems on the atomic scale. Without doubt, this approach has been highly successful, leading to some of the most extraordinarily accurate numerical predictions in any field of science–some in agreement with experiment to as many as 10 significant figures. In addition, quantum-mechanical models have stimulated the development of the technology which dominates life in the 21st Century: lasers, computers and other semiconductor devices, how stars shine, nuclear power, telecommunications (including smartphones), biotechnology, medical instrumentation (such as MRI), and many many other examples.

Still, quantum mechanics, augmented, say, by the Copenhagen interpretation–still the "party line" for most physicists–remains simply a collection of recipes, not a fully-structured physical theory. One might speculate on the possible similarity of the present state of the quantum theory with the laws of thermodynamics, as they were understood in the early 19th Century, at the time of Carnot, Joule and Clausius. Thermodynamics was a model of a successful physical theory which explained the observed behavior of heat and work, for example in steam engines, without invoking any extraneous ideas about the intrinsic nature of matter or energy. No deeper understanding of the fundamental underlying physical reality–molecules and their motions–was necessary for the success of thermodynamics.

It can be argued, a proper scientific theory should tell us what are the actual things it describes and how do these things behave. Two persistent theories which have been around for many years but both of which carry heavy metaphysical baggage are the pilot wave theory of de Broglie and Bohm, and the many-worlds interpretation, first proposed by Hugh Everett.

The de Broglie-Bohm theory, often called *Bohmian mechanics*, is a causal interpretation of quantum mechanics, proposed by Louis de Broglie in 1927 and newly advocated by David Bohm in 1952. This is the simplest example of what is often called a hidden variables interpretation of quantum mechanics. A system of particles is described by its wave function, evolving in accord with the Schrödinger equation. The wavefunction plays the part of a *pilot wave*, which guides the motion of the particles, with trajectories dependent on initial conditions. The particle velocities are determined by the gradient of the wavefunction. Thereby, this formulation of quantum mechanics can be considered a deterministic theory, much like Newtonian mechanics.

According to Everett's many-worlds interpretation, proposed in 1957, the evolution of a quantum system is completely governed by the Schrödinger equation, with no need to augment it by wavefunction collapse. It is proposed that every time a quantum experiment is performed, all possible outcomes are obtained, with each occurring in a different world, even if we are only aware of the one world with the outcome we actually experience. This is accordingly known as the many-worlds interpretation of quantum mechanics, which posits that there must exist a myriad of worlds in the Universe in addition to the world we are aware of. One supporter of this view is the British quantum theorist David Deutsch, who believes that the operation of a quantum computer actually involves simultaneous computations in different worlds. It has also been speculated by some that the many-worlds interpretation might be related to the cosmological idea that we actually inhabit a Multiverse.

Several currently popular formulations of quantum mechanics make use of *decoherence*, the gradual loss of the coherence of quantum superpositions caused by interactions with the macroscopic environment. This idea was first proposed in 1970 by the H. Dieter Zeh, with major contributions by Wojtek Zurek. Decoherence might, in fact, contribute to the mechanism responsible for collapse of the wavefunction. When it has run its course, a quantum superposition such as $\Psi = \sum_n c_n \psi_n$ can be reduced to a classical ensemble of states, such as $\sum_n |c_n|^2 \rho_n$. Decoherence is the principal obstacle to practical realization of quantum computers, since these rely on the unitary evolution of coherent quantum states.

Decoherence is clearly relevant in considerations of the long-controversial *measurement problem*. It might also be evoked to help explain *the arrow of time*–how it is that time flows in one direction despite the fact that the fundamental equations are symmetric wrt time reversal. Decoherence is possibly relevant in consideration of the persistent question of how the "classical world" can emerge from quantum mechanics.

It should also be recognized that quantum mechanics in its present form cannot be the ultimate theory, until it can coherently incorporate gravity. The well-known incompatibility of quantum field theory and general relativity is an indication of the work that remains to be done.

Perhaps the future will eventually see some fuller understanding of quantum mechanics. To quote Viola in *Twelfth Night*: "O time, thou must untangle this, not I. It is too hard a knot for me t'untie."

Chapter 19. Exercises

19.1. A singlet state for two spin-1/2 particles is described by the wavefunction

$$|\Psi\rangle = \frac{1}{\sqrt{2}}\big(|\alpha\rangle_1|\beta\rangle_2 - |\beta\rangle_1|\alpha\rangle_2\big),$$

where $|\alpha\rangle$ and $|\beta\rangle$ are referred to the z-axis. For an axis perpendicular to the z-axis, the corresponding spin-up and spin-down eigenfunctions are given by

$$|\uparrow\rangle = \frac{1}{\sqrt{2}}\big(|\alpha\rangle - |\beta\rangle\big) \qquad |\downarrow\rangle = \frac{1}{\sqrt{2}}\big(|\alpha\rangle + |\beta\rangle\big).$$

Express $|\Psi\rangle$ in terms of $|\uparrow\rangle$ and $|\downarrow\rangle$. You should find the form of the wavefunction to be invariant, a consequence of the spherical symmetry of the singlet spin state.

19.2. Consider an unpolarized light beam incident on an arrangement of polarizers as shown below:

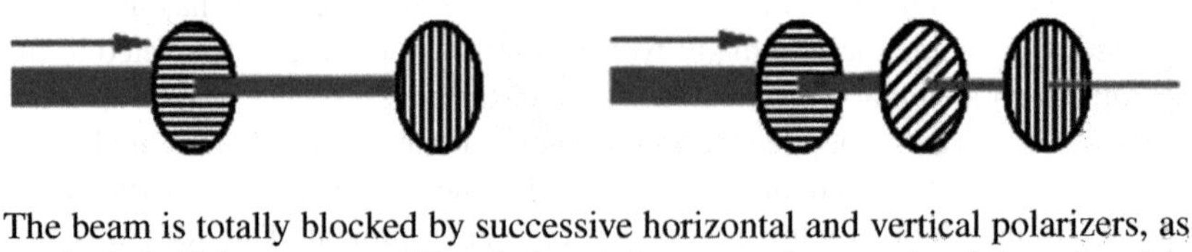

The beam is totally blocked by successive horizontal and vertical polarizers, as on the left. If, however, a 45° polarizer is interposed, as on the right, a fraction of the incident light gets through. Explain this.

Chapter 20

Quantum computers

During the past 50 years, top-of-the-line semiconductor-based computers have doubled in capacity and speed approximately every 18 months, an observation known as "Moore's law." As computer components get smaller, quantum effects become more significant, usually to the detriment of a computer's reliability. But turning adversity to advantage, the two great wonders of the quantum world—superposition and entanglement—suggest powerful methods for encoding and manipulating information, far beyond the capabilities of classical computers. Richard Feynman in 1982 suggested the possibility of constructing a new type of computer taking advantage of quantum principles. As we will show, quantum computers, if they could be constructed, might vastly outperform classical computers. The potential power of quantum computation was first anticipated in a paper by David Deutsch at Oxford in 1985, which described a universal quantum computer as a generalization of a classical Turing machine. The first "killer application" for quantum computers was devised in 1994 by Peter Shor at AT&T Laboratories, an algorithm to perform factorization of large numbers much more efficiently than any classical computer. Encryption techniques such as that of Rivest, Shamir and Adleman (RSA), used for secure electronic transactions, depend on the difficulty of factoring large numbers. Another algorithm, invented in 1996 by Lov Grover, also at AT&T, was a method to speed up searches on large databases. To be sure, no quantum computer yet exists to carry out any such programs. But the motivation is certainly there.

The prospect of applying quantum computers to atomic and molecular problems is highly attractive in view of the relative computational complexities of classical and quantum algorithms. Complexity is a general measure of how computer time and memory resources scale with increasing size. For classical computers, applications to atoms and molecules scale *exponentially* with system size, roughly of the order of 2^n with increasing number of electrons n. However, using quantum-computer algorithms, *in principle*, the increase is only *polynomial*, meaning something like n^3, n^4, etc. Problems which can be solved using polynomially increasing computational resources are commonly called *tractable*, while those that require exponentially growing resources remain *intractable*.

The basic unit of information in computer science is the binary digit or *bit*, whose physical realization is a component capable of two stable states, say $|0\rangle$ and $|1\rangle$. In a conventional computer, these usually correspond to tiny charged and uncharged capacitors within a silicon microchip. On an atomic level, bits

Introduction to Quantum Mechanics. https://doi.org/10.1016/B978-0-12-822310-9.00028-8

might in concept be represented by the two orientations of an electron spin or the two polarization states of a photon. Apart from economy of scale, an atomic two-level system has a capability beyond that of a classical component, namely the possibility of being in a *coherent superposition* such as

$$|\Psi\rangle = a_0|0\rangle + a_1|1\rangle \tag{20.1}$$

Such an entity represents the basic unit of information in a quantum computer—a quantum bit or *qubit*. Unlike a classical bit, which can store only a single value—a 0 or 1—a qubit can store *both* 0 and 1 at the same time. The state of a two-qubit register could be written

$$|\Psi\rangle = a_0|00\rangle + a_1|01\rangle + a_2|10\rangle + a_3|11\rangle \tag{20.2}$$

and contains the equivalent of four classical bits. A quantum register of 64 qubits can store $2^{64} \approx 10^{19}$ values at once. More remarkably, the mutual entanglement of qubits makes it possible to perform computations on all these values at the same time. A quantum computer thus has the capability of operating in a massively-parallel mode. A 300-qubit quantum computer could theoretically store $2^{300} \approx 10^{90}$ bits of information, more than the estimated number of atoms in the known Universe, and also be capable of doing 2^{300} simultaneous calculations. A classical computer can be likened to a solo musical instrument, a quantum computer to a full orchestra. If the music is well played, a symphony is much richer in content than the sum of its parts. A major technical problem in constructing quantum computers is to minimize interactions within the machine and with the environment, which would cause *decoherence*–a breakdown in quantum entanglement.

20.1 Qubits

The fundamental idea of quantum computing is the replacement of the elementary unit of information in digital computers, the bit, with the *qubit*. The qubit can be defined as a normalized quantum state vector belonging to a complex two-dimensional Hilbert space $\mathbb{C}^2$. The basis vectors are conventionally designated, using Dirac notation,

$$|0\rangle = \begin{pmatrix} 1 \\ 0 \end{pmatrix}, \qquad |1\rangle = \begin{pmatrix} 0 \\ 1 \end{pmatrix}. \tag{20.3}$$

A generic element of $\mathbb{C}^2$ can be written as a complex linear combination of two classical states:

$$|\Psi\rangle = \alpha|0\rangle + \beta|1\rangle, \tag{20.4}$$

where α and β are complex numbers. But a qubit is an entirely different type of linear combination, which does not correspond to any well-defined classi-

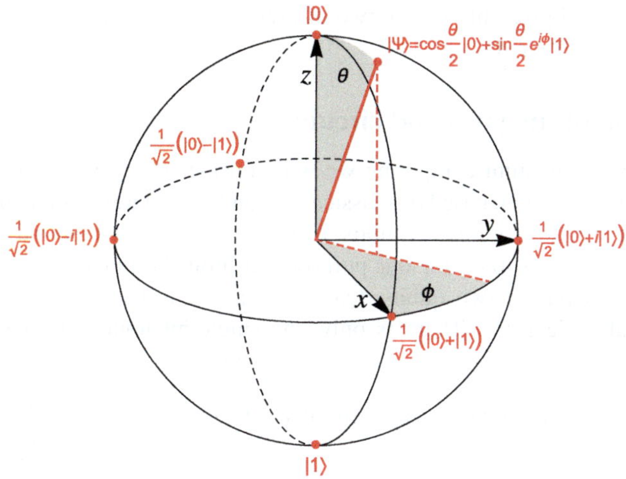

FIGURE 20.1 Bloch sphere. The qubit $|\Psi\rangle$ and the six cardinal points are marked with red dots.

cal entity. The norm $|\Psi|$ is assumed equal to 1; therefore α, β must obey the normalization condition:

$$|\alpha|^2 + |\beta|^2 = 1. \tag{20.5}$$

The *Bloch sphere*, shown in Fig. 20.1, provides a useful geometrical representation of a qubit. The angles θ and ϕ, analogous to latitude and longitude, describe the state of a qubit. Antipodal points represent *orthogonal states*, rather than just negatives of the state vector. On the Bloch sphere, $|0\rangle$ is mapped onto the north pole ($\theta = 0$) and $|1\rangle$ onto the south pole ($\theta = \pi$). The equator, with $\theta = \pi/2$, contains four additional *cardinal points*, with $\phi = 0, \pi/2, \pi, 3\pi/2$, corresponding to $\big(|0\rangle + |1\rangle\big)/\sqrt{2}$, $\big(|0\rangle + i|1\rangle\big)/\sqrt{2}$, $\big(|0\rangle - |1\rangle\big)/\sqrt{2}$, $\big(|0\rangle - i|1\rangle\big)/\sqrt{2}$, respectively. There is a one-to-one correspondence between qubit states and points on the Bloch sphere. A convenient parametrization of a qubit, consistent with the above, is

$$|\Psi\rangle = \cos\frac{\theta}{2}|0\rangle + \sin\frac{\theta}{2}\, e^{i\phi}|1\rangle, \tag{20.6}$$

noting the characteristic spinor half-angles.

Different physical systems can serve as realizations of a qubit: $|0\rangle$ and $|1\rangle$ can represent the two polarization states of a photon, the $\pm\frac{1}{2}$ spin states of an electron or nuclear spin, the direction of a vortex in a superconductor or two different energy levels of an atom or molecule. Since θ and ϕ are real numbers over a continuous range, a qubit can contain a vast amount of information. Quoting Nielsen and Chuang, "Paradoxically, there are an infinite number of points on the unit sphere, so that, in principle, one could store an entire text of Shakespeare in the infinite binary expansion of θ." However, an actual *measurement* of the qubit will give only one of two discrete results, ± 1, which collapses the qubit to either $|0\rangle$ or $|1\rangle$.

20.2 Quantum gates and circuits

The model of quantum computing we will consider is a generalization of the classical circuit model, based on classical bits and logic gates. A quantum circuit carries out transformations on qubits, using gates which correspond to unitary operations. It will suffice for our purposes to limit the unitary gates to those involving just one or two qubits. Quantum gates are the analogs, for qubits, of classical logic gates. There is only one single-bit logic gate, namely, the NOT gate, but several other single-qubit quantum gates are possible. A single-qubit quantum gate transforms the qubit $\begin{pmatrix}\alpha\\ \beta\end{pmatrix}$ into the qubit $\begin{pmatrix}\alpha'\\ \beta'\end{pmatrix}$. To satisfy Eq. (20.5), we must have:

$$|\alpha'|^2 + |\beta'|^2 = |\alpha|^2 + |\beta|^2 = 1, \tag{20.7}$$

so that the *norm* of the state $\begin{pmatrix}\alpha\\ \beta\end{pmatrix}$ is invariant. Since we want to preserve linearity, there is only one class of operators that can serve as quantum gates, namely *unitary* operators. Unitary operators U leave the norm invariant, so that $|U\Psi| = |\Psi|$ for all Ψ. The most important single-qubit gates are shown in Fig. 20.2, along with the 2×2 unitary transformations they produce. For example, application of the Hadamard gate to each of the basis states gives

$$H|0\rangle = \frac{1}{\sqrt{2}}\Big(|0\rangle + |1\rangle\Big), \qquad H|1\rangle = \frac{1}{\sqrt{2}}\Big(|0\rangle - |1\rangle\Big). \tag{20.8}$$

The Pauli-X, Y and Z gates are named after the Pauli spin operators, σ_x, σ_y and σ_z, which have the same matrix representations. The Pauli-X gate is equivalent to the NOT gate, which interchanges basis states

$$X|0\rangle = |1\rangle, \qquad X|1\rangle = |0\rangle, \tag{20.9}$$

just as the classical NOT gate ▷o.

Operators representing rotation of a qubit on the Bloch sphere, about the x, y, and z axes, can be constructed by exponentiation of the Pauli matrices, as the following:

$$R_x(\alpha) = e^{-i\alpha X/2} = \cos\tfrac{\alpha}{2} I - i\sin\tfrac{\alpha}{2} X = \begin{pmatrix}\cos\frac{\alpha}{2} & -i\sin\frac{\alpha}{2}\\ -i\sin\frac{\alpha}{2} & \cos\frac{\alpha}{2}\end{pmatrix},$$

$$R_y(\beta) = e^{-i\beta Y/2} = \cos\tfrac{\beta}{2} I - i\sin\tfrac{\beta}{2} Y = \begin{pmatrix}\cos\frac{\beta}{2} & -\sin\frac{\beta}{2}\\ \sin\frac{\beta}{2} & \cos\frac{\beta}{2}\end{pmatrix},$$

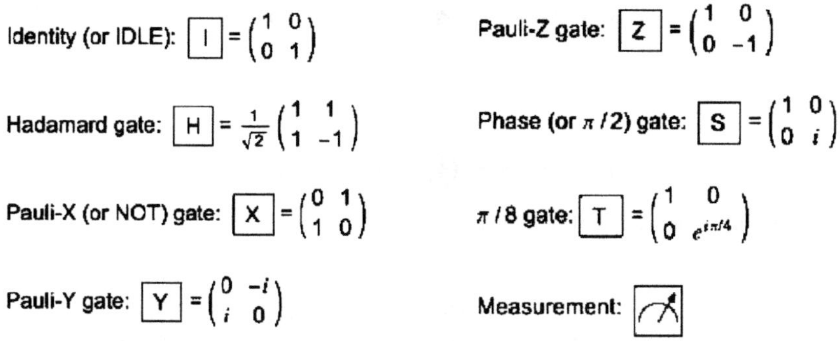

FIGURE 20.2 Single-qubit gates.

$$R_z(\gamma) = e^{-i\gamma Z/2} = \cos\tfrac{\gamma}{2} I - i\sin\tfrac{\gamma}{2} Z = \begin{pmatrix} e^{-i\gamma/2} & 0 \\ 0 & e^{i\gamma/2} \end{pmatrix}. \tag{20.10}$$

A generalization of (20.10), a rotation by θ about a unit vector to the Bloch sphere, $\mathbf{n} = (n_1, n_3, n_3)$ is represented by

$$R_{\mathbf{n}}(\theta) = e^{-i\theta\, \mathbf{n}\cdot\boldsymbol{\Sigma}/2} = \cos\frac{\theta}{2} I - i\sin\frac{\theta}{2}\mathbf{n}\cdot\boldsymbol{\Sigma}, \tag{20.11}$$

where $\boldsymbol{\Sigma} = (X, Y, Z)$. An arbitrary rotation on the Bloch sphere can be described by sequential rotations about three *Euler angles*. From a multitude of possible conventions, we will choose the sequence: (1) rotation about the z-axis by an angle α; (2) rotation about the new x-axis by an angle β; (3) rotation about the new z-axis by an angle γ. This sequence can be represented by the unitary operator

$$U = e^{i\delta} R_z(\gamma) R_x(\beta) R_z(\alpha), \tag{20.12}$$

where the global phase factor $e^{i\delta}$ enables some resulting matrices to be expressed in standard form. Any unitary operation on a single qubit can be written as a combination of rotations in the form (20.12), often in alternative ways. For example, you can show that U becomes the Hadamard gate H with the angles $\alpha = \beta = \gamma = \delta = \pi/2$. Another combination gives $H = i R_x(\pi) R_y(\pi/2)$. As we shall see later, rotation gates are if central importance in the simulation of quantum systems.

Multi-qubit states and operations can be represented as *tensor products* and their matrices constructed as *Kronecker products*. For example, let $A = \begin{pmatrix} a_{11} & a_{12} \\ a_{21} & a_{22} \end{pmatrix}$ and $B = \begin{pmatrix} b_{11} & b_{12} \\ b_{21} & b_{22} \end{pmatrix}$ represent matrices in $\mathbb{C}^2$. Then the matrix for $A \otimes B$ in $\mathbb{C}^2 \otimes \mathbb{C}^2$ is given by

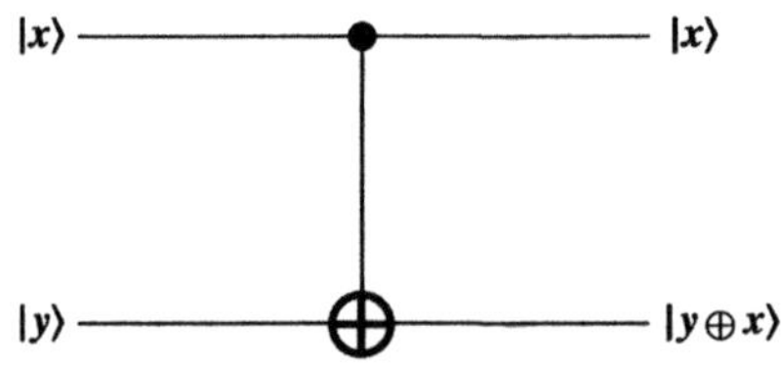

FIGURE 20.3 Controlled-NOT (CNOT) gate.

$$A \otimes B = \begin{pmatrix} a_{11}B & a_{12}B \\ a_{21}B & a_{22}B \end{pmatrix} = \begin{pmatrix} a_{11}b_{11} & a_{11}b_{12} & a_{12}b_{11} & a_{12}b_{12} \\ a_{11}b_{21} & a_{11}b_{22} & a_{12}b_{21} & a_{12}b_{22} \\ a_{21}b_{11} & a_{21}b_{12} & a_{22}b_{11} & a_{22}b_{12} \\ a_{21}b_{21} & a_{21}b_{22} & a_{22}b_{21} & a_{22}b_{22} \end{pmatrix}. \quad (20.13)$$

The two-qubit state with $\Psi_1 = \begin{pmatrix} \alpha_1 \\ \beta_1 \end{pmatrix}$, $\Psi_2 = \begin{pmatrix} \alpha_2 \\ \beta_2 \end{pmatrix}$ is given by

$$\Psi = \Psi_1 \otimes \Psi_2 = \begin{pmatrix} \alpha_1\alpha_2 \\ \alpha_1\beta_2 \\ \beta_1\alpha_2 \\ \beta_1\beta_2 \end{pmatrix}. \quad (20.14)$$

Next we consider two-qubit gates. Two qubits gates operate in the $\mathbb{C}^2 \otimes \mathbb{C}^2$ space. Thus they can be represented by 4×4 unitary matrices. Since the inverse of the unitary matrix U is the matrix $U^\dagger$, all quantum gates are *reversible*, in contrast to classical logic gates. For a quantum gate, the input can be deduced from the output. But for a classical XOR gate $x \oplus y$, for the output 1, we do not know whether the input was $|10\rangle$ or $|01\rangle$. The output does not identify the input. The simplest two-qubit gate is the CNOT (controlled-NOT) gate. We denote the two input qubits by x, y. The first qubit x is called the *control qubit*, the second y the *target* qubit. This target qubit flips if and only if $x = 1$. If $x = 0$ the second qubit remains unchanged. A representation of the CNOT gate is shown in Fig. 20.3. The black dot represents the action of the control qubit. The unitary matrix corresponding to the CNOT gate is given by

$$U_{\text{CNOT}} = \begin{pmatrix} 1 & 0 & 0 & 0 \\ 0 & 1 & 0 & 0 \\ 0 & 0 & 0 & 1 \\ 0 & 0 & 1 & 0 \end{pmatrix}. \quad (20.15)$$

In effect, U_{CNOT} leaves the states $|00\rangle$, $|01\rangle$ invariant, but swaps the states $|10\rangle$ and $|11\rangle$. The CNOT gate is the fundamental creator of entangled qubits, performing, in a sense, the "quantization" of a classical gate to a quantum gate.

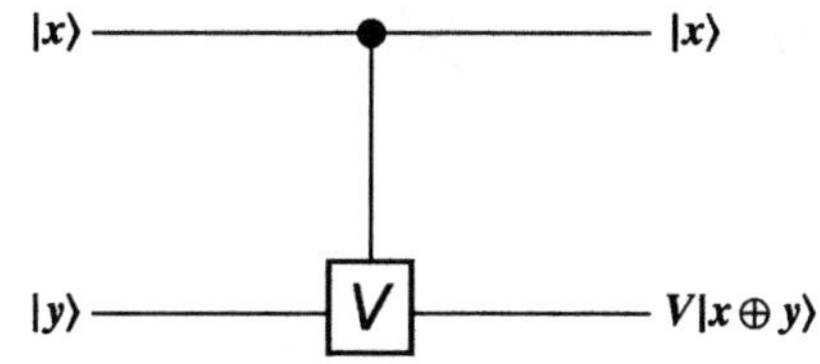

FIGURE 20.4 Controlled-V gate.

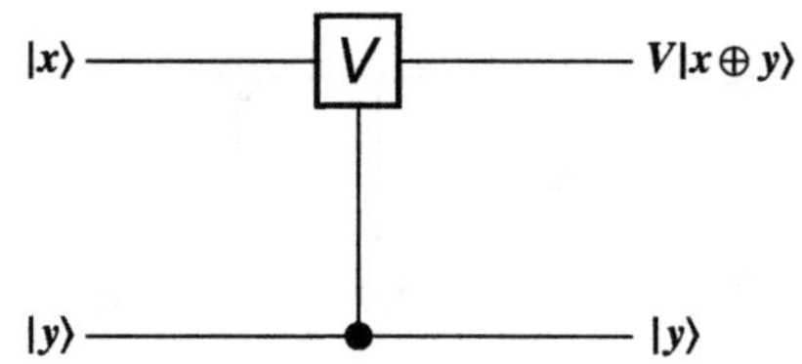

FIGURE 20.5 Inverted controlled-V gate.

A generalization of the CNOT gate is a controlled gates acting on 2 qubits, with the control qubit determining some unitary operation V on the target qubit. If the single qubit gate V is represented by the 2×2 matrix:

$$V = \begin{pmatrix} V_{11} & V_{12} \\ V_{21} & V_{22} \end{pmatrix} \tag{20.16}$$

The corresponding 2-qubit matrix is

$$U_{\text{CV}} = \begin{pmatrix} 1 & 0 & 0 & 0 \\ 0 & 1 & 0 & 0 \\ 0 & 0 & V_{11} & V_{12} \\ 0 & 0 & V_{21} & V_{22} \end{pmatrix}. \tag{20.17}$$

The symbol for the controlled-V gate is shown in Fig. 20.4.

Exchanging the control and target registers gives an *inverted controlled gate*. Fig. 20.5 shows an inverted controlled-V gate, which is represented by the matrix:

$$U_{\text{ICV}} = \begin{pmatrix} 1 & 0 & 0 & 0 \\ 0 & V_{11} & 0 & V_{12} \\ 0 & 0 & 1 & 0 \\ 0 & V_{21} & 0 & V_{22} \end{pmatrix}. \tag{20.18}$$

If V is a phase gate, such as S or T, in general $R_n = \begin{pmatrix} 1 & 0 \\ 0 & e^{i\pi/2^{n-1}} \end{pmatrix}$, then the corresponding controlled and inverted-controlled gates are identical:

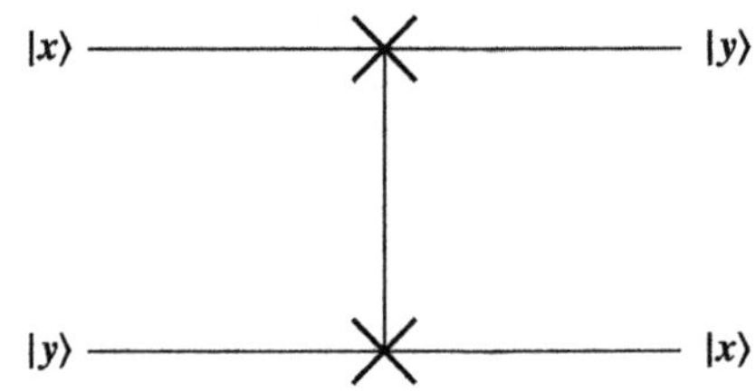

FIGURE 20.6 SWAP gate.

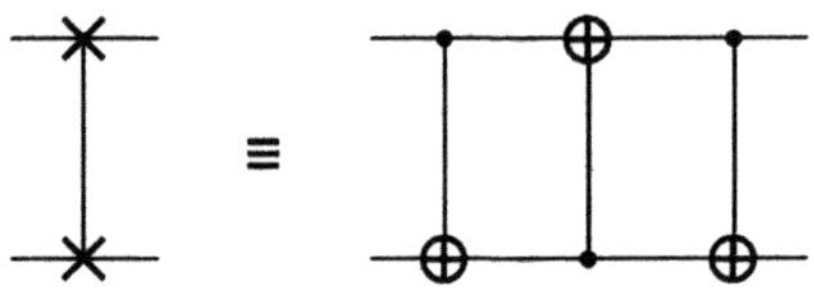

FIGURE 20.7 Sequence of CNOT operations equivalent to SWAP gate.

$$U_{\text{CR}} = U_{\text{ICR}} = \begin{pmatrix} 1 & 0 & 0 & 0 \\ 0 & 1 & 0 & 0 \\ 0 & 0 & 1 & 0 \\ 0 & 0 & 0 & e^{i\pi/2^{n-1}} \end{pmatrix}. \tag{20.19}$$

Another important 2-qubit gate is the SWAP gate, which swaps the states of two input qubits, such that

$$U_{\text{SWAP}} = \begin{pmatrix} 1 & 0 & 0 & 0 \\ 0 & 0 & 1 & 0 \\ 0 & 1 & 0 & 0 \\ 0 & 0 & 0 & 1 \end{pmatrix}. \tag{20.20}$$

This gate is shown in Fig. 20.6. It is easy to verify the equivalence of the SWAP gate with the sequence of CNOT operations shown in Fig. 20.7

In a quantum computer circuit, the path of each qubit is represented by a single horizontal line —— which usually connects a sequence of gates. The qubit is understood to move along the "quantum wire" from left to right. A wire carrying a classical bit, 0 or 1, usually after a measurement, is indicated by a double line $=$. As an example, let us show a quantum circuit which can produce *Bell states*, which are the superpositions $|\Phi^{\pm}\rangle = \frac{1}{\sqrt{2}}\big(|00\rangle \pm |11\rangle\big)$ and $|\Psi^{\pm}\rangle = \frac{1}{\sqrt{2}}\big(|01\rangle \pm |10\rangle\big)$. These are maximally entangled states of two qubits. Bell states can be produced by the circuit shown in Fig. 20.8. The first qubit is passed through a Hadamard gate and then the two qubits are entangled by a CNOT gate. If the input to the system is $|0\rangle \otimes |0\rangle = |00\rangle$, then the Hadamard gate changes the state to

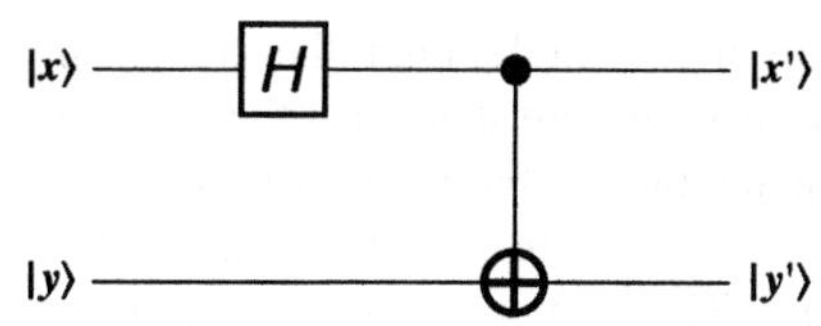

FIGURE 20.8 Circuit producing a Bell state.

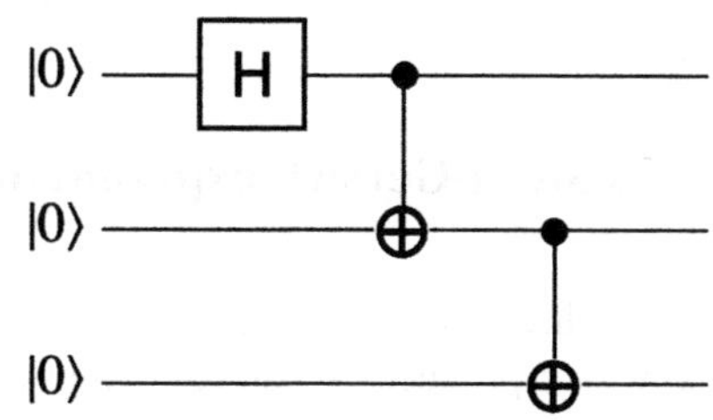

FIGURE 20.9 Circuit producing GHZ state $\frac{1}{\sqrt{2}}\big(|000\rangle + |111\rangle\big)$.

$$\frac{1}{\sqrt{2}}\Big(|0\rangle + |1\rangle\Big) \otimes (|0\rangle = \frac{1}{\sqrt{2}}\Big(|00\rangle + |10\rangle\Big),$$

and the CNOT gate converts this to $\frac{1}{\sqrt{2}}\Big(|00\rangle + |11\rangle\Big) = |\Phi^+\rangle$. For the four possible inputs, we find

$$\begin{aligned} |00\rangle &\to \frac{1}{\sqrt{2}}\Big(|00\rangle + |11\rangle\Big) = |\Phi^+\rangle, \ |01\rangle \to \frac{1}{\sqrt{2}}\Big(|01\rangle + |10\rangle\Big) = |\Psi^+\rangle, \\ |10\rangle &\to \frac{1}{\sqrt{2}}\Big(|00\rangle - |11\rangle\Big) = |\Phi^-\rangle, \ |11\rangle \to \frac{1}{\sqrt{2}}\Big(|01\rangle - |10\rangle\Big) = |\Psi^-\rangle. \end{aligned} \tag{20.21}$$

Bell states can simulate the correlation of measurements in Bohm's version of an Einstein-Podolski-Rosen experiment, described in Sect. 19.5. In the states $|\Phi^+\rangle$ or $|\Phi^-\rangle$, a measurement collapses the two-particle state to either $|00\rangle$ or $|11\rangle$, with equal probability. Thus, if a spin measurement on one particle gives $+1$ or -1, the other particle must necessarily give the same result. Similarly, the states $|\Psi^+\rangle$ or $|\Psi^-\rangle$ collapse in a measurement to either $|01\rangle$ or $|10\rangle$. The spin measurements on the two particle will then, of necessity, give *opposite* results.

Another interesting example is a quantum circuit, shown in Fig. 20.9 which can produce a *GHZ state*, which is an entangled superposition of *three* qubits:

$$|\Psi\rangle_{\rm GHZ} = \frac{1}{\sqrt{2}}\big(|000\rangle + |111\rangle\big). \tag{20.22}$$

GHZ refers to Greenberger, Horne and Zeilinger, who first produced entangled sets of three (and four) photons. Such experiments enabled demonstration of the validity of Bell's theorem in a single run, rather than from statistical analysis of a multitude of runs. Bell's theorem states that the results of quantum mechanics are incompatible with any picture based on local realism.

The quantum algorithms we are considering in this Chapter are based on the *quantum circuit model*. This uses one- and two-qubit gates sequentially connected by wires. We begin with an initial register of qubits to be converted by the action of the circuit into a resultant register of classical bits, on which measurements can be made.

20.3 Simulation of a Stern-Gerlach experiment

In the Stern-Gerlach experiment (see Section 7.6), an unpolarized beam of neutral particles of spin-1/2 is directed through an inhomogeneous magnetic field (blue and red magnets, which produces separated beams of spin-up and spin-down particles). For simplicity, only the outgoing spin-up beam is shown in Fig. 20.10. This beam is then directed through a second magnet, for which the polarization can be rotated by an angle θ from the original. This further splits the beam (except when $\theta = 0$ or π) into spin-up and spin-down beams with respect to the new polarization direction. Again, only the spin-up component is shown. The probability for a particle to emerge with spin-up ($\uparrow$) or spin-down ($\downarrow$) is given by $\cos^2(\theta/2)$ and $\sin^2(\theta/2)$ respectively. The resulting probabilities of $\uparrow$ and $\downarrow$ are shown for five selected angles.

The results of the Stern-Gerlach experiment can be simulated by a quantum computer. The qubits $|0\rangle$ and $|1\rangle$ correspond to the spin states $\uparrow$ and $\downarrow$, respectively. The initial state $|0\rangle$ corresponds to the polarized beam leaving the first magnet. By an appropriate sequence of quantum gates, the results of the beam passing through the second magnet, with polarization angle θ, can be simulated. The statistical results are verified after a large number of runs on the quantum computer, as shown after the measurement symbol.

For example, the $\theta = \pi/4$ rotation is produced by the sequence

$$H\,T\,H|0\rangle = \begin{pmatrix} 1 & 1 \\ 1 & -1 \end{pmatrix} \begin{pmatrix} 1 & 0 \\ 0 & e^{i\pi/4} \end{pmatrix} \begin{pmatrix} 1 & 1 \\ 1 & -1 \end{pmatrix} \begin{pmatrix} 1 \\ 0 \end{pmatrix} = \begin{pmatrix} \frac{1}{2} + \frac{1}{2}e^{i\pi/4} \\ \frac{1}{2} - \frac{1}{2}e^{i\pi/4} \end{pmatrix}.$$

The probability of a result $|0\rangle$ in the subsequent measurement is then given by $\left|\frac{1}{2} + \frac{1}{2}e^{i\pi/4}\right|^2 = 0.8536$, or about 85% spin-up.

20.4 Quantum Fourier transform

The quantum Fourier transform is closely analogous to the well-known *discrete Fourier transform* (DFT), whereby set of N complex numbers x_j ($j = 0, 1, \ldots, N-1$) can be transformed into another set of N complex numbers y_k ($k = 0, 1, \ldots, N-1$) by the relations:

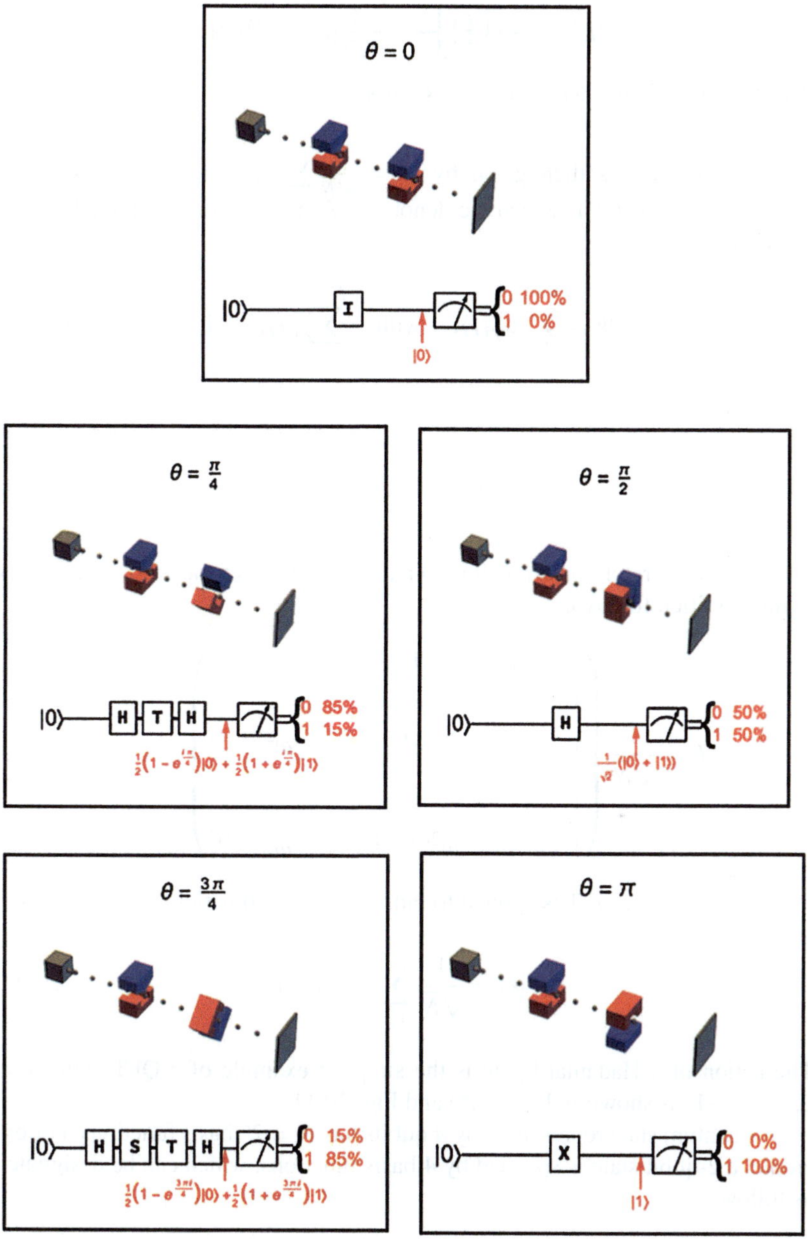

FIGURE 20.10 Stern-Gerlach experiment and quantum-computer simulation.

$$y_k = \frac{1}{\sqrt{N}} \sum_{j=0}^{N-1} x_j \, (\omega_N)^{kj}, \qquad \omega_N = e^{2\pi i/N}. \tag{20.23}$$

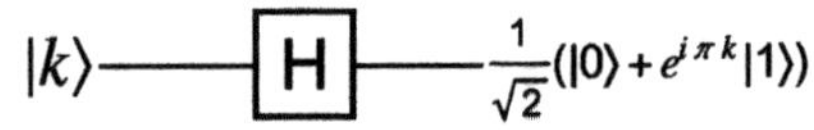

FIGURE 20.11 Circuit with Hadamard gate; $k = 0$ or 1.

The inverse DFT is then given by $x_j = \frac{1}{\sqrt{N}} \sum_{k=0}^{N-1} y_k \, (\omega_N^*)^{kj}$. A *quantum Fourier transform* QFT, which we denote by F, is equivalent to a DFT on the *amplitudes* of a quantum state,

$$\Psi = \sum_{j=0}^{N-1} \alpha_j |j\rangle, \quad \text{with} \quad \sum_{j=0}^{N-1} |\alpha_j|^2 = 1, \tag{20.24}$$

such that

$$F\,\Psi = \sum_{k=0}^{N-1} \beta_k |k\rangle, \qquad \text{with} \qquad \beta_k = \frac{1}{\sqrt{N}} \sum_{j=0}^{N-1} \alpha_j (\omega_N)^{kj}. \tag{20.25}$$

The unitary operator corresponding to a quantum Fourier transform is given by (with ω written for ω_N):

$$F = \frac{1}{\sqrt{N}} \begin{pmatrix} 1 & 1 & 1 & \dots & 1 \\ 1 & \omega & \omega^2 & \dots & \omega^{N-1} \\ 1 & \omega^2 & \omega^4 & \dots & \omega^{2(N-1)} \\ \vdots & \vdots & \vdots & & \vdots \\ 1 & \omega^{N-1} & \omega^{2(N-1)} & \dots & \omega^{(N-1)^2} \end{pmatrix}. \tag{20.26}$$

Most often, the QFT is applied to individual basis functions $|k\rangle$, such that

$$F\,|k\rangle = \frac{1}{\sqrt{N}} \sum_{j=0}^{N-1} (\omega_N)^{kj} \, |j\rangle. \tag{20.27}$$

The action of a Hadamard gate is the simplest example of a QFT, with $N = 2$, $\omega_2 = -1$, as shown in Eq. (20.8) and Fig. 20.11.

A quantum state represented by n qubits has $N = 2^n$ basis functions. For example, a 2-qubit state is spanned by 4 basis functions, which can be designated as follows

$$\begin{aligned} |\mathbf{0}\rangle &= |0\rangle \otimes |0\rangle = |00\rangle, \qquad & |\mathbf{1}\rangle &= |0\rangle \otimes |1\rangle = |01\rangle, \\ |\mathbf{2}\rangle &= |1\rangle \otimes |0\rangle = |10\rangle, \qquad & |\mathbf{3}\rangle &= |1\rangle \otimes |1\rangle = |11\rangle. \end{aligned} \tag{20.28}$$

To summarize:

$$|k\rangle = |k_1\rangle \otimes |k_2\rangle = |k_1 k_2\rangle, \tag{20.29}$$

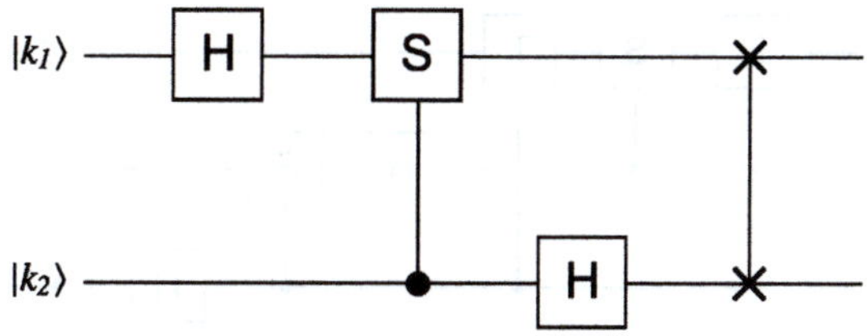

FIGURE 20.12 Circuit for 2-qubit quantum Fourier transform.

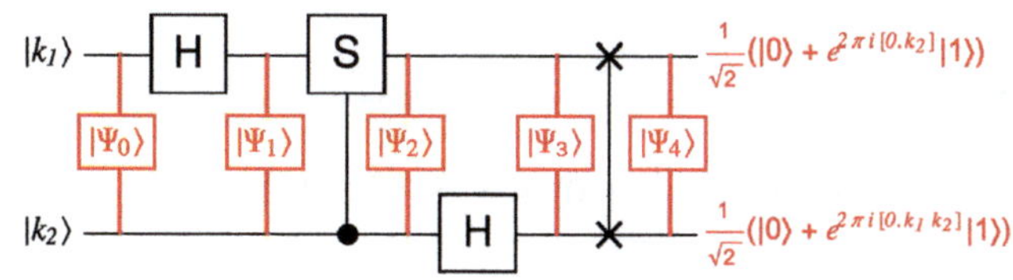

FIGURE 20.13 Details for 2-qubit quantum Fourier transform.

where k is expressed as a binary number. With $N = 4,\ \omega = \omega_4 = e^{i\pi/2} = i$, the QFT of a basis function is given by

$$F\,|k\rangle = \frac{1}{2}\sum_{j=0}^{3}\omega^{kj}\,|j\rangle. \tag{20.30}$$

Noting that $|j\rangle = |j_1\rangle \otimes |j_2\rangle$,

$$F\,|k\rangle = \frac{1}{2}\sum_{j_1=0}^{1}\sum_{j_2=0}^{1}\omega^{2kj_1}\omega^{kj_2}|j_1\rangle \otimes |j_2\rangle = \frac{1}{2}\Big(|0\rangle + \omega^{2k}|1\rangle\Big) \otimes \Big(|0\rangle + \omega^{k}|1\rangle\Big) =$$
$$\frac{1}{\sqrt{2}}\Big(|0\rangle + e^{i\pi k_2}|1\rangle\Big) \otimes \frac{1}{\sqrt{2}}\Big(|0\rangle + e^{i\pi(k_1+\frac{k_2}{2})}|1\rangle\Big). \tag{20.31}$$

It has also been noted that $e^{2\pi i k_1} = 1$.

Fig. 20.12 shows a circuit which implements the 2-qubit QFT. The details of its operation are represented in Fig. 20.13. We begin with the state $|\Psi_0\rangle = |k_1\rangle \otimes |k_2\rangle$. The Hadamard gate on the first register produces the state

$$|\Psi_1\rangle = \frac{1}{\sqrt{2}}\Big(|0\rangle + e^{i\pi k_1}|1\rangle\Big) \otimes |k_2\rangle. \tag{20.32}$$

The subsequent action of the inverted controlled-S gate uses the unitary operator

$$U_{\text{ICS}} = \begin{pmatrix} 1 & 0 & 0 & 0 \\ 0 & 1 & 0 & 0 \\ 0 & 0 & 1 & 0 \\ 0 & 0 & 0 & e^{i\pi/2} \end{pmatrix}, \tag{20.33}$$

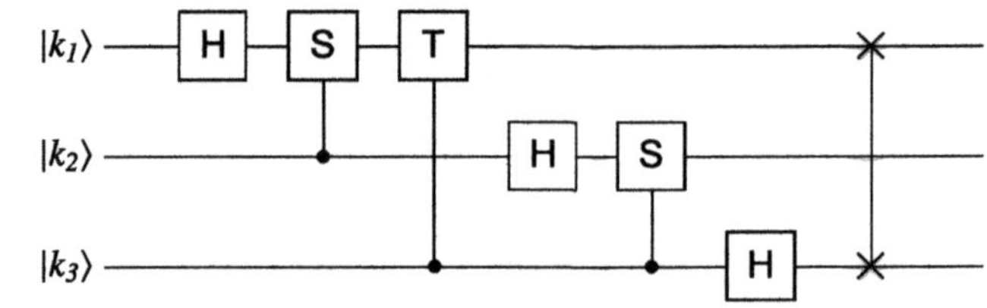

FIGURE 20.14 Circuit for 3-qubit quantum Fourier transform.

which gives

$$|\Psi_2\rangle = U_{\text{ICS}}|\Psi_1\rangle = \frac{1}{\sqrt{2}}\left(|0\rangle + e^{i\pi k_1}e^{i\pi\frac{k_2}{2}}|1\rangle\right) \otimes |k_2\rangle. \tag{20.34}$$

A Hadamard gate in the second register then results in

$$|\Psi_3\rangle = \frac{1}{\sqrt{2}}\left(|0\rangle + e^{i\pi\left(k_1+\frac{k_2}{2}\right)}|1\rangle\right) \otimes \frac{1}{\sqrt{2}}\left(|0\rangle + e^{i\pi k_2}|1\rangle\right). \tag{20.35}$$

The final result is produced by a SWAP gate:

$$|\Psi_4\rangle = \text{SWAP}|\Psi_3\rangle = \frac{1}{\sqrt{2}}\left(|0\rangle + e^{i\pi k_2}|1\rangle\right) \otimes \frac{1}{\sqrt{2}}\left(|0\rangle + e^{i\pi\left(k_1+\frac{k_2}{2}\right)}|1\rangle\right). \tag{20.36}$$

By analogy with the above procedure, we can derive the 3-qubit QFT. Eq. (20.31) generalizes to

$$\begin{aligned} F\,|k\rangle = \frac{1}{\sqrt{2}}\left(|0\rangle + e^{i\pi k_3}|1\rangle\right) \otimes \frac{1}{\sqrt{2}}\left(|0\rangle + e^{i\pi(k_2+\frac{k_3}{2})}|1\rangle\right) \\ \otimes\frac{1}{\sqrt{2}}\left(|0\rangle + e^{i\pi(k_1+\frac{k_2}{2}+\frac{k_3}{4})}|1\rangle\right) = \\ \frac{1}{\sqrt{2}}\left(|0\rangle + e^{2\pi i[0.k_3]}|1\rangle\right) \otimes \frac{1}{\sqrt{2}}\left(|0\rangle + e^{2\pi i[0.k_2k_3]}|1\rangle\right) \\ \otimes\frac{1}{\sqrt{2}}\left(|0\rangle + e^{2\pi i[0.k_1k_2k_3]}|1\rangle\right). \end{aligned} \tag{20.37}$$

In the last line, we have expressed k in binary fractional notation, such that

$$k = k_1 + \frac{k_2}{2} + \frac{k_3}{4} + \cdots = 2[0.k_1k_2k_3\ldots]. \tag{20.38}$$

The corresponding circuit is shown in Fig. 20.14. In all of the above examples, a QFT applied to a single basis qubit transforms it into a equally-weighted linear combination of all N basis qubits, with complex phase factors of unit magnitude.

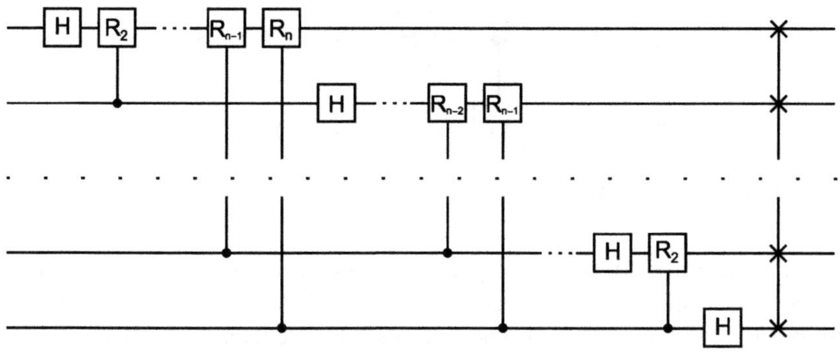

FIGURE 20.15 Circuit for n-qubit quantum Fourier transform.

The inverted controlled-T gate used in the circuit is represented by the unitary matrix

$$U_{\mathrm{ICT}} = \begin{pmatrix} 1 & 0 & 0 & 0 \\ 0 & 1 & 0 & 0 \\ 0 & 0 & 1 & 0 \\ 0 & 0 & 0 & e^{i\pi/4} \end{pmatrix}. \tag{20.39}$$

We can define a generalized *controlled-R_n gate* such that

$$R_n = \begin{pmatrix} 1 & 0 & 0 & 0 \\ 0 & 1 & 0 & 0 \\ 0 & 0 & 1 & 0 \\ 0 & 0 & 0 & e^{i\pi/2^{n-1}} \end{pmatrix}. \tag{20.40}$$

The general n-qubit QFT circuit is shown in Fig. 20.15.

20.5 Phase estimation algorithm

An important application of the quantum Fourier transform is phase estimation. In applications to quantum chemistry, this enables energy eigenvalues to be calculated after the action of the evolution operator e^{-iHt}. Let U be a known $2^n \times 2^n$ unitary matrix and $|\Psi\rangle \in \mathbb{C}^{2^n}$, one of its eigenvectors. We can then write

$$U|\Psi\rangle = e^{i\varphi}|\Psi\rangle, \tag{20.41}$$

since the eigenvalues of a unitary matrix are complex numbers of unit magnitude (sometimes called *eigenphases*). The action of the operator U is performed by some otherwise unspecified "black box," also known as a *oracle*. This can be symbolically represented by $|\Psi\rangle$ —[U]— $e^{i\varphi}|\Psi\rangle$. The basic maneuver leading to the phase-estimation algorithm is known as *phase kickback*. Consider the

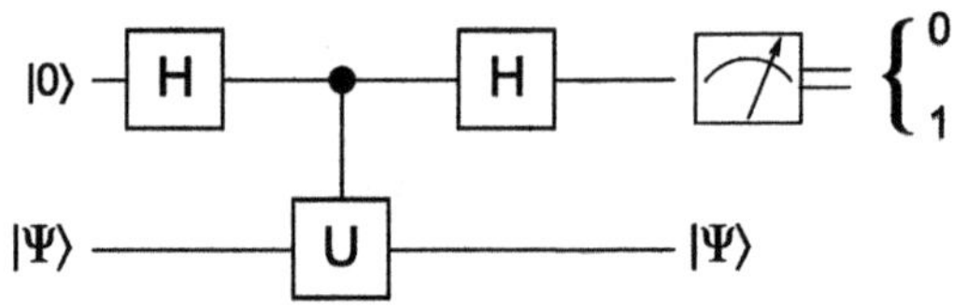

FIGURE 20.16 Phase kickback.

little circuit in Fig. 20.16, involving a single-qubit function $|\psi\rangle$. The initial input to the two registers can be written $|0\rangle \otimes |\psi\rangle$. The Hadamard gate transforms this to $\frac{1}{\sqrt{2}}(|0\rangle \otimes |\psi\rangle + |1\rangle \otimes |\psi\rangle)$. The controlled-U gate then acts only for the $|1\rangle$ component of the control register. The result is

$$\frac{1}{\sqrt{2}}\Big(|0\rangle \otimes |\psi\rangle + |1\rangle \otimes e^{i\varphi}|\psi\rangle\Big) = \frac{1}{\sqrt{2}}\Big(|0\rangle + e^{i\varphi}|1\rangle\Big) \otimes |\psi\rangle, \tag{20.42}$$

with the target register $|\psi\rangle$ emerging unchanged. After action of another Hadamard gate, the control register becomes

$$\begin{aligned} H\frac{1}{\sqrt{2}}\Big(|0\rangle + e^{i\varphi}|1\rangle\Big) &= \frac{1}{2}\Big(|0\rangle + |1\rangle\Big) + \frac{1}{2}e^{i\varphi}\Big(|0\rangle - |1\rangle\Big) \\ &= \left(\frac{1+e^{i\varphi}}{2}\right)|0\rangle + \left(\frac{1-e^{i\varphi}}{2}\right)|1\rangle. \end{aligned} \tag{20.43}$$

Finally, a measurement on the control register results in a bit 0 or 1, with probabilities:

$$p(0) = \left|\frac{1+e^{i\varphi}}{2}\right|^2 = \cos^2\frac{\varphi}{2}, \qquad p(1) = \left|\frac{1-e^{i\varphi}}{2}\right|^2 = \sin^2\frac{\varphi}{2}. \tag{20.44}$$

A statistical analysis then enables the phase φ to be determined.

For a multiqubit function $|\Psi\rangle$, suppose it is desired to estimate the phase φ to an accuracy of n bits. We then need to subject the first n qubits of the eigenvector $|\Psi\rangle$ to a series of controlled operators, involving powers of U, followed by an inverse quantum Fourier transform. The circuit implementing the phase estimation shown in Fig. 20.17. The input of the first register consists of m qubits, all prepared in the state $|0\rangle$. This is an example of an *ancilla*, which is an entangled component of a quantum computer intended to assist in carrying out an algorithm. The second register contains the input of n qubits, which represent the state $|\Psi\rangle$. For concreteness, we consider the particular case $m = 3, n = 3$. An H gate, followed by a controlled U^{2n} (for $n = 0, 1, 2$), maps the qubit $|0\rangle$ into $\frac{1}{\sqrt{2}}\big(|0\rangle + e^{i2^n\varphi}|1\rangle\big)$. In this way the output of the top register is the product state:

$$|\text{OUT}\rangle = \frac{1}{\sqrt{2}}\big(|0\rangle + e^{4i\varphi}|1\rangle\big) \otimes \frac{1}{\sqrt{2}}\big(|0\rangle + e^{2i\varphi}|1\rangle\big) \otimes \frac{1}{\sqrt{2}}\big(|0\rangle + e^{i\varphi}|1\rangle\big). \tag{20.45}$$

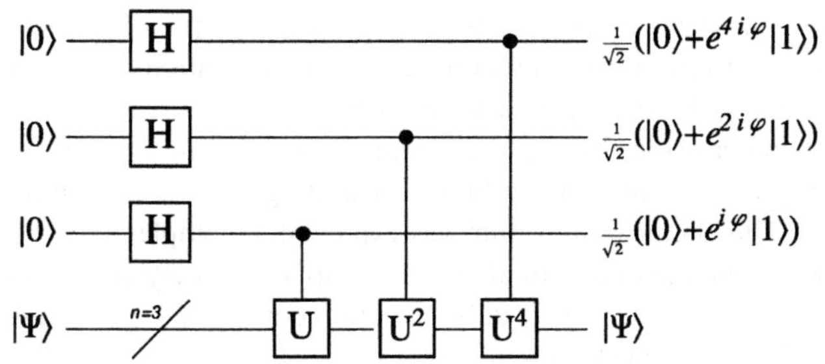

FIGURE 20.17 Circuit implementing phase estimation.

Comparing this with the last line of Eq. (20.37), $|\text{OUT}\rangle$ is seen to represent the Fourier transform of a vector $|k\rangle = |k_1\rangle \otimes |k_2\rangle \otimes |k_3\rangle$. Accordingly, $|k\rangle$ can be determined by the inverse Fourier transform:

$$|k\rangle = F^{-1}|\text{OUT}\rangle. \tag{20.46}$$

The magnitude of k $(0 \leq k \leq 1)$ is given by $k = [0.k_1k_2k_3]$. This represents an approximation to the phase φ, to 3-bit accuracy:

$$\varphi = 2\pi[0.k_1k_2k_3\ldots]. \tag{20.47}$$

Therefore, the circuit of Fig. 20.17 can determine the phase φ to this level of accuracy. The error in this value is bounded by the first neglected bit, numbered $n+1$. Thus the error in the measured phase, say $\delta\varphi$, is of the order of $2\pi/2^{n+1}$. A plot showing a series of repeated determinations of φ might resemble a Gaussian centered at $2\pi k$ with $\sigma \approx \delta\varphi$. Clearly, increasing the number of qubits in $|k\rangle$, improves the accuracy of the computed phase.

20.6 Many-electron systems

Their representation requires systems of multiple qubits. The Hilbert space of D qubits, $\mathbb{C}^{2D}$, is the *tensor product* of D copies of $\mathbb{C}^2$, namely, $\mathbb{C}^2 \otimes \mathbb{C}^2 \ldots \otimes \mathbb{C}^2$, D times. An orthonormal basis in *each* space $\mathbb{C}^2$. The extended basis then consists of 2^D product states:

$$|\delta_1, \delta_2, \ldots, \delta_D\rangle \;=\; |\delta_1\rangle \otimes |\delta_2\rangle \otimes \ldots. \otimes |\delta_D\rangle. \tag{20.48}$$

We use the symbols δ_k, rather than n_k, for the occupation numbers to emphasize that they can equal only 0 or 1 (inspired by the Kronecker delta δ_{ij}). The multi-dimensional systems of particular interest are many-electron atoms or molecules, as represented by Fock states with D spinorbitals, occupied by N electrons in all combinations of occupation numbers. The D-dimensional Fock states will include N occupied spinorbitals and $D-N$ unoccupied (or *virtual*) spinorbitals. Most often, the virtual spinorbitals have positive energies

$\epsilon_k > 0$. On a classical computer, enumeration of all the possible Fock states for a D-dimensional system would require 2^D memory locations. But in a quantum computer, each qubit can be in a superposition of states, containing the same quantity of information. Moreover, computations can, in principle, be run simultaneously on the entire assemblage, exhibiting what is known as *quantum parallelism*. The classical algorithm has exponential complexity wrt increasing system size, while the quantum algorithm is approximately polynomial. As an idea of the scales involved, a 30-qubit quantum computer could simultaneously process $10^{30} \approx 1$ billion states.

Fermion *creation operators* $a_k{}^\dagger$ and *annihilation operators* a_k can add or remove electrons from the Fock state. These obey the anticommutation relations

$$\{a_j, a_k{}^\dagger\} = \delta_{jk}, \qquad \{a_j, a_k\} = \{a_j^\dagger, a_k{}^\dagger\} = 0. \tag{20.49}$$

The *vacuum state* $|\text{vac}\rangle$, the formal starting point for manipulations in Fock space, is defined by

$$|\text{vac}\rangle = |0_1, 0_2, \ldots, 0_D\rangle. \tag{20.50}$$

An electron in a spinorbital k is added to the system using the creation operator $a_k^\dagger$:

$$a_k{}^\dagger|\text{vac}\rangle = |0_1, 0_2, \ldots, 1_k, \ldots\rangle. \tag{20.51}$$

Adding a second electron in the spinorbital j gives

$$a_j{}^\dagger a_k{}^\dagger|\text{vac}\rangle = |0_1, 0_2, \ldots, 1_j, \ldots, 1_k, \ldots\rangle, \tag{20.52}$$

if j comes before k in the Fock space ordering, but

$$a_j{}^\dagger a_k{}^\dagger|\text{vac}\rangle = -|0_1, 0_2, \ldots, 1_k, \ldots, 1_j, \ldots\rangle, \tag{20.53}$$

if k comes before j. The sign change is a consequence of the anticommutation $a_j{}^\dagger a_k{}^\dagger = -a_k{}^\dagger a_j^\dagger$. This accords with the antisymmetry principle for many-electron systems. Clearly, also, $a_k{}^\dagger a_k{}^\dagger = 0$, which is a statement of the Pauli exclusion principle. For the action of a creation operator on an arbitrary Fock state

$$a_k{}^\dagger|\delta_1, \delta_2, \ldots, 0_k, \ldots\rangle = (-1)^{\sum_{j=1}^{k-1}\delta_j}|\delta_1, \delta_2, \ldots, 1_k, \ldots\rangle, \tag{20.54}$$

where the sum counts the number of occupied spinorbitals $j < k$. Of course, $a_k{}^\dagger|\delta_1, \delta_2, \ldots, 1_k, \ldots\rangle = 0$, if the spinorbital k is already occupied.

Analogous relations hold for the annihilation operators a_k. In particular,

$$a_k|\delta_1, \delta_2, \ldots, 1_k, \ldots\rangle = (-1)^{\sum_{j=1}^{k-1}\delta_j}|\delta_1, \delta_2, \ldots, 0_k, \ldots\rangle, \tag{20.55}$$

and $a_k|\delta_1, \delta_2, \ldots, 0_k, \ldots\rangle = 0$.

20.7 Atomic and molecular Hamiltonians

An atomic or molecular N-electrons system can be represented by the Hamiltonian

$$H = \sum_{i=1}^{N} \left\{ -\frac{1}{2}\nabla_i^2 - \sum_A \frac{Z_A}{|\mathbf{r}_i - \mathbf{R}_A|} \right\} + \sum_{i>j=1}^{N} \frac{1}{r_{ij}}, \tag{20.56}$$

in atomic units and with the usual assumption of the Born-Oppenheimer approximation and the neglect of relativistic effects. The Hamiltonian contains only one-body and two-body interactions, which makes it feasible to simulate them with quantum-computer circuits. In second-quantized notation, the Hamiltonian can be expressed as

$$H = \sum_{j,k} h_{j,k} a_j^\dagger a_k + \frac{1}{2} \sum_{j \neq k, p \neq q} v_{j,k,p,q} a_j^\dagger a_k^\dagger a_p a_q \tag{20.57}$$

The coefficients $h_{j,k}$ and $v_{j,k,p,q}$ need to be evaluated in advance on a classical computer. In terms of a selected set of basis functions $\{\phi_k(x)\}$, we have

$$h_{j,k} = \left\langle \phi_j(x) \left| \left\{ -\frac{1}{2}\nabla^2 - \sum_A \frac{Z_A}{|\mathbf{r} - \mathbf{R}_A|} \right\} \right| \phi_k(x) \right\rangle, \tag{20.58}$$

and

$$v_{j,k,p,q} = \left\langle \phi_j(x)\phi_k(x') \left| \frac{1}{|\mathbf{r} - \mathbf{r}'|} \right| \phi_p(x)\phi_q(x') \right\rangle. \tag{20.59}$$

The next step is to assemble a set of quantum gates which will simulate the actions of the creation and annihilation operators in this Hamiltonian. Qubits are often physically realized with the spin-$\frac{1}{2}$ states of electrons or nuclei. The spin-up state is written $|1\rangle$ (or sometimes $|1/2\rangle$), the spin-down state, $|-1\rangle$ (or $|-1/2\rangle$). The analogous qubits are designated $|0\rangle$ and $|1\rangle$, respectively. In Pauli spin algebra, raising and lowering operators are defined by

$$\sigma^{\pm} = \frac{1}{2}\left(\sigma_x \pm i\sigma_y\right), \tag{20.60}$$

such that

$$\sigma^+|-1\rangle = |1\rangle, \qquad \sigma^+|1\rangle = 0, \qquad \sigma^-|1\rangle = |-1\rangle, \qquad \sigma^-|-1\rangle = 0. \tag{20.61}$$

These relations suggest the analogous quantum gates

$$Q_k^\dagger = \frac{1}{2}\left(X_k + iY_k\right), \qquad Q_k = \frac{1}{2}\left(X_k - iY_k\right), \tag{20.62}$$

such that

$$Q_k^\dagger|1_k\rangle = |0_k\rangle, \qquad Q_k^\dagger|0_k\rangle = 0, \qquad Q_k|0_k\rangle = |1_k\rangle, \qquad Q_k|1_k\rangle = 0. \tag{20.63}$$

It is useful to review the multiplicative relations involving Pauli gates on the same qubit:

$$X^2 = Y^2 = Z^2 = I, \;\; XY = -YX = iZ, \;\; YZ = -ZY = iX,$$
$$ZX = -XZ = iY, \;\; QZ = -ZQ = Q, \;\; Q^\dagger Z = -ZQ^\dagger = -Q^\dagger. \tag{20.64}$$

The correspondence between creation/annihilation operators and Pauli gates works very well for actions on a single qubit. However, a problem arises for multiple qubits: operators on different spins commute, but operators on multi-fermion states need to obey the appropriate anticommutation relations. This can be fixed by means of a *Jordan-Wigner transformation*. Essentially, this determines the proper $\pm$ signs, just like the $(-1)^\delta$ factors in Eqs. (20.54) and (20.55). Recall the action of the Pauli-Z gate

$$Z_k|0_k\rangle = +|0_k\rangle, \qquad Z_k|1_k\rangle = -|1_k\rangle. \tag{20.65}$$

This suggests a form for the quantum gate corresponding to the creation operator $a_k^\dagger$, which we designate $A_k{}^\dagger$:

$$A_k{}^\dagger = \otimes_{j=1}^{k-1} Z_j \otimes Q_k^\dagger \otimes_{j=k+1}^{D} I_j. \tag{20.66}$$

Analogously,

$$A_k = \otimes_{j=1}^{k-1} Z_j \otimes Q_k \otimes_{j=k+1}^{D} I_j. \tag{20.67}$$

Note that the gates $A_k{}^\dagger$ and A_k are now *non-local*, since their action depends on the other qubits in the Fock state. For example, for $D = 4, k = 3$, we have

$$A_3{}^\dagger = Z_1 \otimes Z_2 \otimes Q_3^\dagger \otimes I_4, \qquad A_3 = Z_1 \otimes Z_2 \otimes Q_3 \otimes I_4. \tag{20.68}$$

Next, we show how to construct quantum-computer circuits to represent parts of the molecular Hamiltonian. For the diagonal terms in the one-particle sum in (20.57), the operator products $a_k{}^\dagger a_k = N_k$, which represent number operators. The corresponding Jordan-Wigner operators are given by

$$A_k{}^\dagger A_k = \left(\otimes_{j=1}^{k-1} Z_j \otimes Q_k^\dagger \otimes_{j=k+1}^{D} I_j\right)\left(\otimes_{j=1}^{k-1} Z_j \otimes Q_k \otimes_{j=k+1}^{D} I_j\right) =$$
$$Q_k^\dagger Q_k = \frac{1}{2}\left(X_k + iY_k\right)\frac{1}{2}\left(X_k - iY_k\right) = \frac{1}{4}\left(2I_k - i[X_k, Y_k]\right) =$$
$$\frac{1}{2}\left(I_k + Z_k\right), \tag{20.69}$$

noting that $Z_j^2 = I_j$, $X_k^2 = Y_k^2 = I_k$ and $\left[X_k, Y_k\right] = 2i\,Z_k$.

For off-diagonal contributions to the one-particle sum, such as $a_j{}^\dagger a_k$, consider the case $j, k = 2, 3$:

$$A_3{}^\dagger A_2 = \left(Z_1 \otimes Z_2 \otimes Q_3^\dagger \otimes I_4\right)\left(Z_1 \otimes Q_2 \otimes I_3 \otimes I_4\right) =$$
$$Z_1^2 \otimes Z_2 Q_2 \otimes Q_3^\dagger I_3 \otimes I_4 = -I_1 \otimes Q_2 \otimes Q_3^\dagger \otimes I_4, \quad (20.70)$$

while

$$A_2{}^\dagger A_3 = \left(Z_1 \otimes Q_2^\dagger \otimes I_3 \otimes I_4\right)\left(Z_1 \otimes Z_2 \otimes Q_3 \otimes I_4\right) =$$
$$Z_1^2 \otimes Q_2^\dagger Z_2 \otimes I_3 Q_3 \otimes I_4 = -I_1 \otimes Q_2^\dagger \otimes Q_3 \otimes I_4. \quad (20.71)$$

Thus

$$A_2{}^\dagger A_3 + A_3{}^\dagger A_2 = -I_1 \otimes \left(Q_2^\dagger \otimes Q_3 + Q_2 \otimes Q_3^\dagger\right) \otimes I_4 =$$
$$-\frac{1}{2} I_1 \otimes \left(X_2 \otimes X_3 + Y_2 \otimes Y_3\right) \otimes I_4. \quad (20.72)$$

If the indices j, k are not consecutive, then the intervening indices m will contribute factors of Z_m interposed between the j and k factors. Thus

$$A_1{}^\dagger A_3 + A_3{}^\dagger A_1 = -\frac{1}{2}\left(X_1 \otimes Z_2 \otimes X_3 + Y_1 \otimes Z_2 \otimes Y_3\right) \otimes I_4,$$
$$A_1{}^\dagger A_4 + A_4{}^\dagger A_1 = -\frac{1}{2}\left(X_1 \otimes Z_2 \otimes Z_3 \otimes X_4 + Y_1 \otimes Z_2 \otimes Z_3 \otimes Y_4\right), \quad (20.73)$$

and so forth.

Proceeding analogously for the two-particle terms in the Hamiltonian (20.57), we need to evaluate products of the form $A_j{}^\dagger A_k{}^\dagger A_p A_q$ using (20.66) and (20.67). For the sum of two-electron Coulomb plus exchange terms, we find

$$A_j{}^\dagger A_k{}^\dagger A_j A_k - A_j{}^\dagger A_k{}^\dagger A_k A_j = \frac{1}{2}\left(I_j + Z_j\right) \otimes \left(I_k + Z_k\right). \quad (20.74)$$

For the case $j \neq k \neq p \neq q$,

$$A_j{}^\dagger A_k{}^\dagger A_p A_q + A_q^\dagger A_p{}^\dagger A_k A_j =$$
$$-\frac{1}{2}\Big(X_j \otimes X_k \otimes X_p \otimes X_q - X_p \otimes X_q \otimes Y_j \otimes Y_k$$
$$+X_k \otimes X_q \otimes Y_j \otimes Y_p + X_j \otimes X_q \otimes Y_k \otimes Y_p + X_k \otimes X_p \otimes Y_j \otimes Y_q$$
$$+X_j \otimes X_p \otimes Y_k \otimes Y_q - X_j \otimes X_k \otimes Y_p \otimes Y_q + Y_j \otimes Y_k \otimes Y_p \otimes Y_q\Big). \quad (20.75)$$

20.8 Time-evolution of a quantum system

The principal goal of the quantum-computer program we are outlining is to find accurate values of energy eigenvalues, most often just the ground-state energy. The method exploits the fact that the time-evolution of a quantum system exhibits its eigenvalues, in the form of a Fourier superposition of its eigenstates. Let us begin with the time-dependent Schrödinger equation for an arbitrary quantum state $\Psi(x_1, x_2, \ldots, x_N, t)$, abbreviated $\Psi(x,t)$:

$$i\frac{\partial\Psi(x,t)}{\partial t} = H\Psi(x,t). \tag{20.76}$$

The formal solution, with an initial state $\Psi(x,0)$, can be written

$$\Psi(x,t) = e^{-iHt}\Psi(x,0), \tag{20.77}$$

where the exponential, known as the *evolution operator* or *propagator*, is explicitly a representation of the power series

$$e^{-iHt} = \sum_{n=0}^{\infty}\frac{(-iHt)^n}{n!} = 1+(-it)H+\frac{(-it)^2}{2!}H^2+\frac{(-it)^3}{3!}H^3+\ldots. \tag{20.78}$$

The initial state $\Psi(x,0)$ is formally represented by a superposition of eigenstates of the Hamiltonian H:

$$\Psi(x,0) = \sum_m c_m\psi_m(x) = c_0\psi_0(x)+c_1\psi_1(x)+c_2\psi_2(x)+\cdots+\text{continuum}, \tag{20.79}$$

where $H\psi_m(x) = E_m\psi_m(x) = \omega_m\psi_m(x)$. It is useful to express the energies in frequency units $E_m = \hbar\omega_m = \omega_m$. We label the ground state as $m = 0$. Applying the evolution operator, we find

$$\Psi(x,t) = e^{-iHt}\Psi(x,0) = \sum_m c_m e^{-i\,\omega_m t}\psi_m(x) = \\ c_0 e^{-i\,\omega_0 t}\psi_0(x)+c_1 e^{-i\,\omega_1 t}\psi_1(x)+\cdots+\text{continuum}. \tag{20.80}$$

The scalar product of $\Psi(x,t)$ with $\Psi(x,0)$ is an instance of an *autocorrelation function*:

$$F(t) = \langle\Psi(x,0)|\Psi(x,t)\rangle = \langle\Psi(x,0)|e^{-iHt}|\Psi(x,0)\rangle = \\ \sum_m |c_m|^2 e^{-i\,\omega_m t} = |c_0|^2 e^{-i\,\omega_0 t}+\ldots. \tag{20.81}$$

The Fourier transform of the autocorrelation function exhibits the spectrum of eigenvalues

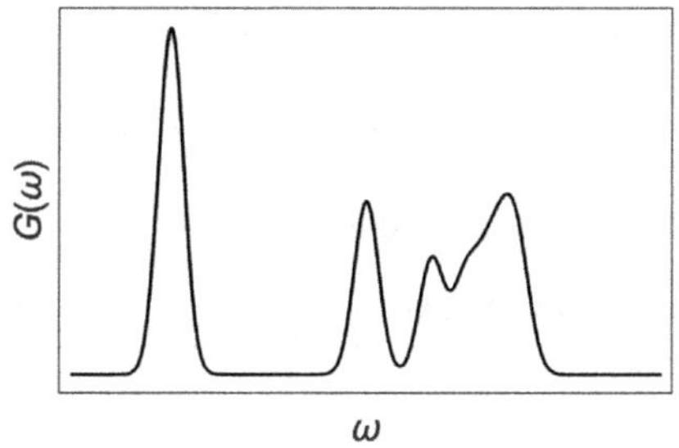

FIGURE 20.18 Simulation of computed spectrum in quantum computation.

$$G(\omega) = \frac{1}{2\pi}\sum_m |c_m|^2\delta(\omega - \omega_m) + \text{continuum}. \tag{20.82}$$

In practice, the peaks of $G(\omega)$ will be finite, because of approximations and inaccuracies, something like the spectrum shown in Fig. 20.18.

The objective pursued in this Chapter is computation of the quantized energy levels of a molecular system. This will be realized by simulation of the dynamics of the system, generated by the time-evolution operator e^{-iHt}. To do this on a quantum computer, we must construct a set of one- and two-qubit gates that implement this exponential transformation on a set of qubits. Approximations to the energy eigenvalues then appear in the phases of qubit states, which are determined using the phase-estimation algorithm.

20.9 Trotter expansions

The time-evolution operator e^{-iHt} is constructed by exponentiation of the second-quantized representation of the molecular Hamiltonian. However, the terms the Hamiltonian, $H = h_1 + h_2 + \ldots$, do not generally commute among themselves. Therefore approximate expansions involving noncommuting exponential operators are necessary.

Consider an exponential operator of the form $e^{(A+B)t}$ where A and B do not, in general, commute.[1] If the time interval t is broken up into n subintervals $\Delta t = t/n$, the *Lie-Trotter product formula* states that[2]

$$e^{(A+B)t} = \lim_{\Delta t \to 0}\left(e^{A\Delta t}e^{B\Delta t}\right)^n. \tag{20.83}$$

---

[1] Recall the Baker-Hausdorff formula $e^{(A+B)t} = e^{At}e^{Bt}e^{-\frac{1}{2}[A,B]t^2} + O(t^3)$.

[2] A simple proof:

$$e^{(A+B)t} = \lim_{n\to\infty}\left(1 + \frac{(A+B)t}{n}\right)^n = \lim_{n\to\infty}\left(1 + \frac{At}{n}\right)^n\left(1 + \frac{Bt}{n}\right)^n = \lim_{n\to\infty}\left(e^{At/n}\right)^n\left(e^{Bt/n}\right)^n.$$

For a finite value of n, this is approximated by

$$e^{(A+B)t} \approx \left(e^{A\Delta t} e^{B\Delta t}\right)^n + O(\Delta t^{n+1}) \tag{20.84}$$

The right hand side can be rearranged to

$$\left(e^{A\Delta t/2} e^{B\Delta t/2} e^{A\Delta t/2} e^{B\Delta t/2}\right)^n = \left(e^{A\Delta t/2} e^{B\Delta t} e^{A\Delta t/2}\right)^n \text{ or}$$
$$\left(e^{B\Delta t/2} e^{A\Delta t} e^{B\Delta t/2}\right)^n. \tag{20.85}$$

This is the form generally used in applications of Trotter approximations for exponential operators, with values of n anywhere from 1 to 10, as appropriate. The propagator e^{-iHt} should act over sufficiently long time to resolve the leading Fourier frequencies to a desired level of precision.

As shown in Eqs. (20.10), exponential operators involving Pauli matrices can be evaluated in closed form.

20.10 Simulations of molecular structure

By applying the Jordan-Wigner transformation and Trotter decomposition, it becomes possible to approximate the energy levels of a fermionic many-body Hamiltonian system on a quantum computer. The energy levels of a quantum system can, in principle, be determined by simulating its time evolution, by action of the operator $U = e^{-iHt}$. This might be done on a quantum computer using a set of one- and two-qubit gates that implement the time evolution on a set of qubits. Recent work by Aspuru-Guzik and coworkers has been able to create a quantum circuit to compute the energy of a molecular system with fixed nuclear geometry using a recursive quantum phase estimation algorithm to simulate a full configuration interaction (FCI) computation.

Let us develop the method beginning with a simple example. A single-particle term in the Hamiltonian, under a Jordan-Wigner transformation, Eq. (20.69) has the form:

$$H = ha^\dagger a \to hA^\dagger A = \frac{h}{2}(I + Z). \tag{20.86}$$

If this were the only term in the Hamiltonian, the time-evolution operator would be given by

$$U = e^{-i\frac{h}{2}(I+Z)t} = e^{-iht/2} e^{-ihtZ/2}. \tag{20.87}$$

The exponential operator reduces to separate factors here since I and Z commute. The second factor has the form of a "rotated" Pauli operator, Eq. (20.10). Therefore

$$e^{-ihZt/2} = \cos\left(\frac{ht}{2}\right) I - i \sin\left(\frac{ht}{2}\right) Z = \begin{pmatrix} e^{-iht/2} & 0 \\ 0 & e^{iht/2} \end{pmatrix} \tag{20.88}$$

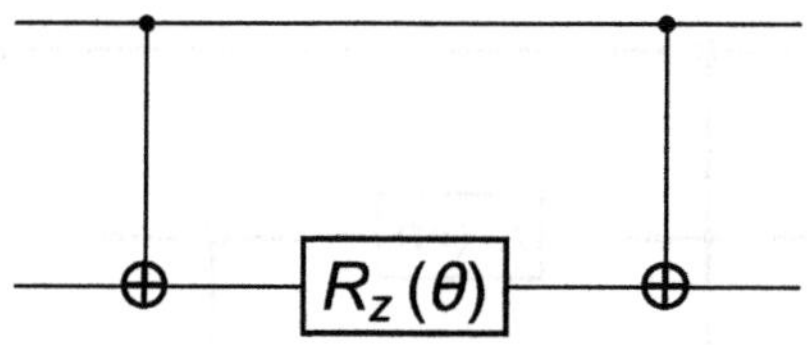

FIGURE 20.19 Gate for two-particle terms.

and

$$U = \begin{pmatrix} e^{-iht} & 0 \\ 0 & 1 \end{pmatrix}. \tag{20.89}$$

Taking $\Psi(0) = \begin{pmatrix} 1 \\ 0 \end{pmatrix}$, the autocorrelation function is given by

$$F(t) = \Psi^{\dagger}(0)U(t)\Psi(0) = e^{-iht}, \tag{20.90}$$

which implies that the system has a single eigenvalue: $E_0 = h$.[3]

Consider next a two-particle contribution to the Hamiltonian, representing the Coulomb plus exchange integrals for spinorbitals 1 and 2. Using (20.74) we have

$$v(a_1{}^{\dagger}a_2{}^{\dagger}a_2a_1 - a_1{}^{\dagger}a_2{}^{\dagger}a_1a_2) \to v(A_1{}^{\dagger}A_2{}^{\dagger}A_2A_1 - A_1{}^{\dagger}A_2{}^{\dagger}A_1A_2) = \frac{v}{2}(I+Z)\otimes(I+Z). \tag{20.91}$$

The propagator corresponding to this Hamiltonian is given by

$$U_{12} = e^{-i\frac{vt}{2}(I+Z)\otimes(I+Z)}. \tag{20.92}$$

The unitary gate that can carry out operations of the form $\exp\left(-i\frac{\theta}{2}Z\otimes Z\right)$ is shown in Fig. 20.19. The CNOT gate first entangles the two qubits. Then the rotation $R_z(\theta) = e^{-i\frac{\theta}{2}Z}$ is applied, followed by a second CNOT gate.

Fig. 20.20 is a highly schematic diagram of a circuit which might be applicable to a simplified two-electron system, such as a helium atom or hydrogen molecule, in which there are two single-particle terms and one interaction term in the Hamiltonian. A minimum basis set of qubits is designated by $|\phi_1\rangle$, $|\phi_2\rangle$. The input parameters are $h_{11} = h_{22} = h$ and $v_{1221} = v$. The gate $T(\theta)$ is represented by the unitary matrix $\begin{pmatrix} 1 & 0 \\ 0 & e^{-i\theta} \end{pmatrix}$. PEA represents the gates for the phase-estimation algorithm, which are shown in full in Fig. 20.17.

3 If this were applied to the harmonic oscillator, with $H = a^{\dagger}a + \frac{1}{2}$, the result would give $E_0 = \frac{1}{2}$.

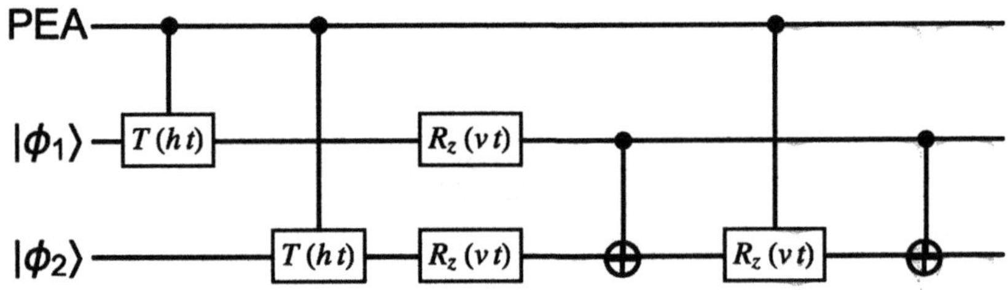

FIGURE 20.20 Simplified circuit for two-particle Hamiltonian.

In summary, we outline the general procedure for a simulation of an atomic or molecular system by quantum computer circuit: (1) The Hamiltonian is expressed as a sum of products of Pauli operators, using the Jordan-Wigner transformation, in conjunction with a set of molecular integrals calculated by conventional computation. (2) Each of these operators is converted into unitary gates such that their sequential execution on a quantum computer can approximate the action of the propagator e^{-iHt}, with application of the Trotter decomposition. (3) The phase estimation algorithm is applied to approximate the eigenvalue of an eigenstate produced by the quantum Fourier transform of the time domain propagation. The output results can serve as an input for an improved approximation, and the procedure can be further iterated, as desired. The rather intricate and extensive details of the circuits which carry out these operations are described more fully in some current references.[4]

[4] A. Aspuru-Guzik, A. Dutoi, P. Love and M. Head-Gordon, Simulated Quantum Computation of Molecular Energies, *Science*, **309**, 1704 (2005); J. D. Whitfield, J. Biamonte and A. Aspuru-Guzik, Simulation of electronic structure Hamiltonians using quantum computers, *Molecular Physics*, **109**, 735-750 (2011); I. Kassal, J. D. Whitfield, A. Perdomo-Ortiz, M.-H. Yung and A. Aspuru-Guzik, Simulating Chemistry Using Quantum Computers, *Ann. Rev. Phys. Chem.*, **62**, 185-207 (2011); P. J. J. O'Malley, et al., Scalable Quantum Simulation of Molecular Energies, *Physical Review X*, **6**, 031007 (2016).

Answers to exercises

NOTE TO STUDENTS: Don't even think of looking here before you attempt to solve the problems yourself!

Chapter 2

2.1. The components of the momentum operator can be expressed

$$\hat{p}_k = -i\hbar\frac{\partial}{\partial x_k}, \quad k = 1, 2, 3.$$

Now extend this relation for $k = 4$ using $p_4 = iE/c$ and $x_4 = ict$:

$$\hat{H} = +i\hbar\frac{\partial}{\partial t},$$

where the energy operator is the Hamiltonian $\hat{H}$. Applying the quantization prescription to the classical energy-momentum relation

$$E = \frac{p^2}{2m} + V(x, y, z), \qquad p^2 = p_1^2 + p_2^2 + p_3^2$$

then leads to the 3-dimensional time-dependent Schrödinger equation (2.30).

2.2. 100 watts = 100 J/sec. The energy of a 550 nm photon is given by

$$E = h\nu = \frac{hc}{\lambda} = \frac{(6.626 \times 10^{-34})(2.998 \times 10^{8})}{550 \times 10^{-9}} = 3.61 \times 10^{-19}\ \text{J}.$$

Thus $100/E = 2.77 \times 10^{20}$ photons/sec.

2.3. Since 1 eV=1.602×10^{-19} J, each electron has a kinetic energy of $(40 \times 10^3)(1.602 \times 10^{-19})$ J. This is equal to

$$E = \frac{1}{2}mv^2 = \frac{p^2}{2m}.$$

The de Broglie relation $\lambda = h/p$, therefore gives

$$\lambda = \frac{h}{\sqrt{2mE}} = \frac{6.626 \times 10^{-34}}{\sqrt{2(9.109 \times 10^{-31})(40 \times 10^{3})(1.602 \times 10^{-19})}}$$

$= 6.13 \times 10^{-12}$ m. This gives sufficient resolution to study the geometric structure of molecules. (Since 40 keV electrons travel at a significant fraction of the speed of light, the relativistic energy-momentum relation must be used. The corrected de Broglie wavelength is actually 6.016×10^{-12} m.)

2.4. Evaluate the partial derivatives

$$\frac{\partial}{\partial x}\Psi(x,t) = \frac{ip}{\hbar}e^{i(px-Et)/\hbar}, \qquad \frac{\partial^2}{\partial x^2}\Psi(x,t) = -\frac{p^2}{\hbar^2}e^{i(px-Et)/\hbar}$$

and

$$\frac{\partial}{\partial t}\Psi(x,t) = -\frac{iE}{\hbar}e^{i(px-Et)/\hbar}.$$

Eq. (2.27) then follows from the relation $E = p^2/2m$.

2.5. Note that $\mathbf{p} \cdot \mathbf{r} = p_x\, x + p_y\, y + p_z\, z$. Then

$$\frac{\partial}{\partial x}\Psi(\mathbf{r},t) = \frac{ip_x}{\hbar}e^{i(\mathbf{p}\cdot\mathbf{r}-Et)/\hbar}, \qquad \text{etc.}$$

and Eq. (2.30), with $V(\mathbf{r})$=0, follows from $E = (p_x^2 + p_y^2 + p_z^2)/2m$.

2.6. Evaluate the derivatives (suppressing A for now):

$$\psi'(x) = e^{-\alpha x} - \alpha x e^{-\alpha x} \qquad \text{and} \qquad \psi''(x) = -2\alpha e^{-\alpha x} + \alpha^2 x e^{-\alpha x}.$$

Then the Schrödinger equation $\hat{H}\psi(x) = E\psi(x)$ becomes

$$-\frac{\hbar^2}{2m}(-2\alpha e^{-\alpha x} + \alpha^2 x e^{-\alpha x}) - \frac{q^2}{x}xe^{-\alpha x} = Exe^{-\alpha x}.$$

Now, cancel out the $e^{-\alpha x}$ and find two independent relations for the terms independent of x and linear in x. The results give $\alpha = mq^2/\hbar^2$, which agrees with the definition and

$$E = -\frac{\hbar^2\alpha^2}{2m} = -\frac{mq^4}{2\hbar^2}.$$

To normalize the function

$$\int_0^\infty |\psi(x)|^2\,dx = 1 = A^2\int_0^\infty x^2 e^{-2\alpha x}\,dx = A^2 \times 2!/(2\alpha)^3,$$

giving $A = 2\alpha^{3/2}$.

Chapter 3

3.1. $y = e^{-kx}$ is a solution of the differential equation $y''(x) - k^2 y(x) = 0$. Note the minus sign.

3.2.

$$P(L/3 \le x \le 2L/3) = \int_{L/3}^{2L/3} |\psi_n(x)|^2 \, dx$$

$$= \frac{2}{L} \int_{L/3}^{2L/3} \sin^2\left(\frac{n\pi x}{L}\right) dx = \frac{2}{L} \frac{L}{n\pi} \left[\frac{\theta}{2} - \frac{\sin 2\theta}{4}\right]_{n\pi/3}^{2n\pi/3}.$$

Note

$$\sin(4n\pi/3) = \sin(4n\pi/3 - 2n\pi) = \sin(-2n\pi/3) = -\sin(2n\pi/3).$$

Thus

$$P = \frac{1}{3} + \frac{1}{n\pi} \sin\left(\frac{2n\pi}{3}\right).$$

As $n \to \infty$, this approaches 1/3.

3.3. Polymethine ion: N^+=C–C=C–C=C–N, 8 electrons (1 from each C, 1 from N^+, 2 from N), $L \approx 7 \times 1.40$ Å.

$$\frac{hc}{\lambda} = \frac{h^2}{8mL^2}(5^2 - 4^2),$$

giving $\lambda = 352$ nm.

3.4. For particle of mass $M = 1.67 \times 10^{-27}$ kg in cubic box with $a = 10^{-14}$ m, ground-state energy is

$$E_{111} = \frac{h^2}{8Ma^2}\left(1^2 + 1^2 + 1^2\right) \approx 6.15 \text{ MeV}.$$

3.5. Energy of 2 electrons in molybox minus that of 2 electrons in cube-atoms:

$$\Delta E = 2 \times \frac{h^2}{8m}\left(\frac{1^2}{(2a)^2} + \frac{1^2}{a^2} + \frac{1^2}{a^2}\right) - 2 \times \frac{h^2}{8ma^2}\left(1^2 + 1^2 + 1^2\right) = -\frac{3}{16}\frac{h^2}{ma^2}.$$

Note that the molybox is more stable (has lower energy). One of the factors promoting formation of molecules from atoms is the increased volume available to valence electrons.

3.6. By analogy with 3-dimensional particle-in-a-box

$$\psi_{n_1 n_2} = \frac{2}{a} \sin\left(\frac{n_1 \pi x}{a}\right) \sin\left(\frac{n_2 \pi y}{a}\right),$$

$$E_{n_1 n_2} = \frac{h^2}{8ma^2}\left(n_1^2 + n_2^2\right), \qquad n_1, n_2 = 1, 2\ldots$$

Ground state $E_{11} = h^2/4ma^2$ is nondegenerate. First excited level, with $E_{21} = E_{21} = 5h^2/8ma^2$, is 2-fold degenerate.

3.7. Six π-electrons occupy E_{11}, E_{12} and E_{21}. Lowest-energy transition is from E_{12} or E_{21} to E_{22}:

$$\frac{hc}{\lambda} = E_{22} - E_{21} = \frac{h^2}{8ma^2}(8-5),$$

$\lambda = 268$ nm when $a = 4.94$ Å.

Chapter 4

4.1.

$$\langle A \rangle = \int \psi^* \hat{A} \psi \, d\tau = a \int \psi^* \psi \, d\tau = a.$$

4.2. Compare the magnitude of $\hbar/2$ with reasonable values for Δx and Δp for a baseball.

4.5. Yes, since $[y, p_x] = 0$.

4.6. Yes, since $[p_z, L_z] = [p_z, xp_y - yp_x] = 0$.

Chapter 6

6.1. The turning points for quantum number occur where the kinetic energy equals 0, so that the potential energy equals the total energy. For quantum number n, this is determined by

$$\frac{1}{2}kx_{\text{max}}^2 = \left(n + \frac{1}{2}\right)\hbar\omega.$$

Recalling that $\omega = \sqrt{k/m}$ and $\alpha = \sqrt{mk}/\hbar$, we find

$$x_{\text{max}}^2 = (2n+1)\frac{\hbar}{\sqrt{km}} = \frac{(2n+1)}{\alpha}.$$

Therefore

$$P(x_{\text{max}} \le x \le \infty) = P(-\infty \le x \le -x_{\text{max}}) = \int_{x_{\text{max}}}^{\infty} |\psi_n(x)|^2 \, dx.$$

[Optional: For $n = 0$,

$$P_{\text{outside}} = 2\int_{1/\sqrt{\alpha}}^{\infty} \left(\frac{\alpha}{\pi}\right)^{1/2} e^{-\alpha x^2}\, dx = \frac{2}{\sqrt{\pi}}\int_{1}^{\infty} e^{-\xi^2}\, d\xi = \text{erfc}(1) \approx 0.158,$$

where erfc is the complementary error function. This result means that in the ground state, there is a 16% chance that the oscillator will "tunnel" outside its classical allowed region.]

6.2. The ground state wavefunction is

$$\psi_0(x) = (\alpha/\pi)^{1/4} e^{-\alpha x^2/2}, \qquad \alpha = (mk/\hbar^2)^{1/2}.$$

Using integrals in Supplement 5A,

$$\langle V\rangle = \int_{-\infty}^{\infty} \psi_0(x)\left(\frac{1}{2}kx^2\right)\psi_0(x)\,dx = \frac{k}{4\alpha} = \frac{1}{4}\hbar\omega = \frac{1}{2}E_0,$$

$$\langle T\rangle = \int_{-\infty}^{\infty} \psi_0(x)\left(-\frac{\hbar^2}{2m}\right)\psi_0''(x)\,dx = \frac{1}{2}E_0.$$

Thus the average values of potential and kinetic energies for the harmonic oscillator are equal. This is an instance of the virial theorem, which states that for a potential energy of the form $V(x) = \text{const}\, x^N$, the average kinetic and potential energies are related by

$$\langle T\rangle = \frac{N}{2}\langle V\rangle.$$

6.3. Taking the square of Eq. (5.46)

$$x^2 = \frac{\hbar}{2m\omega}(aa + aa^\dagger + a^\dagger a + a^\dagger a^\dagger).$$

Using (5.42) and (5.43) we find

$$\langle n|aa|n\rangle = 0, \quad \langle n|a^\dagger a^\dagger|n\rangle = 0, \quad \langle n|a^\dagger a|n\rangle = n, \quad \langle n|aa^\dagger|n\rangle = n+1.$$

Thus

$$\langle V\rangle = \frac{k}{2}\langle x^2\rangle = \frac{m\omega^2}{2}\frac{\hbar}{2m\omega}(2n+1) = \tfrac{1}{2}(n+\tfrac{1}{2})\hbar\omega.$$

The kinetic energy $\langle T\rangle$ works out to the same value.

6.4. The expectation values $\langle x\rangle$ and $\langle p\rangle$ are both equal to zero since they are integrals of odd functions, such that $f(-x) = -f(x)$, over a symmetric range of integration. You have already calculated the expectation values $\langle x^2\rangle$ and $\langle p^2\rangle$

in Exercise 5.2, namely

$$\langle x^2 \rangle = \frac{1}{2\alpha} \qquad \text{and} \qquad \langle p^2 \rangle = \frac{\hbar^2 \alpha}{2}.$$

Therefore $\Delta x \Delta p = \hbar/2$, which is its minimum possible value.

6.5. Since $\langle n|x|n\rangle = 0$ and $\langle n|p|n\rangle = 0$,

$$\Delta x = \sqrt{\langle n|x^2|n\rangle} = \left[\frac{\hbar}{2m\omega}(2n+1)\right]^{1/2}$$

and

$$\Delta p = \sqrt{\langle n|p^2|n\rangle} = \left[\frac{\hbar m\omega}{2}(2n+1)\right]^{1/2},$$

we find $\Delta x \Delta p = (n + \frac{1}{2})\hbar$.

6.6. Using the integrals in Supplement 5A, it is found that $x_{02} = 0$ while x_{01} and x_{12} are nonzero. The general result $\Delta n = \pm 1$ follows from (5.48).

Chapter 8

8.1. De Broglie wavelength $\lambda = h/p$ with $L = r\,p$. Circumference of orbit $2\pi r = n\lambda$, an integer number of wavelengths. This implies $L = nh/2\pi = n\hbar$.

8.2. You can count $n - \ell - 1$ radial nodes, ℓ angular nodes, $n - 1$ total nodes.

8.3. The best formula to use is

$$\frac{1}{\lambda} = Z^2 R\left(\frac{1}{n_1^2} - \frac{1}{n_2^2}\right),$$

where R is the Rydberg constant, 109,678 cm^{-1}. For hydrogen, $1/\lambda = R(1/1^2 - 1/2^2) = 82258.5\ \text{cm}^{-1}$, $\lambda = 121.6$ nm. For helium, $1/\lambda = 4\,R(1/1^2 1/2^2) = 329,034\ \text{cm}^{-1}$, $\lambda = 30.39$ nm.

8.4. Find the maximum of $D_{1s}(r) = 4\pi r^2 \left[\psi_{1s}(r)\right]^2 = \text{const}\, r^2 e^{-2Zr}$. Set $dD/dr = 0$, giving $r_{\text{max}} = 1/Z\ (= a_0/Z)$, same as Bohr radius for $1s$ orbit.

8.5. Define the functions $f_1(r) = e^{-Zr}$, $f_2(r) = (1 + \alpha r)e^{Zr/2}$, $f_3(r) = (1 + \beta r + \gamma r^2)e^{Zr/3}$. Evaluate the integral $\int_0^\infty f_1(r) f_2(r) 4\pi r^2, dr$. The result contains a factor $(Z + 2\alpha)$. Set $\alpha = -Z/2$ so that this vanishes. To find the $2s$ normalization constant, evaluate $\int_0^\infty f_2(r)^2 4\pi r^2, dr$. This equals $8\pi/Z^3$. Thus the normalized $2s$ eigenfunction is given by $\psi_{2s}(r) = \frac{Z^{3/2}}{2\sqrt{2\pi}}(1 - Zr/2)e^{-Zr/2}$.

Proceed analogously for the $3s$ eigenfunction, determining β and γ from the conditions $\int_0^\infty f_1(r)f_3(r)4\pi r^2, dr = 0$ and $\int_0^\infty f_2(r)f_3(r)4\pi r^2, dr = 0$.

8.6. He^{++} and H^+ are bare nuclei so their electronic energies equal zero. He^+ and H are hydrogenlike so their $1s$ energies equal $-Z^2/2$. Thus $\Delta E = -4/2 + 1/2 = -3/2$ hartrees $= -40.8$ eV.

8.7. (i); the other four operators are equal.

8.8.

$$\langle r \rangle = \int_0^\infty \psi_{1s}(r)\, r\, \psi_{1s}(r)\, 4\pi r^2\, dr = \frac{3}{2} \quad \left(= \frac{3}{2} a_0\right),$$

$$\langle r^2 \rangle = \int_0^\infty \psi_{1s}(r)\, r^2\, \psi_{1s}(r)\, 4\pi r^2\, dr = 3 \quad \left(= 3a_0^2\right),$$

$$\langle r^{-1} \rangle = \int_0^\infty \psi_{1s}(r)\, r^{-1}\, \psi_{1s}(r)\, 4\pi r^2\, dr = 1 \quad \left(= \frac{1}{a_0}\right).$$

8.9. Average potential energy:

$$< V > = \int_0^\infty \psi_{1s}(r) \left(-\frac{Z}{r}\right) \psi_{1s}(r)\, 4\pi r^2\, dr = -Z^2.$$

Average kinetic energy:

$$< T > = \int_0^\infty \psi_{1s}(r) \left(-\frac{1}{2}\nabla^2\right) \psi_{1s}(r)\, 4\pi r^2\, dr = Z^2/2.$$

More simply, since total energy $E_{1s} = -Z^2/2$, $\langle T \rangle = E_{1s} - \langle V \rangle$. Note that $\langle V \rangle = -2\langle T \rangle$, consistent with the virial theorem.

8.10. For an easier exercise, do the $2p_z$ orbital instead.

8.11. You should find that this function solves the Schrödinger equation with $E = -Z^2/8$, i.e., $n = 2$. For normalization

$$\text{const} = \frac{Z^{3/2}}{4\sqrt{\pi}}.$$

Noting that $\sin^2(\theta/2) = (1 - \cos\theta)/2$, the function is found to be an s-p hybrid orbital:

$$\psi = \frac{1}{\sqrt{2}} \left(\psi_{2s} + \psi_{2pz}\right).$$

8.12. Solve for R:

$$\int_0^R |\psi_{1s}(r)|^2\, 4\pi r^2\, dr = 0.9,$$

or easier

$$\int_R^{\infty} |\psi_{1s}(r)|^2 \, 4\pi r^2 \, dr = 0.1.$$

We find, using integral table,

$$4\int_R^{\infty} r^2 e^{-2r} \, dr = e^{2R}\,(1+2R+2R^2) = 0.1.$$

Solving numerically, $R = 2.6612 a_0 = 1.41$ Å.

8.13. Let $\psi(r) = e^{-\alpha r}$. Then

$$E(\alpha) = \frac{\int_0^{\infty} e^{-\alpha r}\left(-\frac{1}{2}\nabla^2 - Z/r\right) e^{-\alpha r} 4\pi r^2 \, dr}{\int_0^{\infty} e^{-2\alpha r} 4\pi r^2 \, dr} = \frac{1}{2}\alpha^2 - Z\alpha.$$

$E'(\alpha) = 0$ for minimum, giving $\alpha = Z$. Thus $\psi(r) = e^{-Zr}$ and $E = -Z^2/2$, which, in this exceptional case, gives the exact eigenfunction and eigenvalue.

8.14. In atomic units, $\psi_{1s} = \pi^{-1/2} e^{-r}$, $\psi_{2p_0} = (4\sqrt{2\pi})^{-1} r \cos\theta e^{-r/2}$. Therefore

$$|\psi_{1s}(0)|^2 = \pi^{-1}, \qquad |\psi_{2p_0}(0)|^2 = 0.$$

Integrating over θ

$$\int_0^{\pi} (3\cos^2\theta - 1)\sin\theta \, d\theta = \int_{-1}^{1} (3u^2 - 1) du = 0$$

and

$$\int_0^{\pi} (3\cos^2\theta - 1)\cos^2\theta \sin\theta \, d\theta = \int_{-1}^{1} (3u^2 - 1)u^2 du = \frac{8}{15},$$

so that

$$\left\langle \frac{3\cos^2\theta - 1}{r^3} \right\rangle_{1s} = 0, \qquad \left\langle \frac{3\cos^2\theta - 1}{r^3} \right\rangle_{2p_0} = 2\pi \times \frac{8}{15} \times \langle r^{-3} \rangle_{2p_0} = \frac{1}{30}.$$

Finally

$$\Delta\nu_{1s} = 1420.4 \text{ MHz}, \qquad \Delta\nu_{2p_0} = 17.8 \text{ MHz}.$$

Chapter 9

9.1.

$$\langle T \rangle = \frac{\int_0^{\infty}\int_0^{\infty} \psi(r_1, r_2)\left(-\frac{1}{2}\nabla_1^2 - \frac{1}{2}\nabla_2^2\right)\psi(r_1, r_2)\, 4\pi r_1^2 \, dr_1 \, 4\pi r_2^2 \, dr_2}{\int_0^{\infty}\int_0^{\infty} |\psi(r_1, r_2)|^2 \, 4\pi r_1^2 \, dr_1 \, 4\pi r_2^2 \, dr_2} = \alpha^2$$

and

$$\langle V\rangle = \left\langle -\frac{Z}{r_1} - \frac{Z}{r_2} + \frac{1}{r_{12}}\right\rangle = -2Z\alpha + \frac{5}{8}\alpha.$$

For the optimized variational function, $\alpha = Z - 5/16$, so

$$\langle T\rangle = \left(Z - \frac{5}{16}\right)^2 \quad \text{and} \quad \langle V\rangle = -2\left(Z - \frac{5}{16}\right)^2.$$

Thus $\langle V\rangle = -2\langle T\rangle$, in agreement with the virial theorem.

9.2. Li^+ is He-like with $Z = 3$. Just as for He,

$$E(\alpha) = \alpha^2 - 2Z\alpha + \frac{5}{8}\alpha,$$

with optimal $\alpha = Z - \frac{5}{16} = 2.6875$ and

$$E = -\left(Z - \frac{5}{16}\right)^2 = -7.223 \text{ hartrees}.$$

A more accurate value is −7.280 hartrees.

9.3. For the Li atom with 3 electrons,

$$\hat{H} = \sum_{i=1}^{3}\left(-\frac{1}{2}\nabla_i^2 - \frac{Z}{r_i}\right) + \frac{1}{r_{12}} + \frac{1}{r_{23}} + \frac{1}{r_{31}}.$$

Assuming $\psi(1,2,3) = e^{-\alpha(r_1+r_2+r_3)}$, we find in analogy with helium results,

$$\left\langle -\frac{1}{2}\nabla_i^2\right\rangle = \frac{1}{2}\alpha^2, \quad \left\langle -\frac{Z}{r_i}\right\rangle = -Z\alpha, \quad \left\langle \frac{1}{r_{ij}}\right\rangle = \frac{5}{8}\alpha.$$

The total energy is given by

$$E(\alpha) = \frac{3}{2}\alpha^2 - 3Z\alpha + \frac{15}{8}\alpha,$$

with $Z = 3$. To optimize,

$$E'(\alpha) = 3\alpha - 9 + \frac{15}{8} = 0, \quad \alpha = 2.375, \quad E = -8.4609 \text{ hartrees}.$$

This is *less than* the exact ground state energy −7.478, in apparent violation of the variational principle. But ψ is an "illegal" wavefunction.

Chapter 10

10.1. Spherically symmetrical (S) state whenever valence shell contains only (i) all s-electrons, (ii) half filled shells, (iii) filled shells. Group IA, configuration ns: H, Li, Na, K, Rb. Group IIA, ns^2: Be, Mg, Ca, Sr. Group VB, ns^2np^3: N, P, As, Sb. Group 0: He, Ne, Ar, Kr, Xe. Transition elements: Cr $4s3d^5$, Mn $4s^23d^5$, Mo $5s4d^5$, Tc $5s^24d^5$. Also Cu, Zn, Pd, Ag, Cd, all with d^{10}.

10.2. First excited states: H $2s\ ^2$S, He $1s2s\ ^3$S, Li $1s^22p\ ^2$P, Be $1s^22s2p\ ^3$P, B $1s^22s2p^2\ ^2$S. For C, N and O, the electron configuration is the same as for the ground state but the occupation of degenerate p-orbitals is different: C $2s^22p^2\ ^1$D, N $2s^22p^3\ ^2$D, O $2s^22p^4\ ^1$D. Finally, F $2s2p^6\ ^2$S, Ar $2s^22p^53s\ ^3$P.

10.3. Promote one of the $2s$ electrons to the empty $2p$ orbital. If the four valence electrons have parallel spins, this is a ^{5}S state, which can form four bonds.

10.4. According to $n+\ell$ rule, the ordering of atomic orbitals should be:

$$1s < 2s < 2p < 3s < 3p < 4s < 3d < 4p < 5s < 4d$$
$$< 5p < 6s < 4f < 5d < 6p < 7s < 5f < 6d < 7p.$$

The ordering is sometimes reversed for the entries shown in red (gray in print version). Both rules give the same ground state configuration for Rn ($Z = 86$):

$$1s^22s^22p^63s^23p^64s^23d^{10}4p^65s^24d^{10}5p^66s^24f^{14}5d^{10}6p^6.$$

10.5. The angular momentum in the first Bohr orbit is given by $L = mvr = \hbar$. For atomic number Z, the radius of the lowest orbit is $r = a_0/Z$, where $a_0 = \hbar^2/me^2$. Therefore

$$v = \frac{\hbar Z}{ma_0} = \frac{e^2}{\hbar} Z$$

and $v/c = Z\alpha$. The nonrelativistic energy of this orbit is $E = -Z^2me^4/2\hbar^2 = -13.6Z^2 \approx -87,000$ eV. With the relativistic mass increased by a factor $(1 - v^2/c^2)^{-1/2}$, the energy decreases to approximately $-107,000$ eV.

Chapter 11

11.1. (i) Minimum value of $E(R)$ can be found by setting $E(R) = 0$. It is easy to see from the formula itself that $E(R)$ will have a minimum value of 0 when $R = R_e$. As $R \to \infty$, $E(R)$ approaches D. Thus $D_e = D$, the dissociation energy.

(ii)

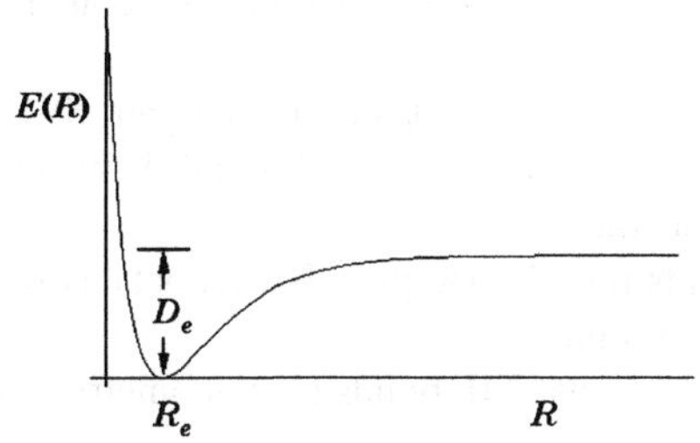

(iii) Remember the expansion for the exponential (In fact, don't ever forget this!)

$$e^x = \sum_{n=0}^{\infty} \frac{x^n}{n!} = 1 + x + \frac{x^2}{2} + \frac{x^3}{6} + \ldots$$

Expanding the Morse function up to terms quadratic in $R - R_e$ gives

$$E(R) = 0 + D\beta^2 (R - R_e)^2 + \ldots$$

This has the form of a harmonic oscillator potential $V(x) = \frac{1}{2}kx^2$ with

$$x = R - R_e \quad \text{and} \quad k = 2D\beta^2.$$

11.2. The central carbon forms two sp-hybrids and two unhybridized p-orbitals, just like acetylene. The sp-hybrids bond to the terminal carbons in a linear arrangement of σ-bonds. Each p orbitals then bonds to a terminal carbon to form a π-bond, as shown below:

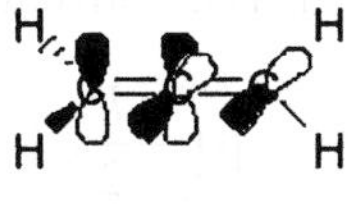

Note that the two CH_2 groups are in perpendicular planes.

11.3. For atomic p-orbitals, the density of a $p_x p_y$ configuration has the form

$$\rho = |\psi_{px}|^2 + |\psi_{py}|^2 = (x^2 + y^2) f(r),$$

which is cylindrically-symmetrical about the z-axis. This remains true for the two π-orbitals formed by linear combination of AO's on the two carbon atoms.

11.4. H_2S: S has 6 valence electrons, 2 form bonds to H leaving 4 electrons or 2 unshared pairs. SH_2E_2 approximately tetrahedral configuration giving two S–H bonds for bent H–S–H molecule. Just like H_2O.
SF_6: 6 S–F bonds, octahedral molecule.

XeF_4: Xe has 8 valence electrons, 4 bonds to F, leaving 2 pairs. XeF_4E_2 octahedral with the two E's on opposite sides to minimize repulsion, so XeF_4 molecule is square planar.
SF_4: 4 S–F bonds, leaving 2 electrons or 1 lone pair. SF_4E trigonal bipyramid with E in one equatorial position. The 4 S–F bonds bend away from the E giving a see-saw shaped molecule.
BrF_5: Br 5 Br–F bonds plus 1 lone pair. BrF_5E octahedral configuration gives geometry of square pyramid.
IF_7: I has 7 valence electrons, 7 IF bonds give geometry of a pentagonal bipyramid.

11.5. CH_2 and SO_2 are bent triatomics. SO_3 and NO_3^- are equilateral triangles. XeO_3 is a triangular pyramid like NH_3.

11.6. Unhybridized oxygen $2p$-orbitals would form two OH bonds 90° apart. According to the valence-shell model, the two bonds would belong to distorted tetrahedra around each oxygen atom.

11.8. $[Fe(H_2O)_6]^{3+}$ $d^5 \to t_{2g}^3 e_g^2$ (high spin). $[Cr(CN)_6]^{4-}$ $d^4 \to t_{2g}^4$ (low-spin). $[Co(NH_3)_6]^{3+}$ $d^6 \to t_{2g}^6$ (low-spin). $[Cu(H_2O)_6]^{2+}$ $d^9 \to t_{2g}^6 e_g^3$.

Chapter 12

12.1. As $\xi \to \infty$, the equation reduces approximately to

$$\xi^2 \frac{d^2\Xi}{d\xi^2} - \frac{R^2|E|\xi^2}{4}\Xi \approx 0.$$

Canceling the ξ^2 and noting that E is negative for bound states,

$$\Xi(\xi) \approx \exp\left(-\frac{R}{2}\sqrt{|E|}\,\xi\right).$$

12.2.

$$\int \psi^2 d\tau = N^2 \int (\psi_A^2 + \psi_B^2 \pm 2\psi_A\psi_B) d\tau = N^2(1 + 1 \pm 2S),$$

thus ψ is normalized when $N = (2 \pm 2S)^{-1/2}$.

12.3.

$$\psi_{1\sigma g} = 1s_A + 1s_B,$$

thus

$$\psi(1,2) = \Big(1s_A(1) + 1s_B(1)\Big)\Big(1s_A(2) + 1s_B(2)\Big) =$$

$$\left\{1s_A(1)1s_B(2) + 1s_B(1)1s_A(2)\right\} + 1s_A(1)1s_A(2) + 1s_B(1)1s_B(2).$$

Term in brackets is the valence bond function for the bond. The remaining terms represent ionic structures H^+H^- and H^-H^+ with both electrons on the same hydrogen atom.

12.4. N_2	$\ldots 1\pi_u^4 3\sigma_g^2$	${}^1\Sigma_g^+$	BO=3
N_2^+	$\ldots 1\pi_u^4 3\sigma_g$	${}^2\Sigma_g^+$	BO=2.5
N_2^-	$\ldots 1\pi_u^4 3\sigma_g^2 1\pi_g$	${}^2\Pi_g$	BO= 2.5

12.5. O_2	$\ldots 3\sigma_g^2 1\pi_u^4 1\pi_g^2$	${}^3\Sigma_g^-$
O_2^-	$\ldots 3\sigma_g^2 1\pi_u^4 1\pi_g^3$	${}^2\Pi_g$
O_2^{2-}	$\ldots 3\sigma_g^2 1\pi_u^4 1\pi_g^4$	${}^1\Sigma_g^+$

12.6. Both excited states have same configuration as ground state, $\ldots 3\sigma_g^2 1\pi_u^4 1\pi_g^2$, but with the following occupancy of $1\pi_g$ orbitals:

$$\uparrow \; \downarrow \quad {}^1\Sigma_g^+ \quad \text{and} \quad \uparrow\!\downarrow \; \_ \quad {}^1\Delta_g.$$

The plus superscript in the first term symbol is rather tricky. Don't worry about it. But if you insist ... two-electron singlet spin state has antisymmetric spin function, thus must have *symmetric* orbital function like $\pi_x(1)\pi_y(2) + \pi_y(1)\pi_x(2)$ which *doesn't* change sign upon transformation $\phi \to -\phi$. Singlet oxygen and other active oxygen species are involved in lipid metabolism.

12.7. Be_2 has configuration ... $2\sigma_g^2 2\sigma_u^2$ ${}^1\Sigma_g^+$. The 2σ orbitals are LCAO's made from hybrids of $2s$ and $2p\sigma$. The $2\sigma_g$ MO probably becomes more strongly bonding while the $2\sigma_u$ becomes more weakly antibonding, with the net effect being weak bonding.

12.8. Setting $dS/dR = 0$ find maximum at $R = 2.1038$ bohrs or $2.1038 \times .593 = 1.115$ Å.

Chapter 13

13.1. Secular determinant:

$$\begin{vmatrix} x & 1 & 0 \\ 1 & x & 1 \\ 0 & 1 & x \end{vmatrix} = x^3 - 2x = 0,$$

where $x = (\alpha - E)/\beta$. Roots $x = 0, \pm\sqrt{2}$, thus $E = \alpha - \sqrt{2}\beta, \alpha, \alpha + \sqrt{2}\beta$. Remember both α and β are negative. Ground state energy (3 electrons) $= 2(\alpha +$

$\sqrt{2}\beta) + \alpha = 3\alpha + 2\sqrt{2}\beta$. One localized π-orbital plus one unpaired electron would have energy $= 2(\alpha + \beta) + \alpha = 3\alpha + 2\beta$. Resonance stabilization energy $= (2 - 2\sqrt{2})\beta = -.828\,\beta = .828|\beta|$. Lowest energy electronic transition given by

$$\frac{hc}{\lambda} = \sqrt{2}\,|\beta|.$$

13.2. For linear H_3, the secular equation is

$$\begin{vmatrix} x & 1 & 0 \\ 1 & x & 1 \\ 0 & 1 & x \end{vmatrix} = x^3 - 2x = 0,$$

with roots $x = 0, \pm\sqrt{2}$. Thus the three MO energies are $\alpha - \sqrt{2}\beta, \alpha, \alpha + \sqrt{2}\beta$. The energy of the three-electron ground state is $3\alpha + 2\sqrt{2}\beta \approx 3\alpha + 2.828\beta$. For triangular H_3,

$$\begin{vmatrix} x & 1 & 1 \\ 1 & x & 1 \\ 1 & 1 & x \end{vmatrix} = x^3 - 3x + 2 = 0.$$

One obvious root is $x = 1$. Division of $x^3 - 3x + 2$ by $x - 1$ gives $x^2 + x - 2$, with roots $x = 1$ and -2. The three MO's are $\alpha + 2\beta, \alpha - \beta, \alpha - \beta$. The energy of the ground state is $3\alpha + 3\beta$. Apparently the triangular form of H_3 has a slightly lower energy.

13.3. The secular equation is

$$\begin{vmatrix} x & 1 & 0 & 1 \\ 1 & x & 1 & 0 \\ 0 & 1 & x & 1 \\ 1 & 0 & 1 & x \end{vmatrix} = x^4 - 4x^2 = 0,$$

with roots $x = 0, 0, \pm 2$. The ground state π-electron energy equals $2 \times (\alpha + 2\beta) + 2\alpha = 4\alpha + 4\beta$. This the same as that for two ethylene molecules, so there is *no* delocalization energy. This system is in fact *antiaromatic*, with $4N$ π-electrons.

13.4. The five occupied π-MO's are $\alpha + 2.303\beta, \alpha + 1.618\beta, \alpha + 1.303\beta, \alpha + 1.000\beta, \alpha + 0.618\beta$. The total energy is $E_\pi = 10\alpha + 13.684\beta$. Since five localized double bonds would have energy $10\alpha + 10\beta$, the resonance energy is approximately $3.684\beta \approx 10$ eV.

13.5. For azulene, the HOMO is $\alpha + 0.477\beta$, the LUMO is $\alpha - 0.400\beta$. Thus $\Delta E = 0.877|\beta|$. Using $|\beta| \approx 2.72$ eV, $|\beta|/hc \approx 21,900\ \text{cm}^{-1}$, we obtain

$$\frac{1}{\lambda} \approx 0.877\,\frac{|\beta|}{hc} \approx 19,200\ \text{cm}^{-1}.$$

This corresponds to an absorption of green light at about 520 nm. If the Hückel computation were quantitatively accurate, the compound should appear red, the complementary color. Actually, azulene is blue, hence its name.

13.6. For hexatriene the HOMO is the 3π, which has its terminal p-orbitals in phase. The thermal reaction should be disrotatory and the photochemical reaction conrotatory.

Chapter 14

14.1. The Hartree-Fock-Slater equation for a $1s$ orbital $\phi(r)$ is given by

$$\left[-\frac{1}{2}\nabla^2 - \frac{Z}{r} + \int \frac{\rho(r')}{|\mathbf{r}-\mathbf{r}'|}d^3\mathbf{r}' - \frac{3\alpha}{2}\left(\frac{3}{\pi}\right)^{1/3}\rho(r)^{1/3}\right]\phi(r) = \epsilon\,\phi(r).$$

The normalized orbital is taken as

$$\phi(r) = \frac{\zeta^{3/2}}{\sqrt{\pi}}\,e^{-\zeta r}.$$

The total electron density is then given by

$$\rho(r) = 2\phi(r)^2 = 2\frac{\zeta^3}{\pi}\,e^{-2\zeta r}.$$

Multiplying the HFS equation by $\phi(r)$ and integrating, we obtain

$$T + V + J + X = \epsilon,$$

with

$$T = \int_0^\infty \phi(r)\left[-\frac{1}{2}\phi''(r) - \frac{1}{r}\phi'(r)\right]4\pi r^2\,dr = \frac{\zeta^2}{2},$$

$$V = \int_0^\infty \phi(r)\left[-\frac{Z}{r}\phi(r)\right]4\pi r^2\,dr = -Z\zeta,$$

$$J = \int_0^\infty \int_0^\infty \frac{\rho(r')\psi(r)^2}{|\mathbf{r}-\mathbf{r}'|}4\pi r'^2\,dr'\,4\pi r^2\,dr = \frac{5}{4}\zeta$$

and

$$X = -\frac{3\alpha}{2}\left(\frac{3}{\pi}\right)^{1/3}\int_0^\infty \rho(r)^{1/3}\psi(r)^2\,4\pi r^2\,dr = -\frac{81\times 3^{1/3}}{64(2\pi)^{2/3}}\alpha\,\zeta.$$

Given the values $Z = 2$, $\zeta = 1.6875$, $\epsilon = -0.9037$, the HFS equation is satisfied when $\alpha = 1.17386$.

Chapter 15

15.1. From the C_{2v} character table we find the following direct products:

$$\mathrm{A}_1 \otimes X = X, \quad X \otimes X = \mathrm{A}_1, \quad A_2 \otimes B_1 = B_2, \\ A_2 \otimes B_2 = B_1, \quad B_2 \otimes B_2 = A_2,$$

where X is any of the four representations. The cartesian coordinates x, y, z transform as B_1, B_2 and A_1, respectively. For a transition to be allowed, the direct product in the dipole integral must contain the totally symmetric representation A_1. Thus x-polarized $A_2 \leftrightarrow B_2$ transitions and y-polarized $A_2 \leftrightarrow B_1$ transitions are allowed. z-polarized transitions are allowed between states belonging to the same representation. All $B_1 \leftrightarrow B_2$ transitions are dipole-forbidden.

15.2. Orbitals of a_1 symmetry are linear combinations of O2s, O2p_z and H1s(1)+H1s(2). Orbitals of b_2 symmetry are linear combinations of O2p_y and H1s(1)-H1s(2). The O2p_x orbital already has b_1 symmetry.

15.3.

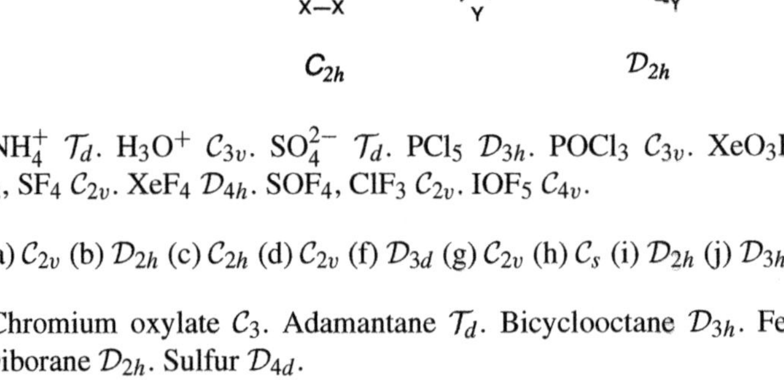

15.4. NH_4^+ $\mathcal{T}_d$. H_3O^+ C_{3v}. SO_4^{2-} $\mathcal{T}_d$. PCl_5 $\mathcal{D}_{3h}$. $POCl_3$ C_{3v}. XeO_3F_2 $\mathcal{D}_{3h}$. PF_3Cl_2, SF_4 C_{2v}. XeF_4 $\mathcal{D}_{4h}$. SOF_4, ClF_3 C_{2v}. IOF_5 C_{4v}.

15.5. (a) C_{2v} (b) $\mathcal{D}_{2h}$ (c) C_{2h} (d) C_{2v} (f) $\mathcal{D}_{3d}$ (g) C_{2v} (h) C_s (i) $\mathcal{D}_{2h}$ (j) $\mathcal{D}_{3h}$ (k) C_{2v}

15.6. Chromium oxylate C_3. Adamantane $\mathcal{T}_d$. Bicyclooctane $\mathcal{D}_{3h}$. Ferrocene $\mathcal{D}_{5d}$. Diborane $\mathcal{D}_{2h}$. Sulfur $\mathcal{D}_{4d}$.

Chapter 16

16.1. The minimum of $V(R)$ occurs where

$$V'(R) = 4\epsilon\,(-12\sigma^{12}R^{-13} + 6\sigma^6 R^{-7}) = 0.$$

This gives $R_e = 2^{1/6}\sigma$. At $R = R_e$, $V(R_e) = -\epsilon$. Since $V(\infty) = 0$, $D_e = \epsilon$. When $R = \sigma$, $V(R) = 0$. This is the point where the potential-energy curve crosses the axis.

16.3. (i)

$$hcE_J = BJ(J+1), \qquad hc(E_1 - E_0) = 2B = 3.410\ \text{cm}^{-1}.$$

(ii)

$$R = 410.6 \Big/ \sqrt{\frac{14\times 16}{14+16}\,B} = 115\ \text{pm}.$$

(iii)

$$k = 58.9\times 10^{-6}\,\tilde{\nu}^2\,\frac{14\times 16}{14+16} = 1590\ \text{N/m}.$$

(iv)

$$hcE_v = (v+\tfrac{1}{2})\tilde{\nu} - (v+\tfrac{1}{2})^2 x_e\tilde{\nu}.$$

$$hc(E_1 - E_0) = \left(\frac{3}{2} - \frac{1}{2}\right)\tilde{\nu} - \left[\left(\frac{3}{2}\right)^2 - \left(\frac{1}{2}\right)^2\right]x_e\tilde{\nu} = 1876\ \text{cm}^{-1}.$$

16.4. The number of bound vibrational levels is equal to

$$n_{\text{bound}} = v_{\max} + 1 \approx \left[\frac{\omega_e}{2\,\omega_e x_e} + \frac{1}{2}\right],$$

where the square bracket means the closest integer. This gives for $n \approx$ 19, 21, 24 for H_2, HD and D_2, respectively.

16.5. For $^1H^{35}Cl$, $\omega_e = 2989.7\ \text{cm}^{-1}$ and $B_e = 10.59\ \text{cm}^{-1}$. The dependence of these parameters on reduced mass can be expressed

$$\frac{\omega_e(1)}{\omega_e(2)} = \sqrt{\frac{\mu(2)}{\mu(1)}}, \qquad \frac{B_e(1)}{B_e(2)} = \frac{\mu(2)}{\mu(1)}.$$

In amu, $\mu(^1H^{35}Cl) \approx \sqrt{1\times 35/(1+35)}$, $\mu(^1H^{37}Cl) \approx \sqrt{1\times 37/(1+37)}$, $\mu(^2H^{35}Cl) \approx \sqrt{2\times 35/(2+35)}$, $\mu(^2H^{37}Cl) \approx \sqrt{2\times 37/(1+37)}$. The results follow:

Molecule	ω_e/cm^{-1}	B_e/cm^{-1}
$^1H^{35}Cl$	2989.7	10.59
$^1H^{37}Cl$	2987.5	10.57
$^2H^{35}Cl$	2143.2	5.442
$^2H^{37}Cl$	2140.1	5.426

16.6. For N_2, the maximum population occurs for the rotational level

$$J_{\text{max}} = \left[\left(\frac{kT}{2hcB_e} \right)^{1/2} - \frac{1}{2} \right] = 7$$

with $B_e = 2.010\ \text{cm}^{-1}$, $T = 300$ K, $hc/k = 1.4388$ cm K.

16.7. For a spherical rotor, the degeneracy increases to $g_J = (2J+1)^2$, the Boltzmann distribution for rotational levels is then

$$N_J = \text{const}\,(2J+1)^2\, e^{-BJ(J+1)hc/kT},$$

with a maximum at the level

$$J_{\text{max}} = \left[\left(\frac{kT}{hcB} \right)^{1/2} - \frac{1}{2} \right].$$

For CH_4 at 300 K, $J_{\text{max}} = 6$.

16.8. The rotational energy of the centrifugally-distorted molecule is given by

$$E_J = \frac{\hbar^2}{2\mu(R+\Delta R)^2} J(J+1) \approx \frac{\hbar^2}{2\mu R^2} J(J+1) - \frac{\hbar^2 \Delta R}{\mu R^3} J(J+1).$$

Equating the Hooke's law force to the centrifugal force

$$k\Delta R = m_1\omega^2 r_1 + m_2\omega^2 r_2 = 2\mu\,\omega^2 R.$$

Use the relations

$$\sqrt{J(J+1)}\,\hbar = \mu R^2\,\omega,$$

$$\omega_e = \frac{1}{2\pi c}\sqrt{\frac{k}{\mu}}$$

and

$$hcB_e = \frac{\hbar^2}{2\mu R^2}$$

to obtain

$$\frac{E_J}{hc} = B_e J(J+1) - \frac{4B_e^3}{\omega_e^2} J^2(J+1)^2,$$

thus identifying

$$D_J = 4B_e^3/\omega_e^2.$$

16.9, 16.10. Symmetry species which transform like x, y or z are IR-active. Those which transform like x^2, xy, etc. are Raman-active. These cartesian combinations can be found in the right-hand columns of character tables.

Chapter 17

17.1. Assuming the molar partition function can be approximated by

$$Q = \Omega e^{-U/kT},$$

we have $\ln Q = \ln \Omega - \frac{U}{kT}$. Therefore

$$S = \frac{U}{T} - k \ln Q = k \ln \Omega.$$

17.2. Approximating $p(0) = 760$ mmHg, and $T = 293$ K, we find

$$p(1600) = 760 \exp(-.029 \times 9.8 \times 1600/8.314 \times 293) \approx 630 \text{ mmHg}.$$

17.3. Using the classical Boltzmann distribution, with $\epsilon(v_x) = mv_x^2/2$,

$$\rho(v_x) = e^{-mv_x^2/2kT} \Big/ \int_{-\infty}^{\infty} e^{-mv_x^2/2kT}\, dv_x.$$

The integral is of the form

$$\int_{-\infty}^{\infty} e^{-\alpha x^2}\, dx = \sqrt{\frac{\pi}{\alpha}},$$

which leads to

$$\rho(v_x) = \left(\frac{m}{2\pi kT}\right)^{1/2} e^{-mv_x^2/2kT}.$$

For the composite 3-dimensional velocity distribution, we can write

$$\rho(\mathbf{v}) = \rho(v_x, v_y, v_z) = \rho(v_x)\rho(v_y)\rho(v_z) = \left(\frac{m}{2\pi kT}\right)^{3/2} e^{-mv^2/2kT}.$$

This is normalized in 3-dimensional velocity space according to

$$\int_{-\infty}^{\infty}\int_{-\infty}^{\infty}\int_{-\infty}^{\infty} \rho(\mathbf{v})\, dv_x\, dv_y\, dv_z = 1.$$

Transforming to spherical polar coordinates in velocity space, noting that $\rho(\mathbf{v})$ is spherically symmetrical, we can write

$$\int_0^{\infty} \rho(\mathbf{v}) 4\pi v^2\, dv = 1.$$

Seeking a normalized function of speed v, independent of direction of velocity, such that

$$\int_0^\infty \rho(v)\,dv = 1.$$

We thus arrive at the Maxwell-Boltzmann-distribution function

$$\rho(v) = 4\pi v^2 \left(\frac{m}{2\pi kT}\right)^{3/2} e^{-mv^2/2kT}.$$

17.4. The probability for a molecule to be in the vibrational state v is given by

$$p_v = \frac{e^{-\epsilon_v/kT}}{q_{\text{vib}}} = e^{-hc\tilde{\nu}v/kT}(1 - e^{-hc\tilde{\nu}/kT}).$$

Using $hc/k = 1.438775$, $\tilde{\nu} = 214.4$, $T = 298$, we find

$$p_0 = 0.645, \quad p_1 = 0.229, \quad p_2 = 0.0813.$$

17.5. The probability for a molecule to be in the rotational level J is proportional to the Boltzmann factor $(2J+1)e^{-J(J+1)hcB/kT}$. As shown in Eq. (16.19), the maximum of this quantity occurs at

$$J_{\text{max}} = \left[\left(\frac{kT}{2hcB}\right)^{1/2} - \frac{1}{2}\right],$$

where the square bracket means the nearest integer.

17.6.

$$J_{\text{max}} = \left[\left(\frac{298}{2 \times 1.438775 \times 1.9987}\right)^{1/2} - \frac{1}{2}\right] = [6.698] = 7.$$

17.7. Taking account of translational, rotational and vibrational contributions,

$$C_{\text{V,m}} = \frac{3}{2}R + R + R\frac{u^2 e^u}{(e^u - 1)^2},$$

where $u = hc\tilde{\nu}/kT = 1.43877 \times 214.5/298 = 1.03562$. We find that $C^{\ominus}_{\text{p,m}} = C_{\text{V,m}} + R = 4.4152R = 36.71\ \text{JK}^{-1}\ \text{mol}^{-1}$.

17.8. Given the parameters

$$M = 253.8,\ \tilde{\nu} = 214.5, u = hc\tilde{\nu}/kT = 1.03562,\ B = 0.03737, \sigma = 2,$$

we have

$$S_m^{\text{trans}} = (1.5 \ln M + 2.5 \ln T - 1.1517)R = 21.3959R,$$

$$S_m^{\text{rot}} = R\ln\left(\frac{ekT}{\sigma hcB}\right) = 8.92704R,$$

$$S_m^{\text{vib}} = R\left[\frac{u}{e^u - 1} - \ln(1 - e^{-u})\right] = 1.00852R.$$

These add up to
$S_m^{\ominus} = 31.3314R = 260.51\ \text{JK}^{-1}\ \text{mol}^{-1}$.

17.9. Calculate the $u_i = hc\tilde{\nu}_i/kT$:

$$u_1 = 16.1113, u_2 = 4.5867, u_3 = 16.4831(\text{two-fold}), u_4 = 7.8577(\text{two-fold}).$$

The vibrational contribution to heat capacity is given by

$$R\sum_i \frac{u_i^2 e^{u_i}}{(e^{u_i} - 1)^2} = (0 + 0.2187 + 2 \times 0 + 2 \times 0.0238)R = 0.2666R.$$

The sum of translational, rotational and vibrational contributions gives

$$C_{\text{V,m}} = \frac{3}{2}R + \frac{3}{2}R + 0.2666R.$$

The result is
$C_{\text{p,m}}^{\ominus} = C_{\text{V,m}} + R = 4.2666R = 35.47\ \text{JK}^{-1}\ \text{mol}^{-1}$.

Chapter 18

18.1. Each CH_2 group is split by the neighboring CH_3 group into a 1:3:3:1 quartet. Correspondingly, each CH_3 group is split by the neighboring CH_2 into a 1:2:1 triplet. Protons in different ethyl groups are too far apart to interact.

18.2. The protons in methane are equivalent and do not exhibit spin-spin splittings. In CD_2H_2 each deuteron has a spin of 1, which by itself would cause splitting into a 1:1:1 triplet. Two deuterons will give a splitting pattern of 1:2:3:2:1, which is what we see for the proton resonances.

18.3. The 3 protons in the methyl group are equivalent with a chemical shift $\delta \approx 2$. The 5 protons on the phenyl group are not strictly equivalent but, evidently, their chemical shifts are nearly equal. Note that the ring protons are significantly deshielded, as shown in Fig. 18.5.

18.4. The methyl protons are split into a doublet by the lone proton on the other carbon atom. The latter proton is itself split into a 1:3:3:1 quartet.

18.5. The integral over solid angle is given by

$$\int_0^{2\pi}\int_0^{\pi}(3\cos^2\theta - 1)\,\sin\theta\,d\theta\,d\phi = 2\pi \times \int_{-1}^{1}(3u^2 - 1)\,du = 0.$$

18.6. A 180° pulse reverses the directions of the precessing nuclei, thus undoing the dephasing after the 90° pulse.

Chapter 19

19.1. The singlet state has the same form

$$|\Psi\rangle = \frac{1}{\sqrt{2}}\Big(|\uparrow\rangle_1|\downarrow\rangle_2 - |\downarrow\rangle_1|\uparrow\rangle_2\Big),$$

with respect to any axis.

19.2. The H-polarized beam is equivalent to an equal mixture of $|+45°\rangle$ and $|-45°\rangle$. Thus half the photons pass through the +45° polarizer. Analogously, the +45°-polarized beam is equivalent to an equal mixture of H and V. Thus half of the remaining photons pass through the V polarizer.

Suggested references

Atkins, P.W., de Paula, J., 2014. Physical Chemistry, 12th edition. Oxford University Press. The best-selling physical chemistry text worldwide. A solid alternative reference for all topics covered in this book.

House, J.E., 2003. Fundamentals of Quantum Chemistry. Elsevier Science & Technology, Amsterdam/New York. A companion volume in the Complementary Science Series. Greater emphasis on numerical problems and worked-out examples than this book.

Claf, A.A., 1957. Calculus Refresher. Dover, New York. A quick tune-up for students whose calculus has become rusty.

Blinder, S.M., 2013. Guide to Essential Math, 2nd edition. Elsevier, Amsterdam. The author's review of the more advanced mathematical requisites for students in chemistry, physics and engineering.

Orear, J., 1979. Physics. Macmillan, New York. Contains all the background in general and modern physics needed for quantum mechanics on the level of this book.

Feynman, R.P., Leighton, R.B., Sands, M.L., 1963. Feynman Lectures on Physics, vols. I, II, III. Addison-Wesley, Reading, MA. Especially vol. III on quantum mechanics. Feynman's unique masterly presentation of the fundamentals of physics. Great reading whenever you are so inclined. Limited overlap with our subject matter but worth skimming for Feynman's quotable pearls of wisdom on physics.

Atkins, P.W., 1989. General Chemistry. Freeman (Scientific American Books), San Francisco. Chapters 7-9 deal with the structure of atoms and molecules.

Porile, N.T., 1993. Modern University Chemistry, 2nd ed. McGraw-Hill, New York. A high-level general chemistry text. Chapter 5 contains a very accessible introduction to quantum theory.

Herzberg, G., 1944. Atomic Spectra and Atomic Structure. Dover, New York. A concise introduction to quantum theory and atomic structure.

Atkins, P.W., 1997. The Periodic Kingdom: A Journey Into the Land of the Chemical Elements. Basic Books, New York. A metaphorical description of the periodic table of chemical elements. Appropriate for any level of scientific background.

Companion, A.L., 1979. Chemical Bonding. McGraw-Hill, New York. A charming little volume covering chemical bonds on an elementary level.

Pauling, L., 1960. The Nature of the Chemical Bond. Cornell University Press, Ithaca, NY. Another science classic by one of the original heroes of quantum chemistry. Heavily slanted towards Pauling's views on resonance and valence-bond theory.

Coulson, C.A., 1961. Valence. Oxford University Press, London. Somewhat dated but a classic introduction to quantum theory of chemical bonding.

McWeeny, R., 1979. Coulson's Valence, 3rd ed. Oxford University Press, London. An update of Coulson's classic, seeking to preserve the flavour of the original.

Pauling, L., Wilson, E.B., 1935. Introduction to Quantum Mechanics. McGraw-Hill, New York. Has long been the definitive exposition of quantum mechanics for chemists. Solutions of Schoödinger equation for harmonic oscillator and hydrogen atom are worked out with step-by-step, giving all mathematical details. Now available as a Dover reprint.

Dirac, P.A.M., 1948. The Principles of Quantum Mechanics, 3rd ed. Clarendon Press, Oxford. A scientific classic by one of the creators of quantum mechanics. Advanced level, but even beginners can enjoy the elegance of Dirac's presentation.

Blinder, S.M., 1974. Foundations of Quantum Dynamics. Academic Press, London. The author's earlier monograph on the fundamental principles of time-dependent quantum mechanics, including transitions and time-dependent perturbation theory.

Levine, I.N., 2014. Quantum Chemistry, 7th edition. Pearson Education, London. A longtime favorite text for undergraduate and graduate courses in quantum chemistry.

Levine, I.N., 1975. Molecular Spectroscopy. Wiley, New York. Originally a companion volume to the above. Detailed coverage of quantum-mechanical foundations of spectroscopy.

Ballhausen, C.J., Gray, H.B., 1965. Molecular Orbital Theory. Benjamin, New York. Contains a detailed account of crystal and ligand field theories for transition metal ions.

Koch, W., Holthausen, M.C., 2001. A Chemist's Guide to Density-Functional Theory, 2nd ed. Wiley-VCH, Weinheim. An up-to-date account of density-functional theory and applications.

Parr, R.G., Yang, W., 1989. Density-Functional Theory of Atoms and Molecules. Oxford University Press, New York and Oxford.

Cotton, F.A., 1990. Chemical Applications of Group Theory. Wiley, New York. The standard reference on group theory for chemists.

Derome, A.E., 1987. Modern NMR Techniques for Chemistry Research. Pergamon, New York. A practical introduction for chemists using NMR.

Stryer, L., 1989. Molecular Design of Life. Freeman, New York. A lucid introduction to the chemical and physical basis of molecular biology.

Herbert, N., 1985. Quantum Reality: Beyond the New Physics. Doubleday, Garden City, NY. A patient exposition dealing with the "weirdness" of quantum mechanics.

Gribbin, J., 1996. Schrödinger's Kittens and the Search for Reality. Little, Brown & Co., Boston, MA. A stimulating nontechnical account of the perplexing paradoxes of quantum theory.

Aczel, A.D., 2001. Entanglement: The Greatest Mystery in Physics. Four Walls Eight Windows, New York. A popular account of recent developments on the wonders of the quantum world.

Johnson, G., 2003. A Shortcut Through Time: The Path to the Quantum Computer. Knopf, New York. An engaging popular-level introduction to the possibilities of quantum computation.

Karplus, M., Porter, R.N., 1970. Atoms and Molecules: An Introduction for Students of Physical Chemistry. Benjamin, New York. Valuable for computational details on atomic and molecular orbitals and several key topics in quantum chemistry.

McQuarrie, D.A., 1983. Quantum Chemistry. University Science Books, Mill Valley, CA. A student-friendly introduction to the subject.

Ratner, M.A., Schatz, G.C., 2001. Introduction to Quantum Mechanics in Chemistry. Prentice-Hall, Upper Saddle River, NJ. A recent quantum chemistry text for both undergraduate and graduate students. Keyed to approximately the same level as this book.

Fitts, D.D., 1999. Principles of Quantum Mechanics as Applied to Chemistry and Chemical Physics. Cambridge University Press, Cambridge, UK. Another alternative, as above.

Nielsen, M.A., Chuang, I.L., 2010. Quantum Computation and Quantum Information, 10th anniversary edition. Cambridge University Press, Cambridge, UK. The definitive reference on this subject.

Fano, G., Blinder, S.M., 2017. Twenty-First Century Quantum Mechanics: Hilbert Space to Quantum Computers. Springer International Publishing, New York. Mathematical methods and conceptual foundations of quantum mechanics from a modern point of view.

Blinder, S.M., House, J.E. (Eds.), 2018. Mathematical Physics in Theoretical Chemistry. Elsevier Science Publishing, Amsterdam. A collection of fairly advanced articles on various mathematical methods applied to theoretical chemistry.

McQuarrie, D.A., 1973. Statistical Thermodynamics. Harper and Row, New York. The standard reference on the subject.

Index

D

E

F

G

H

O

P

Q

R

T

U

9780128223109